Proceedings of the I-ESA Conferences

Volume 9

This series publishes the proceedings of the IESA conferences which began in 2005 as a result of cooperation between two major European research projects of the 6th Framework R&D Programme of the European Commission, the ATHENA IP (Advanced Technologies for Interoperability of Heterogeneous Enterprise Networks and their Applications, Integrated Project) and the INTEROP NoE, (Interoperability Research for Networked Enterprise Applications and Software, Network of Excellence). The I-ESA conferences have been recognized as a tool to lead and generate an extensive research and industrial impact in the field of interoperability for enterprise software and applications.

More information about this series at http://www.springer.com/series/10534

Keith Popplewell · Klaus-Dieter Thoben ·
Thomas Knothe · Raúl Poler
Editors

Enterprise Interoperability VIII

Smart Services and Business Impact of Enterprise Interoperability

Springer

Editors
Keith Popplewell
Future Manufacturing Applied Research Centre
Coventry University
Coventry, UK

Thomas Knothe
Business Process and Factory Management
Fraunhofer IPK
Berlin, Germany

Klaus-Dieter Thoben
BIBA
Universität Bremen
Bremen, Germany

Raúl Poler
Higher Polytechnic School of Alcoy
Polytechnic University of Valencia
Alcoy, Alicante, Spain

ISSN 2199-2533 ISSN 2199-2541 (electronic)
Proceedings of the I-ESA Conferences
ISBN 978-3-030-13692-5 ISBN 978-3-030-13693-2 (eBook)
https://doi.org/10.1007/978-3-030-13693-2

Library of Congress Control Number: 2019932622

This Springer imprint is published by the registered company Springer Nature Switzerland AG.
The registered company address is: Gewerbestrasse 11, 6330 Cham, Switzerland

Organizers and Sponsors

Organizers

InterOP - VLab

DFI

Fraunhofer
IPK
INSTITUTE
PRODUCTION SYSTEMS AND
DESIGN TECHNOLOGY

Sponsors

DFG Deutsche Forschungsgemeinschaft

ifip

SGIMRI
中德智能制造研究院

Preface

Industry 4.0, Internet of things, block-chain technologies and digital transformation require a foundation for seamless, clear and secure communication called "interoperability". Moreover, the cooperation between different organizations such as manufacturers, service providers and government requires "enterprise interoperability" between applications and systems. Consequently, interoperability of organizations is a major issue in successfully building enterprise networks. Interoperability becomes a key factor when smart services inside organizations and autonomous factories have to cooperate. Self-organizing production networks will have an enormous impact upon companies and their networks and require new business models for modelling, testing, executing and managing smart service infrastructures. Interoperability permeates all aspects of a business, from strategic and tactical planning to operational processes of the company and horizontally in the many facets of business cooperation. Accordingly, I-ESA'18 (Interoperability for Enterprise Systems and Applications) joins new business models, smart services, IoT and cloud technologies. Connecting the world's leading researchers and practitioners of enterprise interoperability and related domains, including interoperability aspects of enterprise systems and applications, I-ESA'18 presents an outstanding opportunity to exchange experiences and business ideas between researchers, service providers, entrepreneurs and industrial stakeholders.

I-ESA'18 is the ninth of a series of conferences: Genève (2005), Bordeaux (2006), Madeira (2007), Berlin (2008), Coventry (2010), Valencia (2012), Albi (2014), Guimarães (2016) and a special edition in Beijing (2009), this time under the motto "Smart Services and Business Impact of Enterprise Interoperability". The I-ESA'18 conference was hosted by Fraunhofer IPK and jointly promoted by DFI (Deutsches Forum für Interoperabilität e.V.) and INTEROP-VLab (European Virtual Laboratory for Enterprise Interoperability—http://www.interop-vlab.eu).

World-leading researchers and practitioners in the area of enterprise interoperability contributed to this book. You will find integrated approaches from different disciplines: Computer Science, Engineering and Business Administration.

The I-ESA'18 program included several keynotes presented by high-level renowned experts from industry, government and academia:

- Prof. Dr.-Ing. Dieter Wegener, Siemens AG, Vice President, Germany
- Dr. Nenad Ivezic, Systems Integration Division, Engineering Laboratory, National Institute of Standards and Technology (NIST), USA
- Mr. Gerald Santucci, European Commission (retired 2017), Ambassador of the INTEROP-VLab, Belgium
- Mr. Stefan Zimmerman, Head of COE Industrie 4.0 at Atos Global B&PS, Germany.

This book is organized into twelve parts addressing the major research in the scope of Interoperability for enterprise systems and applications:

Part I	Security
Part II	I 4.0 and Industrial Automation
Part III	Platforms and Infrastructures for Enterprise Interoperability
Part IV	Semantic Interoperability
Part V	Interoperability Testing
Part VI	Ontology Modeling
Part VII	Block Chain and Decentralized Approaches
Part VIII	Interoperability Application Scenarios
Part IX	Interoperability in Manufacturing and Repair and Operation (MRO)
Part X	Modelling and Frameworks
Part XI	Entities in IoT
Part XII	Interoperability in M2M Interaction.

Coventry, UK — Keith Popplewell
Bremen, Germany — Klaus-Dieter Thoben
Berlin, Germany — Thomas Knothe
Alcoy, Spain — Raúl Poler

Acknowledgements

We wish to thank all the authors, invited speakers, reviewers, senior program committee members and participants of the conference that made this book a reality and the I-ESA'18 conference a success. We express our gratitude to all organizations, which supported the I-ESA'18 preparation, especially INTEROP-VLab and its German Pole DFI. We are deeply grateful to the local organization support notably Thomas Knothe, Frank-Walter Jaekel and Andrea Koch for their excellent work for the preparation and the management of the conference. We are proud and thankful to the Sponsors of I-ESA'18, namely the Deutsche Forschungsgesellschaft and the Sino German Intelligent Manufacturing Institute.

Contents

Part XII Interoperability in M2M Interaction

About the Editors

Keith Popplewell started his career in operational research and specialised in computer-aided engineering and production planning systems design with Raleigh Industries and Boots Company plc. During this time, he took a doctorate in production engineering and production management at the University of Nottingham. Subsequently, he became technical director in a software house specialising in the design, development and implementation of CAE systems, before joining the Department of Manufacturing Engineering at Loughborough University in 1985.

In 2000, he became Jaguar Cars Professor of Engineering Manufacture and Management and Head of Department at Coventry University, before, in 2006, he accepted the post of Director of Coventry University's Future Manufacturing Applied Research Centre (FMARC). His research interests centre on design, modelling and operation of global and network manufacturing enterprises, and in particular on providing intelligent knowledge oriented support for virtual organisations, especially focusing on the needs of SMEs. In this context, he is President of the INTEROP-VLab AISBL.

Klaus-Dieter Thoben is Professor for Integrated Product Development in the Faculty of Production Engineering at the University of Bremen, Germany. At the same time, he is Director of BIBA (Bremen Institute of Production and Logistics) and Spokesman of LogDynamics (Bremen Research Cluster of Dynamics in Logistics).

His main research activities include but are not limited to the application of information and communication technologies to support cooperative processes and environments with a specific focus on product development (smart/intelligent products engineering), product and systems lifecycle management, collaborative process chains, and enterprise networks. He has more than 25 years of experience in coordination and management of European and national funded research projects including EU-funded Networks of Excellence and large integrated projects (IPs). He has published some 500 research papers in national as well as international conferences and journals.

He is active on various levels and with different roles in various scientific as well as professional communities such as DFI e.V. (Deutsches Forum für Interoperabilität),

IFIP (International Federation of Information Processing) TC 5 Working Groups, WiGeP (Wissenschaftliche Gesellschaft für Produktentwicklung), ForWind, and VDI (Verein Deutscher Ingenieure).

Prof. Dr.-Ing. Thomas Knothe is Head of the Department of Business Process and Factory Management at Fraunhofer IPK, Corporate Management Division, and is responsible for the appliance methodology of the Enterprise Modelling Tool MO^2GO.

After finishing his study in information technology for production, he started his career at Fraunhofer IPK in 1998. He is lecturing at several universities in Germany and abroad, and since 2017, he holds an honorary Professorship at University of Applied Science in Wildau.

He coordinates several national and international research projects according to Industry 4.0 in Germany. His industrial background comes from leading various business and process development projects in IT service, aerospace, automotive and process industry. He holds several inventions in intelligent manufacturing, e.g. self-organized planning and control for maintenance repair and overhaul of transportation systems like aircraft.

He is Chairman of the German Association for Interoperability (DFI e.V.) and as Member of ISO involved into International Standardization regarding information processing for production.

Raúl Poler is Professor in Operations Management and Operations Research at the Universitat Politècnica de València (UPV), València, Spain. He received his Ph.D. in Industrial Engineering from UPV in 1998.

He is Director of the Research Centre on Production Management and Engineering (CIGIP). He is Founding Partner of the Spin-off UPV EXOS Solutions S.L. He is Director of the Master in Industrial Engineering and Logistics (MUIOL) at Alcoy Campus UPV.

He has led several Spanish Government and European R&D Projects. He has published more than 300 research papers in a number of leading journals and in several international conferences.

He is Member of the Board of Director of the INTEROP-VLab and Chair of its Education Committee. He is Secretary of INTERVAL. He is Member of the Executive Board of the Association for the Development of Organization Engineering (ADINGOR). He is Chair of the Education Activity of the IFIP WG 5.8 Enterprise Interoperability.

Part I
Security

Cybersecurity Requirements for Supporting Enterprise Interoperability of Multi-sided Platforms

Violeta Damjanovic-Behrendt

Abstract We report on work in capturing cybersecurity requirements for cloud-based and IoT-enabled multi-sided platforms (MSPs). Our approach is designed to capture security aspects related to business rules and constraints of MSPs, thus shaping the platform's behaviour and the participants' interaction and leading towards safer enterprise interoperability. We design the MSPs Privacy Requirements Framework and the MSPs Security Architecture, in order to cater for specific use case-centric and platform-centric cybersecurity requirements. To ensure compliance with the upcoming GDPR, we discuss the mapping between elicited cybersecurity requirements and GDPR rules. The new GDPR is expected to have significant implications on businesses in the EU, and our approach is designed to achieve full compliance with it.

Keywords Cybersecurity · Platforms · Multi-sided platforms · Enterprise interoperability · Privacy · GDPR

1 Motivation

Enterprise interoperability of multi-sided platforms (MSPs) enables separately developed enterprise systems, which incorporate different business models and domains to effectively share their data, exchange information through negotiation, interact and perform customized business processes and maximize their efficiency through interworking. MSPs support interaction and interoperability between two or more sides; for example, a two-sided platform manages the interaction between two distinct groups (e.g. consumers and software providers), while MSPs facilitate the recurring interaction between more than two distinct groups [1]. Some examples of multi-sided platforms are Uber, Xbox, eBay, Airbnb, to name a few.

This paper presents our approach to capturing cybersecurity requirements supporting the design and development of the NIMBLE platform's infrastructure

V. Damjanovic-Behrendt (✉)
Salzburg Research, Jakob Haringer Str. 5/II, 5020 Salzburg, Austria
e-mail: violeta.damjanovic@salzburgresearch.at

K. Popplewell et al. (eds.), *Enterprise Interoperability VIII*, Proceedings of the I-ESA Conferences 9, https://doi.org/10.1007/978-3-030-13693-2_1

and its services. The NIMBLE platform is designed as an MSP, enabling multi-sided B2B trade and enterprise collaboration. The platform development is funded through the EU H2020 research and innovation programme (for more details, see: https://www.nimble-project.org/).

The NIMBLE's MSP business model involves contractual relationships between buyers and suppliers, as well as logistics and other services affiliated with the platform. Designed for cloud environments, the NIMBLE models support a federation of platform instances, each providing a set of core services and offering additional, specifically tailored services enabling interoperation at regional, sectorial or topical levels. The flexibility of such an MSP increases the complexity of enterprise interoperability and opens issues related to cybersecurity and possible serious harm that can be caused to the participating companies, e.g. revealing sensitive information (personal data, design and operational information), losing customers, facing a host of legal and financial penalties, putting businesses at risk through cyber jamming communication, spoofing and data manipulations affecting the decision-making process.

Paper organization. Section 2 discusses related work on MSPs and platform evolution models, which are used to analyse the overall growth, effectiveness and performances of MSPs in supporting enterprise interoperability. A related work on cybersecurity frameworks and strategies for MSPs is also presented in this section. Section 3 discusses cybersecurity methods used as a baseline to create the MSPs Privacy Requirements Framework in the NIMBLE. Section 4 describes our approach to cybersecurity requirements elicitation for MSPs, encompassing several aspects targeting enterprise interoperability. For example, we firstly create our MSPs Security Architecture, from which we further derive core platform-centric security controls and specify a set of related cybersecurity requirements. Secondly, we map these security controls into the new General Data Privacy Regulation (GDPR) requirements, which came into effect in the EU in May 2018. Section 5 concludes the paper.

2 Related Work

2.1 Multi-sided Platforms and Their Evolution Models

Enterprise MSPs contribute to frictionless access to markets through reduction in both search costs and shared transaction costs among multiple participating sides [2]. The direct interaction over MSPs occurs with high frequency between participants affiliated with the same platform. Such frequent interaction expands network effects and fosters an extensive platform adoption. In turn, this positively affects the value of the platform to all affiliated sides [1], while the complexity of economic and technology factors that drive the strategic design of MSPs increases accordingly.

MSPs prove to be systems that evolve gradually over time, and the phenomenon of the platform's evolution is examined in several platform evolution models:

- Hagiu [2] observes platform evolution as *gradual transition* from being one-sided to two-sided and multi-sided platforms;
- Evans [3] recognizes the importance of a *critical mass of users*, which can be achieved by following a zigzag strategy for attracting new platform participants;
- Tiwana [4] focuses on concrete *evolutionary metrics* to estimate the speed and effectiveness of a platform's evolution.

Tiwana's model is the only model that recognizes the importance of aligning platform architecture, governance and business strategy in order to progress the MSPs. However, the above models do not provide coherent decisions on either architecture or governance and business strategies for MSPs that would practically drive platform owners and developers. Thus, the authors in [5] design the *Reach and Range Framework* for MSPs as an analytical tool providing in-depth understanding of the MSP's key mechanisms (such as *reach* and *range*), which are used to address the main strategic challenges during platform evolution. In our work, we follow the Tiwana's model for the design of our security requirements methodology in order to elicit more complete cybersecurity requirements and better support their management and further evaluation.

2.2 Cybersecurity Frameworks and Strategies Affecting Multi-sided Platforms

Capturing functional and non-functional cybersecurity requirements for MSPs puts a strong emphasis on an early integration of security and privacy with software development, which is ensured through the key concepts of Information Security, as defined in the ISO/IEC 27000:2009 standard [6]. This standard ensures that the information is neither violated nor compromised through possible critical situations, i.e. device malfunctions, threats (software attacks, ransomware, viruses and the like), identity theft, hazards, natural disasters, etc.

In our work, we select several privacy requirements frameworks which are important either because they are influential with regulations or because they have been designed to provide practical advice for developers:

- *The Fair Information Practices (FIP) framework* opens a list of privacy elements which are useful to be discussed in the system design phase. The current EU *Directive on the Protection of Personal Data* (1995) is based on FIP, which brings the following eight principles on personal data [7, 8]: *Collection Limitation Principle, Data Quality Principle, Purpose Specification Principle, Use Limitation Principle, Security Safeguards Principle, Openness Principle, Individual Participation Principle* and *Accountability Principle.*
- *The Seven Laws of Identity framework* improves usable security topics (user-centric privacy) through the definition of seven principles: *User Control and Consent, Minimal Disclosure for a Constrained Use, Justifiable Parties, Directed*

Identity, Pluralism of Operators and Technologies, Human Integration and Consistent Experience Across Contexts.

- *The Data Minimization strategy* is designed with the aim to reduce privacy risks and provide users with maximum control over their sensitive information [9]. Specific mechanisms can be applied to validate the integrity of algorithms, demonstrate compliant handling of data, prove that data collectors and processors respect privacy policies, etc. The Data Minimization principles are as follows: *minimize collections of data in the system; minimize disclosure by constraining the flow of information to parties other than the entity to whom the data relates; minimize replication of data; minimize centralization by avoiding single point of failure in the system; minimize linkability of data;* and *minimize retention of data in the system.*
- *Microsoft Privacy Guidelines for Software Products and Services.* Despite recent privacy issues with Windows 10 [10, 11], this document offers practical guidelines for creating notice and consent experiences, providing sufficient data security, maintaining data integrity, supplying controls for developing software products, etc. One of the core principles in this document is about user's consent, related to what personal data will be collected, with whom it will be shared and how it will be used.
- Finally, the new GDPR (see: http://www.eugdpr.org/) is a complex regulation created to enhance personal privacy rights, increase duty for protecting personal data, provide mandatory personal data breach reporting, etc. [12]. The GDPR will become enforceable in May 2018 and will have significant implications on businesses in the EU.

3 Cybersecurity Considerations for Multi-sided Platforms

Our approach to security and privacy requirements elicitation for the purpose of engineering and delivering secure platform solutions for a variety of MSP's users (e.g. suppliers, logistic operators, service providers, cloud providers, retailers and platform providers) combines several views on MSPs:

- A platform-centric view resulting in the design of the MSPs Security Architecture and its alignment to the technical NIMBLE platform architecture;
- A use case-centric view resulting in security and privacy control services for MSP's users;
- A data-centric view that designs and implements data security and privacy services, governance models and necessary GDPR compliance models.

In our work, we follow the formulation of security as a property to prevent *unauthorized access to and modification of information and data, as well as unauthorized use of resources* [13], while privacy is seen as a common application of security technologies, with a significant intersection with data provenance that adds security controls for preserving both data integrity and confidentiality [14, 15]. Privacy ensures the development of platform services that satisfy user's requirements related to privacy protection and disclosure of both personal information and corporate information.

For the development of privacy-preserving mechanisms for MSPs, we combine the FIP framework, the Seven Laws of Identity, the Data Minimization principles and methods for the GDPR implementation as described in [16].

Table 1 presents an excerpt from our work showing the convergence of the proposed privacy frameworks into privacy requirements for MSPs.

3.1 MSPs Privacy Requirements Framework

Table 2 illustrates the proposed MSPs Privacy Requirements Framework, which addresses three views: user-centric, platform-centric and data-centric views that incorporate a set of GDPR requirements.

4 Cybersecurity Requirements Capturing for MSPs

In our work, the process of cybersecurity requirements elicitation for MSPs includes the following steps:

- **Use case-centric security and privacy requirements elicitation**, which is based on requirements collected from platform's participants;
- **Platform-centric security and privacy requirements elicitation**, which is based on the problem context of the platform's system, its architecture and design of components and services;
- **Mapping between use case-centric and platform-centric cybersecurity requirements** with the aim of eliminating possible inconsistencies and repetitions between requirements; and
- **Security and privacy requirements evaluation** that is based on STRIDE (Spoofing, Tampering, Repudiation, Information Disclosure, Denial of Service, and Elevation of Privilege) threat analysis [19]. STRIDE analyses assets of the platform and its data flows and identifies possible threats and vulnerabilities that can affect the platform. The threats point to what the attacker can do to harm the platform, while vulnerabilities are weaknesses of the platform that could be easily exploited by the attacker.

Figure 1 illustrates our approach to requirements capturing for MSPs. Firstly, we apply the ISE/IEC 27000 series of Information Security standard for capturing security requirements related to the platform's participants and their use cases and to the MSP's technical architecture and components. In this way, we identify and specify both use case-centric and platform-centric security requirements related to MSPs. For the privacy requirements, we use the MSPs Privacy Requirements Framework

Table 1 Conversion of adopted privacy frameworks into privacy requirements of MSPs

Fair Information Practices	Laws of identity	Data min. principles	GDPR	Conversion into privacy requirements
Collection limitation principle	User control and consent	–	Implementing privacy procedures for seeking, recording and managing consent	Does the user give consent for the collection and use of his sensitive data? Do we have established privacy methods and tools for seeking, recording and managing consent?
Purpose specification principle	–	–	Documenting all personal and sensitive personal data that the organization hold	Is the purpose of collecting sensitive data clearly defined? Are personal data and sensitive personal data adequately documented in the system?
Individual participation principle	Human integration	–	Deletion and modification of personal data by the data subject; privacy procedures for access requests	Does the user have rights to his sensitive data (e.g. obtaining data from a data controller, having the data erased, rectified, completed or amended, when required)? Is there a way for the user to request access to his data?

Table 2 MSPs Privacy Requirements Framework

MSP Privacy Requirements Practices	Conversion into privacy requirements
User-centric approach	
User consent w.r.t. data collection	Does the user give consent for the collection and use of his sensitive data? Do we have established privacy methods and tools for seeking, recording and managing consent?
User rights and controls w.r.t. data collection	Does the user have rights to his sensitive data (e.g. obtaining data from a data controller, having the data erased, rectified, completed or amended, when required)? Does the user have control over his sensitive data? Is there a way for the user to request access to his data?
Data-centric approach	
Purpose specification	Is the purpose of collecting sensitive data clearly defined?
Data quality	Is sensitive data accurate, complete and up to date? Is user's age verified?
Minimize data collection	Is the amount of collected data minimal?
Minimize disclosure	Does the system disclose the least amount of sensitive data and with its limited use?
Minimize replication	Is the amount of entities where data is stored or processed, minimal?
Minimize centralization	Is the number of single point of failure in the system minimal? Is sensitive data minimally centralized in the system?
Minimize linkability	Is the amount of linked sensitive data minimal?
Minimize data retention	Is retention of data in the system minimal?
Security safeguards	Is sensitive data protected?
Openness and interworking of providers and technologies	Does open developments, practices and policies respect sensitive data? Does interworking technologies have privacy mechanisms taking into account respect for sensitive data?
GDPR requirements	
Reviewing and keeping up-to-date privacy notices	Are privacy notices updated? Are privacy notices clearly explained? Do privacy notices follow laws for processing personal data?
Detecting, reporting and investigating a personal data breach	Are the mechanisms for detecting personal data breaches established? Are the mechanisms for reporting personal data breaches defined? Do we have defined procedures for investigating a personal data breach?

(continued)

Table 2 (continued)

MSP Privacy Requirements Practices	Conversion into privacy requirements
Assessing privacy impact	Is the ICO's code of practice on Privacy Impact Assessment implemented? Are we using another method to assess privacy impact?
Implementing the guidance from the Article 29 Working Party	Is the guidance from the Article 29 Working Party (EU-GDPR) which is about processing under the authority of the controller or processor implemented?

(presented in Sect. 3.1). Secondly, we map the use case- and platform-centric security and privacy requirements in order to eliminate inconsistencies and repetitions between requirements. Finally, we use STRIDE-based methods for the evaluation of security requirements (for more details, see [18]).

The objective of our approach to cybersecurity elicitation is to identify the most important security controls and measures to be implemented and maintained during the MSP's security life cycle. In addition, we design an MSPs Security Architecture (see Fig. 2) for the analysis of platform-centric cybersecurity requirements.

4.1 MSPs Security Architecture

MSPs Security Architecture is derived from the NIMBLE Microservice Architecture [17]. Platform-centric security and privacy requirements for each of the identified security controls are elaborated in detail in [18]. The basic security controls have been addressed by the following core components:

- *The FrontEnd component* is designed to ensure an easy-to-use interaction with the users and to handle authentication, load balancing and related security controls ensuring that only authenticated users can access the platform's services and data. Security monitoring methods must be in control of provenance data, revealing information about the platform's connection parameters.
- *The Open API component* implements access control management, authorization methods and security monitoring applications for monitoring unauthorized attempts to invoke an API.
- *The Data Store and Data Management components* ensure various security methods for protecting against front-end threats, log threats, tampering, ensuring data integrity and preventing information disclosure.
- *Core Services, Service Registry and Service Discovery components* include security controls for protecting Product Catalogues against unauthorized access and distribution, SQL injection attack against Product Catalogues, etc.
- *Core Security and Privacy controls* implement Identity Management, Access Control Management, Authorization and Authentication, Data Provenance Manage-

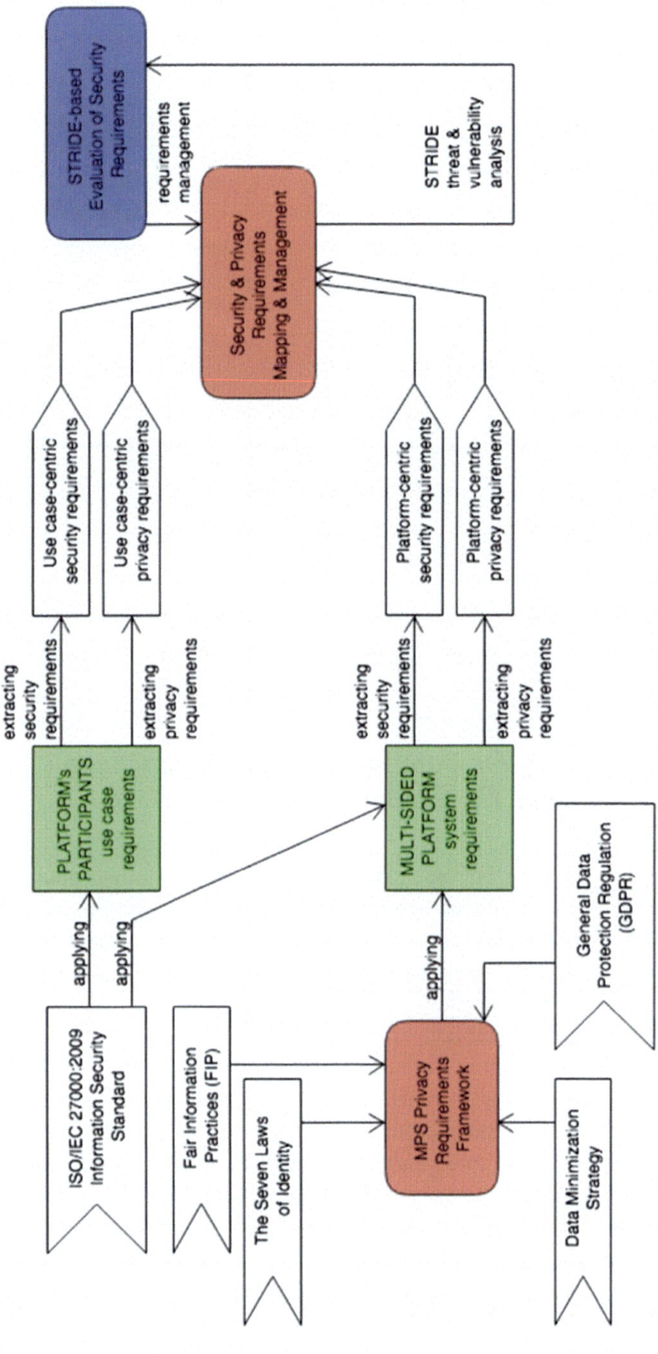

Fig. 1 Cybersecurity approach to requirements capturing for MSPs

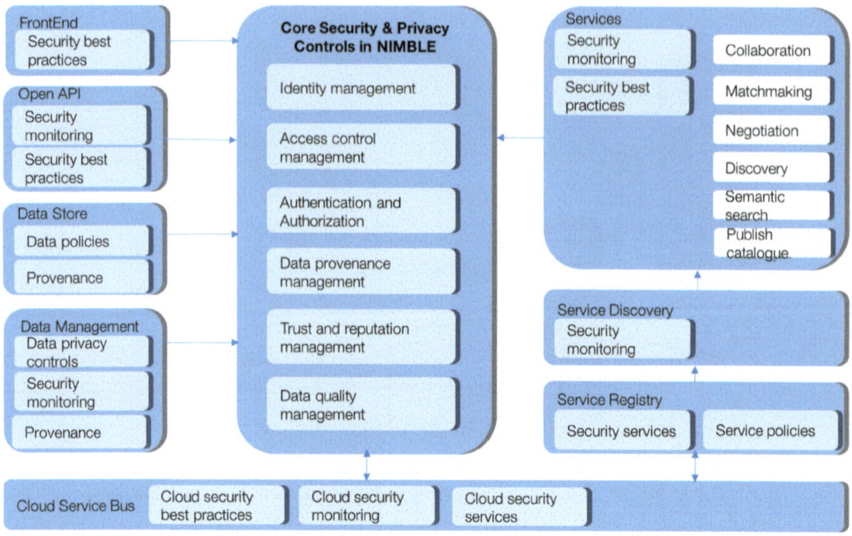

Fig. 2 MSPs Security Architecture

ment, Trust and Reputation Management, and Data Integrity and Data Quality Management.

4.2 *Security Controls for Data Integrity and Data Quality Management and Their Mapping with the GDPR*

A failure to control the distribution of data, data integrity and data quality often leads to data breaches, loss of sensitive information and data manipulation, which need to be prevented using adequate security controls, e.g. only authenticated users can access data in a controlled manner. Data manipulation related to MSPs enables comparison of products and suppliers, filtering and ordering information in a way that forces unfair trade and monopolies.

Provenance information about access to the system needs to be kept in audit logs, while security controls for anomaly detection are performed to capture unusual behaviour. Provenance information matters in cybersecurity as a measure for preventing data manipulation that can cause harmful changes in product specifications (e.g. power outages, data sabotage). The secure exchange of business information through file sharing, email and messaging system for negotiation is another big concern for platform participants interacting over MSPs.

To demonstrate mapping compliance of the MSPs Security Architecture and its security controls with the GDPR requirements, in Table 3, we discuss security con-

Table 3 Security controls supporting Data Integrity and Data Quality Management

Security control name	Security control description and its compliance with the GDPR
Data Integrity and Data Quality Policy	Data Integrity and Data Quality Policy must be clearly defined and based on the MSPs Privacy Requirements Framework (in order to incorporate different approaches: use case-centric view, data-centric view and the GDPR requirements)
Data input validation	Controls over various factors: predictable behaviour, manual override, timing, etc. must be integrated, which corresponds to the Data Quality Principle and the GDPR requirement for verifying sensitive data for its accuracy, completeness and for being up to date
Data and metadata protection	Protection against unauthorized access and manipulation; Automated restricted access; Cryptographic protection; GDPR requirement for deletion of personal data and/or personal data modification by the data subject; GDPR requirement for supporting subject's requests to access personal data
Data protection at rest	Cryptographic protection, offline storage; GDPR requirement for deletion and/or modification of personal data by the data subject
Data protection in shared resources	Cryptographic protection; GDPR requirement for deletion of personal data and/or personal data modification by the data subject
Notification of data integrity violations	Monitoring services must be provided; GDPR requirement for detecting, reporting and investigating personal data breaches; GDPR requirement for reviewing existing privacy notices and keeping them up to date
Informed consent by design	User must have an informed consent on the data usage, which prevents the use of data in a way that is not according to the user wish; GDPR requirement for implementing privacy procedures for seeking, recording, and managing user's consent

trols supporting Data Integrity and Data Quality Management (more details are given in [18]).

5 Conclusion

The ultimate role of MSPs in digital automation is to increase speed to market, minimize costs and optimize manufacturing and logistic processes through enterprise interoperability. The overall growth, effectiveness and performances of MSPs can be analysed using the platform's evolution models. We consider the Tiwana's platform evolution model, which aligns with the architecture of MSPs, governance and business strategies to further progress evolution of MSPs, as an approach that can be extended towards cybersecurity. Therefore, in this paper, we discussed the MSPs Privacy Requirements Framework and its specific MSPs Security Architecture featuring security controls that are specifically designed to ensure compliance with the GDPR. In that way, the role of the MSPs can be seen as a "regulatory role", which shapes enterprise interaction and online behaviour and which is expected to lead towards safer enterprise interoperability.

References

1. Hagiu, A., & Wright, J. (2011). *Multi-sided platforms*. Working Paper 12-024. Boston, MA: Harvard Business School.
2. Hagiu, A., & Wright, J. (2008). *Multi-sided platforms: From microfoundations to design and expansion strategies*. Working Paper 09-115. Boston, MA: Harvard Business School.
3. Evans, D. S. (2009). How catalysts ignite: The economics of platform-based start-ups. In A. Gawer (Ed.), *A platform, markets and innovation* (pp. 99–130). Cheltenham and Northampton, US: Edward Elgar.
4. Tiwana, A. (2014). *Platform ecosystems, aligning architecture, governance, and strategy*. Amsterdam: Morgan Kaufmann.
5. Staykova, K., & Damsgaard, J. (2016). Adoption of mobile payment platforms: Managing reach and range. *Jounal of Theoretical and Applied Electronic Commerce Rese*arch. ISSN 0718-1876.
6. ISO/IEC 27000:2009 (E). (2009). Information technology—Security techniques—Information security management systems—Overview and vocabulary. ISO/IEC.
7. OECD Guidelines on the Protection of Privacy and Transborder Flows of Personal Data (1980). Online: http://bit.ly/2xfYFv3. Last Access September 2017.
8. World Privacy Forum. (2008). *A brief introduction to fair information practices*. Online: http://bit.ly/2xmslsb. Last Access September 2017.
9. Gürses, S., Tronsoco, C., & Diaz, C. (2011). *Engineering privacy by design*. COSIC 2011. Online: https://www.esat.kuleuven.be/cosic/publications/article-2589.pdf.
10. Warren, T. (2017). *Microsoft finally reveals what data Windows 10 really collects*. Online available from: http://bit.ly/2EJRrC2. Last Access October 2017.
11. Boomi, D. (2017). How *Windows 10 data collection trades privacy for security*. Online: http://bit.ly/2gQyhAK. Last Access October 2017.
12. Microsoft whitepaper. (2017). *Beginning your General Data Protection Regulations (GDPR) Journey for Windows 10. (2017)*. Online available: http://bit.ly/2iRu1p2.
13. Amini, M. (1993). *Formal methods for information security*. Online: http://bit.ly/2mxvb6Y.
14. Martin, A., Lyle, J., & Namilkuo, C. (2012). Provenance as a security control. In *Proceedings of the 4th USENIX conference on theory and practice of provenance*, USA.
15. Sultana, S., Bertino, E., & Shehab, M. (2011). A provenance based mechanism to identify malicious packet dropping adversaries in sensor networks. *ICDCSW, 2011*, 332–338.

16. Overview of the GDPR. (2017). Online: http://bit.ly/29lxF0U.
17. Innerbichler, J., Gonul, S., Damjanovic-Behrendt, V., Mandler, B., & Strohmeier, F. (2017). NIMBLE collaboration platform: Microservice architectural approach to federated IoT. In *Proceedings of the 1st Global IoT Summit 2017 (GIoTS'17)*, Switzerland.
18. NIMBLE D6.1. (2017). *Security and privacy requirements*. Project Deliverable. (To appear: https://www.nimble-project.org/deliverables/).
19. Shostack, A. (2014). *Threat modelling. Designing for security*. USA: Wiley.

Improving the Semantic Interoperability of IoT Early Warning Systems: The Port of Valencia Use Case

João Moreira, Luís Ferreira Pires, Marten van Sinderen, Roel Wieringa, Prince Singh, Patrícia Dockhorn Costa and Miguel Llop

Abstract An early warning system (EWS) is a distributed system that monitors the physical world and issues warnings if it detects abnormal situations. The Internet of Things (IoT) offers opportunities to improve monitoring capabilities of EWS and to realize (near) real-time warning and response. This paper presents the development of an interoperable IoT-based EWS to detect accident risks with trucks that deliver goods at the Valencia port area. Our solution addresses the semantic integration of a variety of data sources with processing in safety-critical applications for effective emergency response. The solution considers existing domain-specific ontologies and standards, along with their serialization formats. Accident risks are assessed by monitoring the drivers' vital signs with ECG medical wearables and the trucks' position with speed and accelerometer data. Use cases include the detection of health issues and vehicle collision with dangerous goods. This EWS is developed with the SEMIoTICS framework, which encompasses a model-driven architecture that guides the application of data representations, transformations, and distributed

J. Moreira (✉) · L. Ferreira Pires · M. van Sinderen · R. Wieringa · P. Singh
University of Twente, Enschede, Netherlands
e-mail: j.luizrebelomoreira@utwente.nl

L. Ferreira Pires
e-mail: l.ferreirapires@utwente.nl

M. van Sinderen
e-mail: m.j.vansinderen@utwente.nl

R. Wieringa
e-mail: r.j.wieringa@utwente.nl

P. Singh
e-mail: p.m.singh@utwente.nl

P. D. Costa
Federal University of Espírito Santo (UFES), Vitória, Brazil
e-mail: pdcosta@inf.ufes.br

M. Llop
Valencia Port, Valencia, Spain
e-mail: MLlop@fundacion.valenciaport.com

K. Popplewell et al. (eds.), *Enterprise Interoperability VIII*, Proceedings of the I-ESA Conferences 9, https://doi.org/10.1007/978-3-030-13693-2_2

software components. This framework enables an EWS to act as a semantic broker for situation-aware decision support.

Keywords Semantic interoperability · Early warning system · IoT

1 Introduction

Disaster risk reduction (DRR) is a systematic approach to analyze potential disasters and reduce their occurrence rate and potential impact. The main DRR component is an early warning system (EWS), which is a distributed information system that is able to monitor the physical world and issue warnings if it detects abnormal situations [1]. EWSs can benefit from the Internet of Things (IoT) technologies to realize (near) real-time data acquisition, risk detection, and message brokering between data sources and warnings' destinations [2]. Three major challenges in the development of IoT-based EWS are: (i) semantic integration of a variety of data sources that adhere to different standards, ontologies and data models; (ii) near-real-time processing in time- and safety-critical applications; and (iii) data analysis for effective situation awareness and decision support [2]. In this paper, we describe the SEMIoTICS framework [3], which has been designed to address these challenges. We discuss how SEMIoTICS is being used to develop an interoperable IoT EWS (INTER-IoT-EWS) to detect accidents with trucks delivering goods at the port of Valencia, which is a scenario of the H2020 INTER-IoT project [4]. This project aims to enable semantic integration among IoT platforms at the device, network, middleware, application, and semantic layers. The INTER-IoT-EWS integrates health and logistics data provided by different devices, made available through different IoT platforms and represented with different syntactic and semantic standards. INTER-IoT-EWS use cases include the early detection of a vehicle collision, health issues with drivers, and accidents involving dangerous goods. The use cases' validation plan is presented and lists the performed and current activities. This paper is further structured as: Sect. 2 presents the motivation of our research, Sect. 3 presents the SEMIoTICS framework, Sect. 4 presents the INTER-IoT case study, and Sect. 5 the lessons learned, limitations, and the future work.

2 Motivation

2.1 Early Warning System (EWS)

An EWS is a system for "the provision of timely and effective information, through identified institutions, that allows individuals exposed to a hazard to take action to avoid or reduce their risk and prepare for effective response" [1].

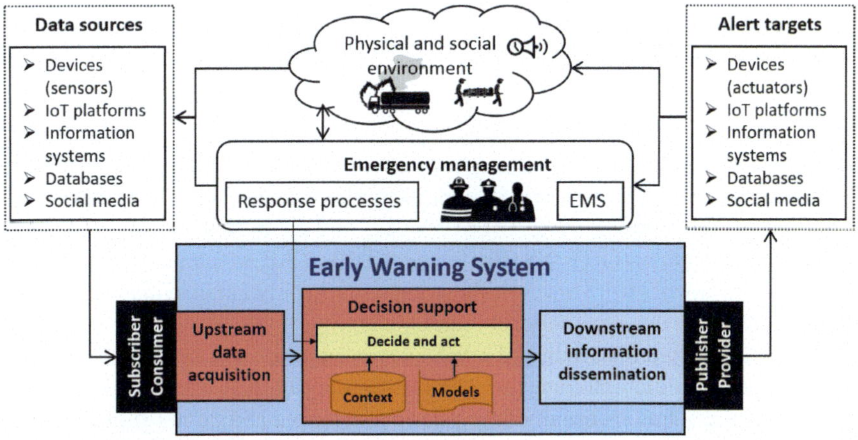

Fig. 1 Typical EWS architecture interacting with the environment

An effective EWS must be people-centered and integrate knowledge about the risks, risks' monitoring and warning, dissemination of meaningful warnings, and public awareness [5]. Modern EWSs comprise software and hardware for data acquisition, situation awareness, decision making, and information dissemination. Some experimental prototypes incorporate IoT technology to improve their functionality [2]. The conceptual architecture of EWS typically consists of three parts [1, 2, 6] (Fig. 1):

- *Upstream data acquisition*: Distributed sensor systems transform observations into digital signals, preprocess the associated data values to ensure that they contain relevant information for decision making, and transmit these data values to a message- and/or event-oriented middleware (broker).
- *Decision support*: Data are stored in data storage and are subject to rules to detect situations of interest. These rules are represented as models, which can be deterministic (e.g., rule-based approach) and/or non-deterministic (e.g., machine learning approach). Once a situation is detected, the EWS considers the requirements of the alert targets to assess the risk and determine the emergency response.
- *Downstream information dissemination*: Different target groups, comprising humans (e.g., the public) and machines (e.g., sirens), receive adequate notifications.

Interoperability is an important feature of effective EWSs for the integration of internal components and interworking of different EWSs. The level of interoperability depends on the standardization of interfaces, data exchange formats, and protocols [6]. The design problem addressed by our research is the improvement of IoT EWSs' interoperability among different data sources and targets, including other EWSs, enabling risky situation identification, and early warning emergency notifications.

2.2 Problem Definition

IoT standards have been defined to improve EWSs syntactic interoperability in multi-agency sensor information integration [7–9], such as the OGC's Sensor Web Enablement (SWE),[1] the OASIS Emergency Data Exchange Language (EDXL)[2], and Health Level Seven (HL7) standards. For example, the FEMA's (USA) Integrated Public Alert and Warning System (IPAWS) and the German Indonesian Tsunami Early Warning System (GITEWS) implement EDXL-CAP, which is a common alert data format protocol [10]. However, these approaches only target syntactic interoperability while we also need semantic interoperability.

The semantic interoperability of EWSs has been addressed by approaches that apply domain-specific ontologies to support meaningful data integration [11, 12]. These semantic solutions usually have poor performance and do not support effective response preparation [13]. In contrast, the *Semantic IoT EWS* approach [2] targets the challenges of scalable time-sensitive data handling from heterogeneous sources, enabling effective responses. This approach balances lightweight and heavyweight semantics: the former for upstream and the latter for downstream data. Moreover, this approach introduces an ontology, the *Decision Support Ontology* (DSO),[3] which is extended with the W3C Semantic Sensor Network (SSN) and OGC SWE terms. Although DSO's goal is "to aggregate and align multiple ontologies to support compound EWS semantics and ontology commitments," it lacks the support for multiple domain ontology alignments at runtime, i.e., it does not provide a mechanism for describing and executing ontology alignments at runtime. Furthermore, the DSO was serialized as XML with the Web Ontology Language (OWL), which inherits the verbosity of RDF/XML and the complexity of OWL, affecting the performance on data exchange and processing.

Our research goal is *to improve the semantic interoperability of emergency services for IoT EWSs, i.e., improve the semantic integration capacity of components of an IoT EWS and enable seamless integration with other IoT EWSs*. We identified the following challenges to achieve this goal:

(C1) *Semantic integration of a variety of data sources:* Avoid loss of semantics when multiple ontologies, standards and data models from different and overlapping domains are involved, considering their syntactic and semantic alignments.

(C2) *Processing in time- and safety-critical applications:* Provide the required performance for upstream data acquisition, emergency risk detection and brokering messages, in terms of scalability and total transaction time.

(C3) *Data analysis for effective responses:* Enable high-quality situation awareness (perception, comprehension, and projection) to avoid false positives and improve decision support based on emergency procedures.

[1] http://www.opengeospatial.org/ogc/markets-technologies/swe.

[2] https://www.oasis-open.org/committees/tc_home.php?wg_abbrev=emergency.

[3] http://tridec.server.de/ontologies/TRIDEC5.2.2.owl#.

3 The SEMIoTICS Framework

The "SEmantic Model-driven development for IoT Interoperability of emergenCy serviceS" (SEMIoTICS) framework aims at improving semantic interoperability among EWSs and their components [3, 14].

SEMIoTICS consists of an architecture (Fig. 2), technologies, and guidelines that are based on model-driven engineering (MDE), inspired by the mediation information system approach [15] and the semantic healthcare lifecycle [16]. SEMIoTICS uses the Endsley's situation awareness theory [17], which is harmonized with the Unified Foundational Ontology [3], enabling high-quality situation awareness (challenge *C3*). The framework also applies MDE transformations to integrate models and generate code for each component of the architecture, according to use case requirements. More details of SEMIoTICS can be found in [18].

The framework architecture has six elements that address the three main functions of an EWS: (1) input handler: upstream data acquisition; (2) abstraction: foundational ontology; (3) context model: domain ontology; (4) situation model: complex event processing; (5) situation awareness: data flows; and (6) output handler: downstream emergency notification. It follows the publisher/subscriber pattern and has RESTful services using JSON and XML, addressing challenge *C1* by enabling web services' syntactic interoperability. JSON for Linked Data (JSON-LD) was adopted in SEMIoTICS to support semantic interoperability and upstream data acquisition performance. JSON-LD is a structured way of using JSON, designed to be a lightweight syntax to serialize RDF, providing interoperability to JSON data at web scale. JSON-LD is a W3C standard recommended by schema.org and Google.

JSON-LD does not fully address challenge *C1* because data can still be represented with multiple different ontologies. To tackle this issue, the architecture supports the identification of functional components that reflect possible decentralized control of EWS functions, recommending interoperability standards to connect these components, and identifying adaptor components to bridge different standards or standards and proprietary solutions. The framework separates adaptors for syntactic interoperability from adaptors for semantic interoperability, allowing adaptor solutions that focus on one particular interoperability problem, and mix and match syntactic and semantic standards with a minimum of different adaptors. Adaptors

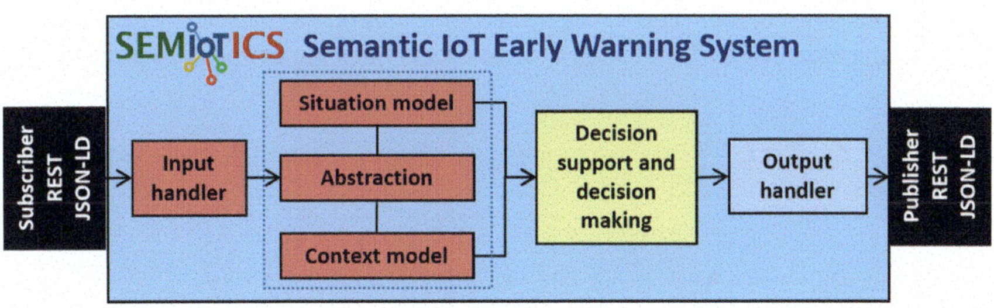

Fig. 2 SEMIoTICS framework for semantic IoT EWS

are implemented as syntactic and semantic translations [19]. The input handler is responsible for message translation, which relies on the syntax of each ontology being used and, therefore, also requires semantic as well as syntactic translations, e.g., from RDF/XML to JSON-LD and from HL7 to EDXL. Messages are translated from the original ontologies to our context model (core ontology) [18], which is aligned to W3C SSN and incorporates terms from EDXL and HL7. This approach aims at facilitating the data and semantics maintenance when integrating distinct domains, measuring, and tracking data quality.

The abstraction component refers to foundational ontologies, which are designed to maximize the support for interoperability of high-level categories, e.g., event, process, physical object, and system. The core ontology and SSN are grounded on the UFO (through OntoUML) and DOLCE Ultralite (DUL), respectively [3]. UFO and DOLCE share the same definitions for some conceptualizations, facilitating the alignment between the ontologies extended with them. This component benefits from the harmonization and axiomatization of terms related to "situation awareness" in UFO [3] and its impact on languages for context and situation modeling, discussed in [18].

The situation model is responsible for the situation identification mechanism, i.e., the formalization of the emergency risk detection [18]. Our approach allows the specification and implementation of complex event processing (CEP). CEP is a common component of IoT platforms to correlate data using temporal predicates (events' relations). For example, Cepheus[4] is the CEP engine of FIWARE IoT platform, based on Java ESPER[5] and, therefore, the event processing language (EPL). The SEMIoTICS guidelines describe how CEP technologies can implement the situation models in ESPER/EPL or SCENE (Drools Fusion) [20]. The situation model addresses *C2*, by incorporating adequate technology, and *C3*, by enabling situation awareness.

Decision support is enabled by the adoption of a workflow management system that enables the end user to design business processes as data flows, e.g., emergency plans. Big data integration tools for workflow development can generate code and are able to deploy data flows at runtime, e.g., Node-RED.[6] We cover the deployment and execution of the data flows for decision making by adopting such tool, addressing challenge *C3*.

The output handler is responsible for brokering the emergency risk notifications to the proper targets, according to the emergency procedures defined in the decision support component. For each predetermined risk, targets are enumerated with their information requirements. The data format of the notifications follows EDXL standards serialized as JSON-LD. Risk notification services are exposed as publishers.

[4]https://catalogue.fiware.org/enablers/iot-data-edge-consolidation-ge-cepheus/.

[5]http://www.espertech.com/.

[6]https://nodered.org/.

4 Case Study: EWS to Detect Accidents at the Port of Valencia

We proposed SEMIoTICS from our research in interoperability of IoT-based emergency systems, and, to validate it, we need to develop an IoT EWS and test in within an emergency scenario (i.e., a case study) composed of use cases. For demonstration and validation purposes, the INTER-IoT project described a scenario to decrease the risk of fatal accidents at the port of Valencia, improving health prevention and enabling quick reaction by reducing time response [21]. The goal is to exploit how e-Health can use IoT platforms dedicated to logistics to prevent the occurrence of accidents and to support evacuation or attention in case of emergency situations.

4.1 Requirements and Use Cases

The requirements of the scenario are:

(FR1) IoT platforms should be able to coordinate with emergency systems by detecting risks of accidents and accidents with trucks within the port area (collision and drivers' health issues), alerting their urgency and severity. The acceptance criterion is to check whether the port IoT platform is able to coordinate with emergency systems located in the vicinity.

(FR2) The hauler IoT platform and the port IoT platform should be able to share health information about the driver, monitored in real time through an electrocardiography (ECG) device. This device should be used for real-time ECG monitoring of drivers, transmitting data to a smartphone, which should act as a gateway, transmitting data to the cloud, both raw and calculated data, e.g., ECG sequence and heart rate (HR). These data need to be integrated with the port emergency control system.

(NFR1) IoT platforms should be semantically and syntactically interoperable. The acceptance criterion is the existence of a mechanism to translate data format and semantics of exchanged message to achieve communication with a common understanding on both sides.

(NFR2) E-Health and logistics should be integrated at the application and semantics level, including primitives for data interpretation of medical and transportation data.

(NFR3) The energy consumption (battery level) of the devices being used for the situation identification mechanism should be monitored.

Five use cases were defined to validate the achievement of these requirements:

(UC01) Vehicle collision detection: Uses accelerometer data of the truck from mobile phone and health device;

(UC02) Hazardous health changes: Detect occurrences of stress and arrhythmia (e.g., bradycardia and tachycardia);

(UC03) Temporal relations between UC01 and UC02: Detect if a health issue occurred before, during, or after a vehicle collision;
(UC04) Wrong-way driving: Integrates the trucks location data and the streets' direction within the port;
(UC05) Accidents with dangerous goods: Monitor dangerous goods being transported (according to the UN list of dangerous goods) in all use cases (1–4), adding adequate information regarding emergency procedures for effective response.

UC03 is particularly interesting because it requires the integration of data from both domains (health and logistics) and represents complex behaviors. For example, there is a possibility that bradycardia is detected followed by a continuous decrease in the heart rate after a collision. This situation reflects an accident where the driver is injured, classified as extremely severe with immediate urgency. In this situation, the vehicle collision is identified with both accelerometers from the ECG device and from the smartphone, considering device features as accuracy and energy consumption.

4.2 INTER-IoT-EWS: EWS Developed with SEMIoTICS

Our solution prototype (Fig. 3) includes the Shimmer ECG 3 device[7] to collect ECG data from drivers. This device has high accuracy and usability, being able to transmit data from a TinyOS application (running within the device) to a mobile phone application (Android) through Bluetooth. This mobile application receives and forwards the data to the cloud, acting as a gateway. Data are sent to the cloud and published in a broker as RDF/XML messages following the European Telecommunications Standards Institute (ETSI) Smart Appliances REFerence ontology (SAREF)[8] ontology extended with HL7 aECG (Annotated ECG), supported by the UniversAAL IoT platform.[9]

Similarly, the *MyDriving* mobile application for logistics (open use case of the Azure IoT platform[10]) transmits the data about the truck position, speed, accelerometer, and goods information to the cloud infrastructure. These logistics data are serialized as JSON messages, following the structure of SAREF ontology aligned to LogiCO ontology.[11] SAREF was chosen because of its capabilities for tracking devices' energy consumption. The IoT Platform Semantic Mediator (IPSM) module [22] is responsible for syntactically and semantically translating these data: from JSON and RDF/XML to the INTER-IoT JSON-LD syntax, which is structured JSON-LD (two @graph) with middleware information, and from SAREF to the INTER-IoT

[7]http://www.shimmersensing.com/products/ecg-development-kit.
[8]http://ontology.tno.nl/saref/.
[9]http://www.universaal.info/.
[10]https://azure.microsoft.com/en-us/campaigns/mydriving/.
[11]http://ontology.tno.nl/logico/.

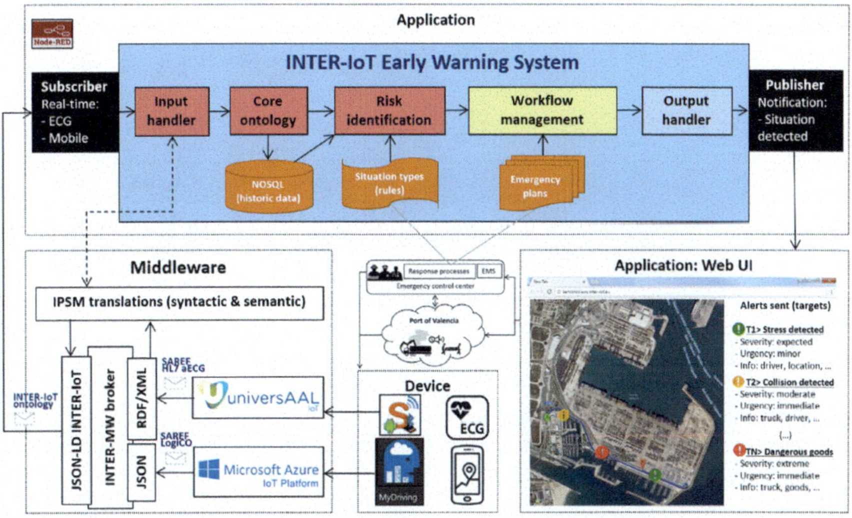

Fig. 3 EWS to detect accident risks and accidents at the port of Valencia

core ontology semantics, which is aligned with SSN. These translations are configured a priori in IPSM by the developer through a REST service.

The data represented as INTER-IoT JSON-LD syntax and INTER-IoT core ontology semantics are published in the broker in a topic, to which the EWS subscribes to receive real-time data. Then, the EWS input handler certifies whether new translations to harmonize the data in the SEMIoTICS core ontology are necessary and if so, the input handler requests the translations to IPSM.

Data are annotated with the core ontology and stored in a NoSQL database (MongoDB) as historical data. Both real-time data and historical data are used by the risk identification component, i.e., the SCENE CEP engine [20]. Situation types are defined a priori, as rule sets, describing the risky situations of interest based on emergency plans. Each situation type is linked to a response process, i.e., the specific workflow to be executed once a situation is identified. Therefore, the risk identification component triggers a workflow, which executes the linked processes.

The workflow component is responsible for checking the information requirements of each alert target, passing this information to the output handler, which is responsible for transforming the data to EDXL compliant semantically enriched messages. Therefore, the output handler enables the brokering of notifications of the situations detected, following the JSON-LD syntax and the EDXL data model, which is able to link to the used semantics. A web UI application shows each alert

Table 1 Data sources

External	Health	Logistics
Data	Driver's ECG, HR, accelerometer	Position, speed, accelerometer, goods
Device	Shimmer ECG 3 (capture), mobile	Mobile (MyDriving Android or iOS)
IoT platform	UniversAAL	MS Azure IoT
Ontologies	ETSI SAREF, HL7/aECG, FHIR	ETSI SAREF, LogiCO

sent by the EWS with its severity and urgency, and other information, including the targets that received the notification and the message sent to each target. The EWS is developed with Node.js and Node-RED. Table 1 summarizes the data sources used in the prototype.

4.3 Validation Plan

The validation plan of our solution follows the challenges *C1*, *C2*, and *C3* listed in Sect. 2.2 and is given in Table 2. It is organized as (a) factory acceptance tests (FAT): In a laboratory environment, the EWS is deployed in the cloud and the components integration tested through mock objects; and (b) site acceptance tests (SAT): a pilot in the port, where accidents will be simulated in accordance with the port emergency exercises. Both FAT and SAT assess whether the system works for the intended risks' detection and warning.

Since our approach is based on semantic translations, semantic loss at runtime will be used to calculate semantic interoperability, which will also be based on the semantic expressiveness of the EWS models (on specification level).

A comparison is included between our solution and a non-semantic approach for upstream data acquisition (from multiple devices), risks detection, and brokering. Thus, the plan includes the performance evaluation of data transfer, process, and brokerage. This plan includes data management according to the "Findable, Accessible, Interoperable and Reusable" (FAIR) data principles,[12] which enables research data to be reused.

Currently, the INTER-IoT-EWS is under implementation and testing phase. The initial execution and first results of A2 are presented in [19] and for A3 in [18]. Activities A1, A4, and A5 are ongoing. The first results of A4 show that using

[12]https://www.force11.org/group/fairgroup/fairprinciples.

Table 2 Validation activities

#	Activity	Description	Addresses
A1	Functional evaluation	Test cases with different levels of severity and urgency, checking the adherence with emergency procedures (pragmatic interoperability)	C1, C2, C3, FR1, FR2
A2	Semantic interoperability tests: semantic loss	Transformations in sequence from ontology A to ontology B and from B to A, i.e., check how x differs from $T(T(x)_{A>B})_{B>A}$, where $T(x)_{A>B}$ represents the semantic translation function from A to B [19]	C1, NFR1, NFR2
A3	Semantic interoperability tests: expressiveness	Specification level, i.e., how the models describe reality from different points of view [18]	C1, NFR1, NFR2
A4	Performance evaluation: data transfer	Compare JSON and JSON-LD as payload, measuring the impact of using JSON-LD rather than JSON, following the structure of the involved ontologies	C2, NFR3
A5	Performance evaluation: data process	Total time to translate; annotate data and insert into the database; access and process data for risk identification; and create alert messages (serialize as EDXL)	C2, NFR3
A6	Performance evaluation: data brokering	Scalability and resilience measured for single cluster and multi-broker, e.g., semantic IoT EWS approach [2], with sensor throughputs of up to 700 msg/sec	C2, C3, NFR3

JSON-LD brings an irrelevant burden to total transaction time when compared to JSON, and thus, migrating from JSON to JSON-LD is viable for the majority of the IoT solutions.

5 Conclusions

Current proposals for IoT-based EWS only partially address the semantic integration of a variety of data sources along with processing in time-critical applications and data analysis for effective responses. Our SEMIoTICS framework addresses this problem by applying different modeling languages, ontologies, and technologies toward the improvement of interoperability within and between IoT EWSs. To validate this framework, we are developing an EWS prototype, and we are currently applying it for detecting accidents at the port of Valencia.

Preliminary results include the INTER-IoT-EWS solution architecture, the required syntactic and semantic translations, and a validation plan guiding factory and site acceptance tests for measuring the interoperability support of the SEMIoTICS framework through the INTER-IoT-EWS. Initial tests indicate that the solution is adequate to cover the challenges, but this is an ongoing work to be reported in the near future.

The SEMIoTICS framework has been designed to be general enough to be applicable to other types of emergencies. However, the framework still lacks a mechanism to cope with the quality of information (QoI) at the network level, such as a Grubbs' test for outlier detection, and a statistical algorithm that can classify anomalous or invalid sensor values. Future work includes the development of a QoI mechanism and the completion of the execution of the validation plan.

Acknowledgements This work has been carried out under the CAPES funding BEX 1046/14-4 and EU-H2020-ICT grant INTER-IoT 687283.

References

1. UN. (2006). *Global survey of early warning systems: An assessment of capacities, gaps and opportunities toward building a comprehensive global early warning system for all natural hazards*. United Nations Report.
2. Poslad, S., et al. (2015). A semantic IoT early warning system for natural environment crisis management. *IEEE Transactions on Emerging Topics in Computing*.
3. Moreira, J. L. R., et al. (2015). Towards ontology-driven situation-aware disaster management. *Journal of Applied Ontology*.
4. Ganzha, M., et al. (2016). Semantic interoperability in the internet of things: an overview from the INTER-IoT perspective. *Journal of Network and Computer Applications*.
5. Comes, T., Mayag, B., & Negre, E. (2014). Decision support for disaster risk management: Integrating vulnerabilities into early-warning systems. In *Information systems for crisis response and management in mediterranean countries (ISCRAM-med)*.

6. Wächter, J., & Usländer, T. (2014). The role of information and communication technology in the development of early warning systems for geological disasters: The Tsunami show case. In *Early warning for geological disasters: Scientific methods and current practice*.
7. Alamdar, F., Kalantari, M., & Rajabifard, A. (2016). *Towards multi-agency sensor information integration for disaster management*. Environment and Urban Systems: Computers.
8. Raape, U., et al. (2010). Decision support for Tsunami early warning in Indonesia: The role of OGC Standards. In *Geographic information and cartography for risk and crisis management: Towards better solutions*.
9. Zambrano, A. M., et al. (2015). Sensor web enablement applied to an earthquake early warning system. In *International Conference on Internet and Distributed Computing Systems*.
10. Chronaki, C. E., et al. (2011). Interoperability in disaster medicine and emergency management. *Journal of Health Informatics*.
11. Ramar, K., & Mirnalinee, T. T. (2012). An ontological representation for Tsunami early warning system. In *IEEE-advances in engineering, science and management (ICAESM)*.
12. Barros, R., et al. (2015). EDXL-RESCUER ontology: An update based on faceted taxonomy approach. In *CEUR Workshop Proceedings*.
13. Middleton, S., et al. (2013). The seven main challenges of an early warning system architecture. In *Information systems for crisis response and management (ISCRAM)*.
14. Moreira, J. L. R., et al. (2016). Improving semantic interoperability of big data for epidemiological surveillance. In *I-ESA, BDI4E Workshop*.
15. Fertier, A., et al. (2016). Use of big data for continuous interoperability in crisis management. In *Enterprise Interoperability VII: I-ESA Proceedings*.
16. Mačinković, D., & Aničić, N. (2016). The systems development life cycle to facilitate progression towards semantic and organizational interoperability for healthcare system. In *Enterprise Interoperability VII: I-ESA Proceedings*.
17. Endsley, M. R. (1995). Toward a theory of situation awareness in dynamic systems. *Journal of Human Factors*.
18. Moreira, J. L. R., et al. (2017). Ontology-driven conceptual modeling for early warning systems: Redesigning the situation modeling language. In *MODELSWARD*.
19. Moreira, J. L. R., et al. (2017). Towards IoT platforms' integration: Semantic translations between W3C SSN and ETSI SAREF. In *Semantics. Workshop SIS-IoT*.
20. Costa, P. D., et al. (2016). Rule-based support for situation management. In *Fusion methodologies in crisis management: Higher level fusion and decision making*.
21. INTER-IoT. (2016). *INTER-IoT deliverable: D2.4*. Use cases manual.
22. Ganzha, M., et al. (2017). Streaming semantic translations. In *2017 21st international conference on system theory, control and computing (ICSTCC)*.

Software-Defined Networking to Improve Cybersecurity in Manufacturing Oriented Interoperability Ecosystems

Francisco Fraile, José Luis Flores, Raúl Poler and Eduardo Saiz

Abstract Industry 4.0 is reshaping the manufacturing industry. Through Industrial Internet of Things (IIoT) and cloud computing, manufacturing orientated interoperability ecosystems allow companies to reinvent the use of manufacturing data in value creation. Cybersecurity is a critical aspect in the design of these interoperability platforms to ensure safety in manufacturing operations. Software-Defined Networking (SDN) allows to control the network architecture and behavior in a programmatic way, thus enabling innovative cybersecurity concepts that can be applied to manufacturing orientated interoperability ecosystems. The Virtual Factory Open Operating System (vf-OS) is an innovative multi-sided platform designed to enable collaboration between manufacturing companies. The vf-OS Holistic Security and Privacy Concept incorporates the latest standards and technologies to enable interoperability with industrial control systems (ICS) in the vf-OS ecosystem. This paper uses this state-of-the-art security concept to describe the role of SDN in securing manufacturing oriented interoperability ecosystems and presents an innovative proposal to further improve cybersecurity using SDN technology.

Keywords Privacy and security in enterprise interoperability · Platforms and infrastructures for enterprise interoperability · Interoperability in industry 4.0

1 Introduction

After the three industrial revolutions brought by steam engines, electricity, and automation, the concepts and technologies of the fourth revolution, coined Industry 4.0, are helping manufacturing companies to reinvent themselves, increasing productivity or shifting to new business models in order to remain competitive [1]. Among these concepts and technologies that are reshaping the manufacturing

F. Fraile (✉) · R. Poler
Universitat Politècnica de València, 46023 Valencia, ES, Spain
e-mail: ffraile@cigip.upv.es

J. L. Flores · E. Saiz
IK4-Ikerlan Technology Research Centre, 20500 Guipuzkoa, Spain

K. Popplewell et al. (eds.), *Enterprise Interoperability VIII*, Proceedings of the I-ESA Conferences 9, https://doi.org/10.1007/978-3-030-13693-2_3

Fig. 1 vf-OS Platform concept

industry, Industrial Internet of Things (IIoT) [1] and cloud computing allow manufacturing companies to leverage the potential of production data far beyond the possibilities of any traditional manufacturing execution system (MES) [2]. These technologies enable the creation of data streams along the entire production process and among supply chains that can be used to monitor manufacturing assets and/or value streams in real time. In combination with data analytics, machine learning, and artificial intelligence, they provide the basis for gaining predictive insights into any supply chain process [1], from customer relationship management, through manufacturing flow management, to returns management. Moreover, bringing together these cybernetic systems and the physical world through actuators, cobots, and other smart devices that can sense and interact with the physical world—comprising what is known as cyber-physical systems [3]—provides endless possibilities for manufacturing. Some of the applications of these technologies are advanced fault detection for maintenance operations, waste reductions for lean manufacturing operations or support to collaborative manufacturing.

The Virtual Factory Open Operating System (vf-OS) Platform [4] is a multi-sided platform orientated to manufacturing and logistics companies that exploit these technologies through a range of services to integrate better manufacturing and logistics processes within organizations and among supply networks. These services are instantiated by applications (vApps) developed by independent software developers and available within a given vf-OS Platform installation, which can be hosted in the cloud. The vf-OS Platform concept is shown in Fig. 1.

The value proposition for the different customer groups of the multi-sided platform is clear:

- Manufacturing users can select and use vApps from the marketplace to integrate manufacturing and logistics processes, enabling collaboration in the value chain. If they do not find any applications suitable for their needs they can demand the development of new solutions to software developers.
- Software developers gain access to a new and high-growth potential market of applications for Industry 4.0 and the Factories of the Future.
- Manufacturing providers delivering products and services to manufacturing users have new ways to collaborate and interact with their customers and provide added value.
- Service providers can provide new services (hosting, storage, cloud services, etc.) to realize the vf-OS ecosystem.

Within the vf-OS system architecture, there are certain components designed to interact with all kinds of manufacturing assets, both physical devices (e.g., PLCs or sensors) and business software applications (e.g., ERPs or CRMs). These components, known as Input–Output (IO) Components, confirm the Virtual Factory I/O (vf-IO) and implement interoperability mechanisms specifically oriented to manufacturing processes.

In order to materialize this, it is necessary to implement holistic cybersecurity and privacy concept covering all the interactions in the multi-sided platform shown in Fig. 1. Securing the use of vApps that interact with industrial control systems (ICS) is a particularly sensitive issue, given the criticality of industrial control in manufacturing processes. This paper describes the specific security standards, specifications, and technologies used in the design of the vf-OS holistic cybersecurity concept to enable secure interoperability with manufacturing assets. Later, it describes a series of security solutions based on Software-Defined Networking (SDN) [5] that can be used to implement and enhance the proposed network layer security mechanisms. Furthermore, SDN technology allows to control the behavior and the topology of the network in a programmatic way and represents a new paradigm in networking that can bring many benefits to manufacturing oriented interoperability ecosystems such as vf-OS.

2 vf-OS Holistic Security and Privacy Concept

2.1 Thread Model and Response

vf-OS provides a flexible infrastructure to develop and deploy applications that act as interoperability mechanisms between manufacturing assets within organizations and among supply networks. Flexibility is achieved thanks to the microservices [6] architecture of the vf-OS Platform: vApps are built on top of a series of heterogeneous, potentially distributed, components that provide well-defined web services.

Protecting and preserving this infrastructure is a significant challenge since it concentrates all classical security requirements (confidentiality, integrity, and availability) in a challenging environment with a multi-fold threat model:

- Public Services: The vf-OS marketplace is a public web service on the Internet providing applications, components, identities, and services. Every web service on the Internet is subjected to any kind of threat, and the attack surface comprises a significant list of elements (information gathering, configuration management, data in transit, authentication, etc.). Moreover, attackers have the target permanently available to test different attack techniques. In order to address this threat, the main reference for vf-OS is the Open Web Application Security Project (OWASP) [7] initiative security guidelines and recommendations to secure web services, which are the core of vf-OS technology.
- The vf-OS internal microservices need to be deployed in factories, implementing interoperability mechanisms with industrial control systems (ICS) as well as with other services. This represents a very critical environment from a security perspective, since attacks to ICS may not only cause great economic losses to manufacturing companies but also put in risk the safety of operators. The security architecture presented in the next section has been designed according to the ISA/IEC-62443 [8] security standard for ICS. In the terminology of IEC 62443, vf-OS is installed as an industrial device, following the same flow as any certificated industrial device. In addition to this, vf-OS implements a centralized security management system based on the Security Content Automation Protocol (SCAP) [9] to enable automated vulnerability management, measurement and policy compliance evaluation of heterogeneous and complex systems, facilitating the alignment with OWASP and the management of IEC-62443.
- Software developers use a public development kit, i.e., the vf-OS Open Application Kit (vf-OAK) to develop vApps. This implies the possibility of introducing bugs and malware which can affect public services and/or internal services. The OWASP Secure Coding Practices Reference Guide and Checklist [7] are implemented in the vf-OAK to ensure that security aspects are taken into consideration in the development process. Additionally, as described in the next section, the security architecture implements a public key infrastructure (PKI)-based system to identify developers and several mechanisms to guarantee security during the application life cycle.
- Personal data protection: Multi-party architectures need to ensure the privacy of user personal data. The recent adoption of the GDPR [10] and the future launching (May 25, 2018) of the new regulation, implying mandatory enforcement, has a direct impact in how vf-OS manages personal data. vf-OS privacy has been designed with data protection by design and by default. vf-OS implements mechanisms to protect personal information (such as pseudonymization), inform of possible data breaches, ask for explicit consent when required, and provide records of processing activities from the vf-OS permanent logging service.

2.2 Security Architecture

Figure 2 illustrates the vf-OS Security Architecture which provides interoperability mechanisms compliant with the ISA/IEC-62443 secure ICS network architecture. This standard applies a defense-in-depth strategy where the network is divided into zones or segments according to the functionality of the systems connected to the network. The Enterprise Level 1 zone contains the vApps and other services that can be accessed through the Internet, like the enterprise content management system (ECM).

The Enterprise Level 1 is considered a second demilitarized zone. The Enterprise Level 0 zone connects other corporate services that can be accessed across the entire corporate network, like the customer relationship management (CRM) or enterprise resource planning (ERP) software. The Industrial Zone Level 2 interconnects the operative systems like the SCADA or MES Systems. The Industrial Zone Level 1 and 0 interconnect the most critical industrial control network components, PLCs, and field components like IO modules and sensors. Redundant switches interconnect systems in each zone. Firewalls implement filtering rules to limit network access between the different segments so that only allowed connections between levels can be established. This way, an attacker willing to penetrate the ICS will meet the different Firewalls preventing further access across the network and protecting critical industrial control components.

The microservice architecture adopted in vf-OS allows to deploy each vf-OS component at the optimal network segment according to the ISA/IEC-62443 defense-in-depth strategy. The Enterprise Level 1 hosts vApps that provide interoperability mechanisms to enable collaboration among supply chain networks, as well as external services, which are components that exchange information with cloud services. The Enterprise Level 0 connects the vf-OS Kernel Services, proving the core vf-OS services and the API connectors which provide interoperability mechanism with corporate software systems and applications. On the other hand, device drivers are located within the Industrial Zone Levels, depending on how they interact with automation equipment. Device drivers can implement industrial communication protocols like OPC UA [11] to interact with control and field components from the operation segment or can be embedded into PLCs or smart sensors.

vf-OS introduces a new segment in the Enterprise Level for the vf-OS Security Command Center, which is the vf-OS component that implements the authentication and authorization services that orchestrate the entire security concept. The Command Center implements a role-based access control–attribute–based access control (RBAC-ABAC) [12] model to implement access control and restrict access to data. The RBAC-ABAC model combines the best features of RBAC with attribute-based systems to improve security for distributed and rapidly changing applications. In this model, security policies [8] determine which manufacturing assets can be accessed

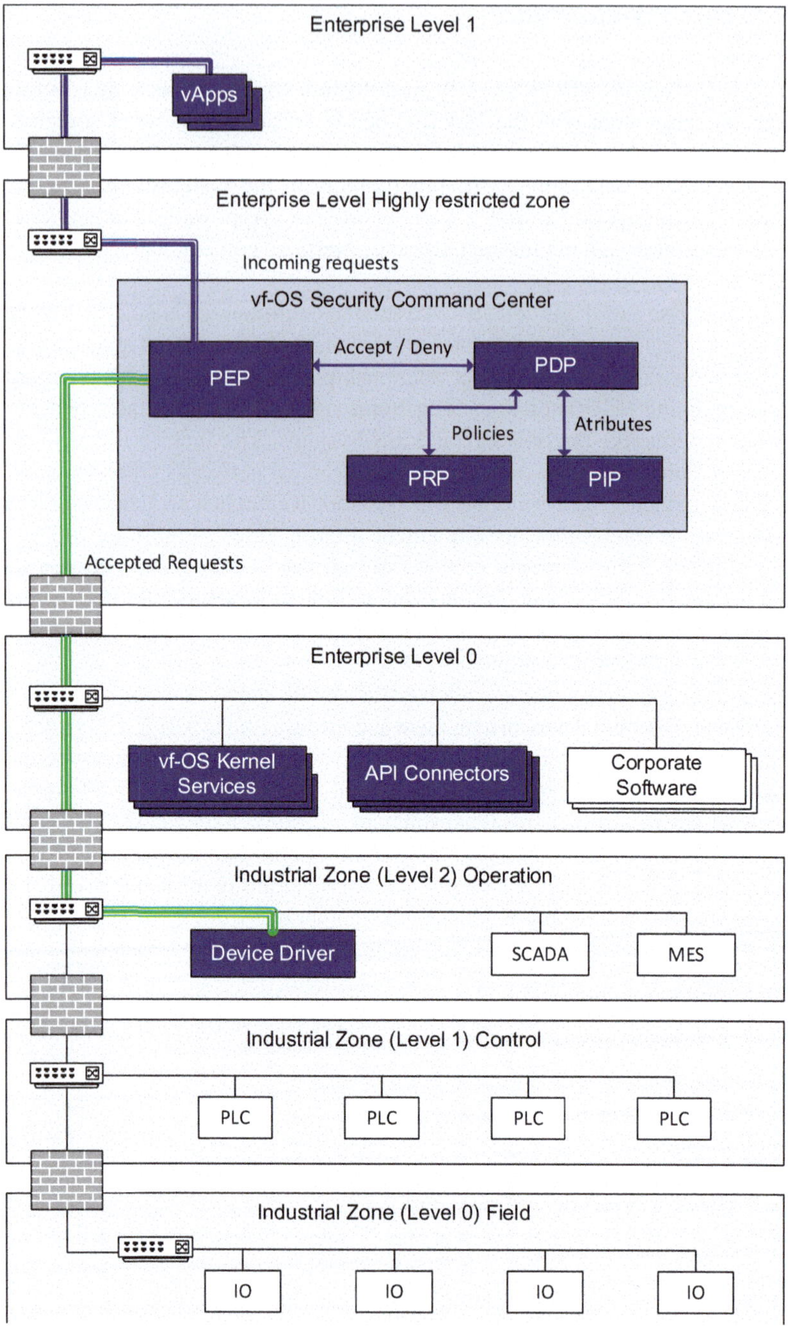

Fig. 2 vf-OS Secured Architecture

and what are the operations allowed (e.g., read or read/write). Defined security policies are later assigned to user groups in the RBAC-ABAC model. The Security Command Center acts as a security proxy. The internal Policy Enforcement Point (PEP) intercepts every request made by vApps and forwards it only when it complies with the access rules established by the security profile. The forwarding decision is made by the Policy Decision Point (PDP) which retrieves the policies and additional attributes, respectively, from the Policy Retrieval Point (PRP) and Policy Information Point (PIP) components. The Security Command Center also provides RESTful APIs to control the security model so that it can be flexibly adapted to every use case. The vf-OS Holistic Security and Privacy Concept is compatible with other open platforms such as FIWARE [13], and it is possible to securely integrate components from these external platforms (enablers).

Firewalls, on the other hand, restrict network access based on rules that determine which connections are allowed or restricted in each network segment. State-of-the-art application Firewalls implement application control by means of in-depth packet inspection to apply filters based on application layer rules. For instance, they are able to determine HTTP traffic and block connections, but they have no knowledge of ad hoc REST APIs, and therefore, they cannot apply more sophisticated security restrictions. This means that in the security architecture, vf-OS implements application layer security mechanisms, the underlying protocols (HTTPS) implement communication layer mechanisms, whereas Firewalls implement network layer security mechanisms. However, it is important to note that these mechanisms at the different layers are loosely coupled.

This is not the case if SDN technology is introduced in the architecture, since it allows to integrate switches and Firewalls in the holistic security concept, coordinating security decisions at the different layers to provide enhanced security, performance, and flexibility. Next section describes the SDN technology in the context of security for interoperability platforms as a means to achieve this integration.

3 SDN Network Layer Security

As explained above, the secured network architecture of ISA/IEC 62443 standard in Fig. 2 proposes the use of Firewalls to limit the exposure of industrial control devices to Internet attacks with a defense-in-depth strategy.

With this architecture, an attacker pretending to intrude the Industrial Zone (Level 1) Control network segment needs to attack each Firewall from the Enterprise Level 1 to gain access to the successive network segments. Basically, this consists of figuring out how to generate traffic that meets the security filtering rules programmed into each Firewall and find means to use this information in order to perform another attack to the next level. Application control allows to deploy application layer filtering rules so that only traffic matching specific application level rules can access the network. This makes it possible to integrate network security rules with application layer security mechanisms to some extent. The main limitation is the processing capacity of the Firewall to read complex context information. For instance, Firewalls are able to detect HTTP request for every method but they are not able to implement dynamic filtering rules adapted to the business logic of a specific REST API.

Conversely, SDN technology allows to implement this kind of advanced filtering techniques in order to mitigate security risks. SDN Firewalls were introduced in [5] SDN Firewalls, whereas [14] presents a SDN Firewall specifically designed for Industry 4.0. Basically, a SDN Firewall allows an external controller to modify the forwarding tables that define the behavior of the network. This not only means that the controller can configure the bridges between network segments but also to control the security configuration of every network interface. This takes the defense-in-depth strategy of ISA/IEC 62443 to a new level, since each network boundary faces other system boundaries through secured interfaces.

This concept of protective network structure for manufacturing systems with a specialized SDN Firewall based on the OPC UA and OpenFlow standards is presented in [14]. The SDN has two main functionalities: first, to detect automation devices automatically and group them into the appropriate network segment, and second, to integrate OPC UA application layer security mechanism into the SDN Firewall controller. This way, administrators need only to allow access between OPC UA server and clients. The SDN Firewall controller uses this information to set up the filtering rules applied at each network interface in the industrial zone.

The same concept can be applied in vf-OS to extend the protective network structure to the entire network and to integrate all the security mechanisms at the different layers. This concept is shown in Fig. 3. When a vApp wants to connect to a vf-OS component, the request is forwarded to the vf-OS Security Command Center. The PDP determines whether the connection is accepted or not depending on the defined security policies and attributes. The PEP enforces the decision, but this time, by controlling the approved physical connections in the SDN network, instead of acting like a proxy. Furthermore, more attributes can be defined at the device driver level to integrate specific industrial protocol security rules into the RBAC-ABAC model.

This way, the configuration and control of all the security mechanisms are integrated and centralized in the vf-OS Security Command Center. The network security configuration is dynamic so that connections are only allowed when vApps require them (temporal filtering) and only between the required network components (spatial filtering) which makes it harder for attackers to learn what kind of traffic meets the security rules.

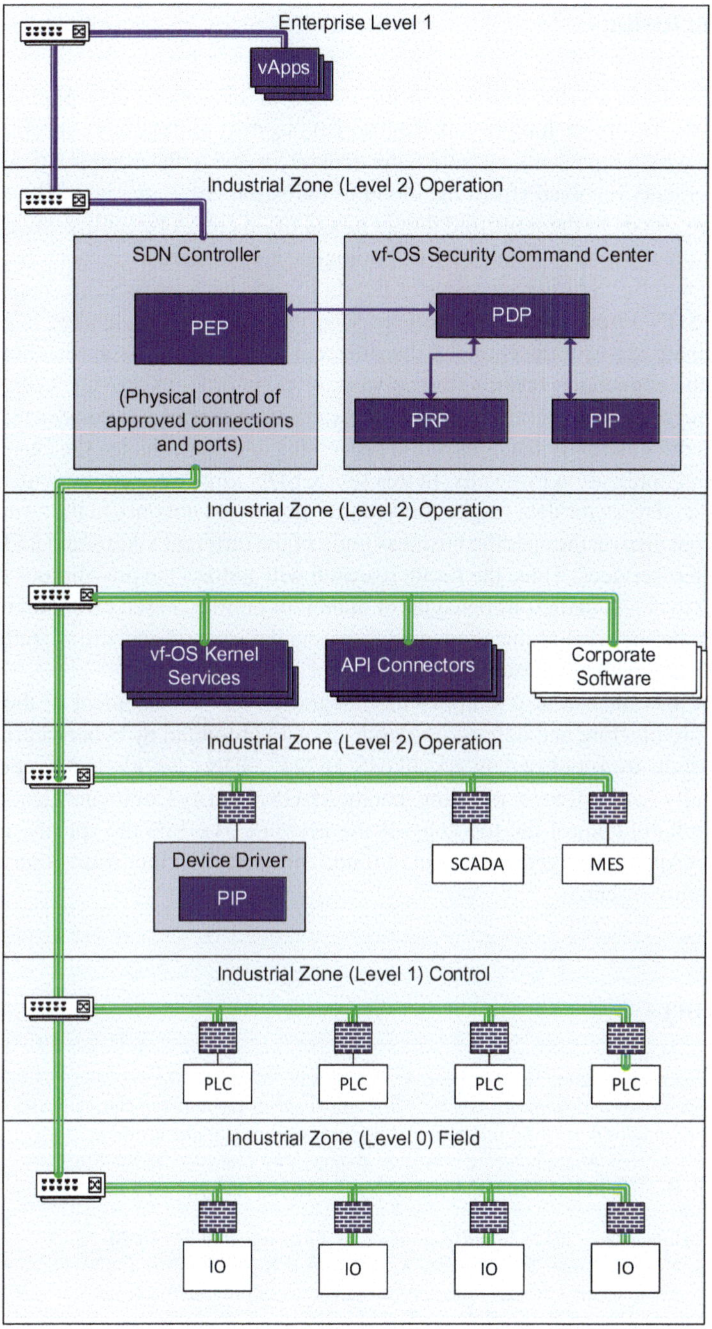

Fig. 3 vf-OS Security Concept with SDN Firewalls

4 Conclusions

Cybersecurity is a critical aspect in the design of interoperability platforms and ecosystems. This paper has presented the holistic security and privacy concept applied in vf-OS, which represents a challenging ecosystem due to the complexity and variety of elements involved (from the cloud to industrial environments). The security concept responds to the multifold thread model of this interoperability environment applying the most modern security technologies and standards.

In order to further improve security, flexibility, and performance, this paper introduces a SDN Firewall into the holistic security concept. This makes it possible to coordinate security decisions at the different layers, from the application layer, through the connection layer, to the network layer, in order to provide a centralized security response to cyberattacks in interoperability platforms and ecosystems. SDN Firewalls are inherently faster at traffic processing and can enhance the current possibilities of application Firewalls. In this sense, SDN Firewalls can implement more sophisticated rules for data inspection and filtering, based not only in the structure of packets, but also on the specific business logic of the different vApps and underlying REST microservices. Thus, the future research will address the possibilities of SDN Firewalls to detect malicious behavior or malicious entities, based on context knowledge (e.g., knowledge of installed manufacturing devices, vApps, users), rather than just analyzing isolated network data packets.

SDN Firewalls can also simplify the integration and management of the secure network architecture and defense-in-depth strategy promoted by cybersecurity standards for industrial control networks like ISO 62443. SDN networks can be controlled dynamically with software, meaning that the vf-OS Security Command Center could also potentially control the topology of the network based on the specific requirements of vApps at any given moment of time. This is another interesting line of work for the future research.

References

1. Kagermann, H., Helbig, J., Hellinger, A., & Wahlster, W. (2013). *Recommendations for implementing the strategic initiative INDUSTRIE 4.0: Securing the future of German manufacturing industry; final report of the Industrie 4.0 Working Group*. Forschungsunion.
2. Kletti, J. (2013). *Manufacturing execution system–MES* (1st ed.). Berlin: Springer.
3. Baheti, R., & Gill, H. (2011). Cyber-physical systems. *The Impact of Control Technology, 12*, 161–166.
4. vf-OS Homepage. (2017). http://www.vf-os.eu. Last Accessed 31 Oct 2017.
5. Nunes, B., Mendonca, M., Nguyen, X., Obraczka, K., & Turletti, T. (2014). A survey of software-defined networking: Past, present, and future of programmable networks. *IEEE Communications Surveys & Tutorials, 16*(3), 1617–1634.
6. Nadareishvili, I., Mitra, R., McLarty, M., & Amundsen, M. (2016). *Microservice architecture: Aligning principles, practices, and culture* (1st ed.). Sebastopol: O'Reilly Media Inc.
7. OWASP Homepage. (2017). https://www.owasp.org/index.php/Main_Page. Last Accessed 31 Oct 2017.

8. International Electrotechnical Commission (IEC) TS 62443-1-1 ed1.0. (2009). *Industrial communication networks—network and system security—part 1-1: Terminology, concepts and models* (pp. 7–83).
9. Radack, S. M. (2010). Security content automation protocol (SCAP): Helping organizations maintain and verify the security of their information systems. *ITL Bulletin*.
10. GDPR Homepage. (2018). https://www.eugdpr.org/. Last Accessed 15 Jan 2018.
11. OPC Foundation. (2010). *OPC unified architecture specification part 2: security model*.
12. Kuhn, D. R., Coyne, E. J., & Weil, T. R. (2010). Adding attributes to role-based access control. *Computer, 43*(6), 79–81.
13. FIWARE Homepage. (2017). https://www.fiware.org/. Last Accessed 31 Oct 2017.
14. Tsuchiya, A., & Fraile, F., Poler, R., & Ortiz, A. (2017). Software defined networking firewall for OPC UA MES systems. In *11th International Conference on Industrial Engineering and Operations Management, Springer, Valencia*.

Part II
I 4.0 and Industrial Automation

Industry 4.0 Out of the Box

Patrick Gering and Patrick Drange

Abstract The digital revolution results in changes in the production and value-added processes, which is a challenge especially for SMEs who cannot keep up with the rapid pace of technology development. This is mainly due to the lack of know-how and lack of strategies to lead one's own company into the fourth industrial revolution as well as high investment costs in industry 4.0 solutions with unclear benefits. The increased amount of data in production systems together with the increasing networking of all systems offer new possibilities for quality assurance and quality management, though. The main objective of Industry 4.0 out of the box is to enable manufacturing SMEs to carry out the individual tracking of value creation with minimal planning and implementation costs. Industry 4.0 out of the box contains all necessary hardware and software for flexible tracking of production and testing parameters at part level and the collection of order-specific parameters without disrupting normal production.

Keywords Industry 4.0 · Production monitoring · Quality management

1 Industry 4.0 for SMEs—Industrial Demands

No other topic currently determines German industry as intensively as Industry 4.0 and exerts greater influence. Industry 4.0 was launched by the Federal Government to secure and, at best, expand the leading position of German industry in the world's leading industrialized countries in the medium and long term [1].

The digital revolution results in changes in the production and value-added processes, which is a challenge especially for SMEs. This is mainly due to the lack of know-how and lack of strategies to lead one's own company into the fourth industrial revolution.

P. Gering (✉) · P. Drange
Fraunhofer IPK, Pascalstraße 8-9, 10587 Berlin, Germany
e-mail: patrick.gering@ipk.fraunhofer.de

K. Popplewell et al. (eds.), *Enterprise Interoperability VIII*, Proceedings of the I-ESA Conferences 9, https://doi.org/10.1007/978-3-030-13693-2_4

SMEs cannot keep up with the rapid pace of technology development. Further obstacles are high investment costs (34%) and lack of technology standards (31%), as well as a lack of know-how and resources (19%) [2].

The increased amount of data in production systems together with the increasing networking of all systems offer new possibilities for quality assurance and quality management, though. Starting with the acquisition, analysis and evaluation of process data in real time, assistance systems offer the possibility of intuitive and efficient control of production processes based on networked data sources [3, 4]. In the long term, it will be possible to optimize production machines in real time based on a fully networked society and production chain.

Existing manufacturing process management and control systems like manufacturing execution systems (MES), often promised as the solution towards Industry 4.0, are widely applied. The implementation of a manufacturing execution system incorporating business processes and manufacturing processes is still challenging and expensive. Implementation often takes more than half a year, even with model-based specifications defined by Bajric and Jäkel [5] and is affected by organizational as well as technical interoperability barriers [6, 7].

Experiences from different research projects like MetamoFAB [8] and JUMP 4.0 [9] have shown that existing production planning systems need too much time to implement and are, therefore, too capital-intensive for SMEs. They are too rigid with regard to new individual processes and the adaptation is time-consuming.

Furthermore, old machinery parks do not offer state-of-the-art interfaces or interfaces at all to evaluate machine date. Standards, technology maturity, robustness, opportunities and benefits as well as consequential costs of industry 4.0—technologies are not transparent for companies, especially SMEs. Even more, different expectations regarding the degree of system support and its autonomy exist within companies. Too much information technology and autonomous systems are seen as a limitation in flexibility and the benefits are unclear. The lack of skilled personnel in SMEs for selection and mastering of new IT systems is another hurdle. Therefore, they are not willing to invest. The aim of 'Industry 4.0 out of the box' is to network quickly and ad hoc machines and systems via external sensors to plan and track individual orders.

This paper focuses on the demand of Industry 4.0 solutions for SMEs, the Idea of Industry 4.0 out of the box as well as the quality services for sensor data evaluation.

2 The Idea of Industry 4.0 Out of the Box

Industry 4.0 out of the box offers manufacturing SMEs a solution that enables them to deploy cyber-physical systems with minimal prior knowledge and low planning, implementation and investment costs. Industry 4.0 out of the box contains all necessary hardware and software for flexible tracking of production and testing parameters at part level and the collection of order-specific parameters without disrupting normal production. A model-based approach ensures easy and flexible implementation and configuration, no previous knowledge of industry 4.0 is required.

In detail, Industry 4.0 out of the box aims to solve the following problems relevant for SMEs:

1. **No fast reaction to customer requests possible**: SMEs are often not able or just with substantial effort to provide requested data such as a detailed status within the production process or specific production parameters. Industry 4.0 out of the box is developed as a modular-based system and easily expandable with additional sensors to collect production parameters and adaptable to new production processes. The follow-up costs are transparent, because Industry 4.0 does not need interfaces to external systems.
2. **No flexible tracking of the production process possible**: Industry 4.0 out of the box supports customer order-specific production processes. The production process can be modelled customer order based. The relevant sensors and corresponding key figures can be assigned individually for each customer request. This ensures a fast response and query of key figures at any time and leads to a reduction in production, tracking and monitoring costs.
3. **No meaningful key figures for quality assurance are available**: A networked productions system with production and machine data acquisition is rarely available. Even in fully automated, industry-capable production factories, there is a lack of information on the processes and machine/plants for completeness. However, these are urgently needed for further optimization and the elimination of errors. Industry 4.0 out of the box collects data from included or added sensors and provides the data for monitoring and tracking as a dashboard as well as data evaluation with quality assurance services.
4. **Lack of expertise in digitally integrated production**: Industry 4.0 out of the box enables companies to gain experience with the new methods without prior knowledge. On this experience, own qualified requirements for industry 4.0 solutions can be formulated and professional competence expanded. SMEs do not lose the connection to technological developments and can recognize opportunities for their own company.
5. **Existing solutions are too expensive, extensive and complex**: The upgrade to a fully networked production system is expensive, extensive and complex for SMEs, especially without a clear understanding of the benefits. Industry 4.0 out of the box is an easy to install add-on to existing production systems. It does not interfere with existing systems and promises minimal planning, implementation and investment costs.
6. **Training costs are too high**: Industry 4.0 out of the box includes wizard-based installation and configuration instructions without dependencies to existing systems. This ensures a low learning curve, but still demonstrates basic principles of a networked production system.

3 Architecture

The architecture of Industry 4.0 out of the box as seen in Fig. 1 consists of four main components:

1. MO^2GO Model Database (MO^2GO Model DB)
2. Cyber-Physical-Connector (CP-Connector)
3. OpenIoTFog
4. Industry Cockpit.

With the integrated enterprise modelling (IUM) approach, MO^2GO is a comprehensive modelling solution that has already been tried and tested in practice. In comparison to BPMN, entire company structures can be mapped in detail using simple means, in addition to business and production processes. For this purpose, all relevant objects of a company as well as their properties and relations are mapped in the information model and divided into the IUM object classes 'Product', 'Order' and 'Resource' according to the principles of object-oriented modelling [10]. In addition, the IUM conforms to the international standard ISO 19440 (constructs for enterprise modelling) [11], which ensures compatibility with other methods and tools. In Industry 4.0 out of the box, MO^2GO is used to model the production process to be monitored. To achieve that, the separate production steps as well as product states are modelled with MO^2GO. The included sensors in Industry 4.0 out of the box are connected as resources to the different process steps accordingly (see Fig. 2). Industry 4.0 out of a box comes with a pre-defined set of standard production steps in as a model library and can be extended easily. All the sensors available in Industry out of the box are present as a resource and can be assigned to a process steps. A stepwise installation instruction guides the user through the installation and configuration process of the sensors.

As a middleware for sensor data collection, the Internet of Things platform OpenIoTFog is used. Based on concepts from consortia such as the Industrial Internet Consortium (IIC) [12] and the Industry 4.0 platform [13], the plug-and-play framework represents a so-called asset administrative shell to integrate existing physical, industrial systems in factories of the future as a digital twin. The plug-and-play framework collects data from various sensors via industrial fieldbus systems and various (industrial) wireless technologies and standardizes, communicates and aggregates them via secure standards-based interfaces [14]. In Industry 4.0 out of the box, OpenIoTFog Collects the sensor data in a database for further evaluation.

The CP-Connector is the core software component of Industry 4.0 out of the box and aims to bring Industry 4.0/Internet of Things capabilities to existing production processes and machines. Information acquisition, enrichment and presentation are the goal of the software architecture. The middleware is responsible for recording of actual values of the sensor data. The CP-Connector is the control unit for the coordination of sensor data. The CP-Connector enriches the recorded data by linking them with the associated processes, product states and customer orders. It is the binding element of the physical and cyber world. In Industry 4.0 out of the box, the

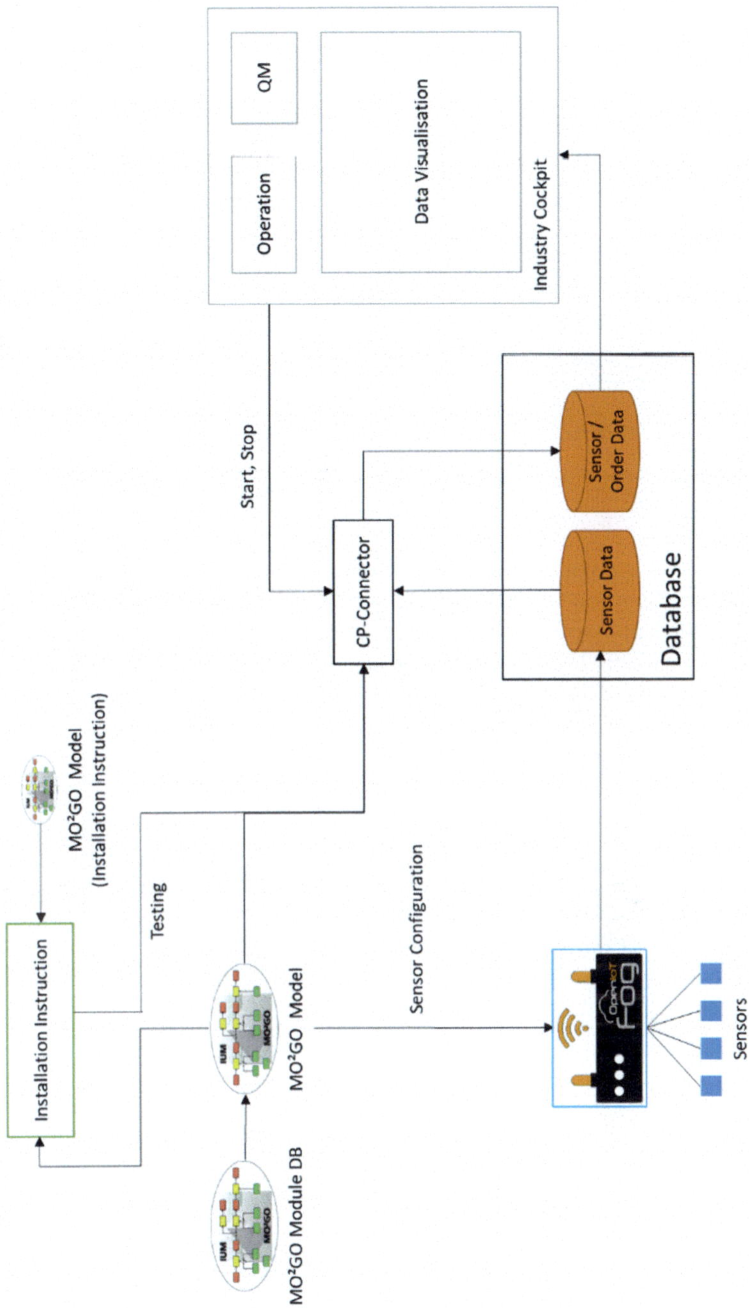

Fig. 1 Architecture

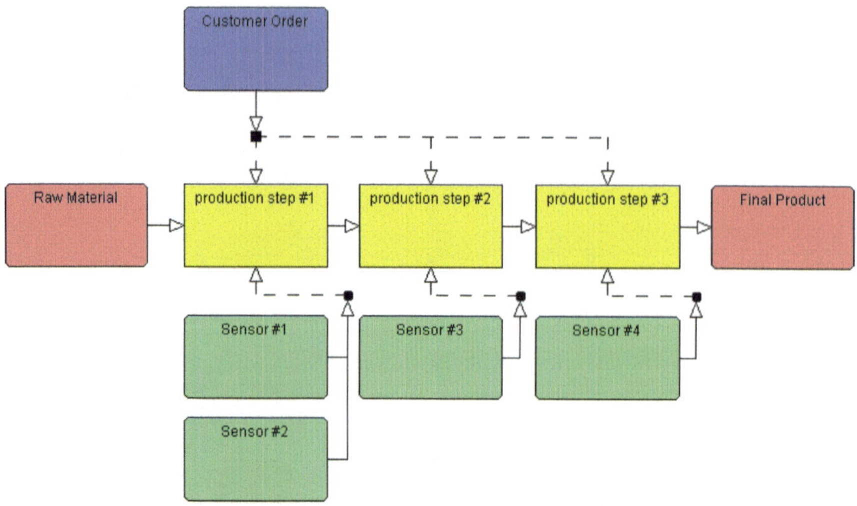

Fig. 2 Production process

CP-Connector is used without controlling functions, only information acquisition and enrichment. It generates a tuple for unambiguous assignment of information: (SensorId, SensorTypId, sensor value, sensor unit, time, process, sub-process and order/product, machine). The CP-Connector separates the acquisition of the actual values from the setting of the controller's setpoint values. This has the advantage that different real-time conditions can be set. This also results in independence, which is to some extent desirable in production.

The industry cockpit presents an overview of collected sensor data and order status and is used to control the CP-Connector. Via a Web-based user interface developed for mobile devices with different screen sizes order, numbers for customer orders can be generated or assigned if generated in already available order management systems. Within the industry cockpit, the order states can be set process step based (i.e. production step #1 fulfilled) on the modelled production process in MO^{2}GO.

4 QS Services

With the help of QS Services, a first step towards the vision of a self-optimizing production process is made possible. On the basis of production-relevant data, which can be collected with the help of sensors from industry 4.0 out of the box, it is possible to carry out statistical and quality-relevant analyses. After each input, knowledge is generated, archived and included in future decision-making processes. This results in a self-learning system that is constantly evolving.

Accordingly, QS Services is to be understood as an assistance system, which assists the user, e.g., the operator (a quality engineer), in the analysis of process data. The application of the services will be possible with the help of different technologies (devices) so that the use of these services at different locations and machines will not be an obstacle. With the help of QS Services, it will be possible to efficiently read out, analyse and understand process data based on all inputs (type of error on the product, limit values, cutting values, etc.) and to suggest measures. This gives the responsible quality engineer or machine operator the opportunity to adjust the necessary process settings at an early stage and actively influence quality in a positive way.

The methodology of QS Services is presented below on the basis of an example.

A company regularly detects qualitative defects in the product. This has already led to increased complaints in the past. Despite intensive root cause research carried out by the quality management department and changes in process parameters, the problem continues to occur continuously. In this situation, QS Services can help the company identify causes and solutions.

After the implementation of all process-related sensors from the 'Industry 4.0 out of t he box', one or more products are remanufactured. Physical quantities such as temperature, speed, force or pressure are measured. The sensor data thus determined is processed and made available to the QS Services. For efficient analysis, the person responsible is asked to provide information on the nature and extent of the existing quality problem after starting the program. By entering further parameters, such as tolerance limits, cutting values or similar parameters, QS Services will be positive influenced on the algorithms, so it is, therefore, possible to propose specific measures for improvement.

During the development of QS Services, experimental tests were conducted to determine and correlate significant influencing factors on product quality. The algorithms developed with the aid of statistical methods form the basis of the analyses in QS Services.

Once the problem has been successfully described with regard to product quality and additional process parameters have been entered, the analysis of the relevant process data starts. Within a few seconds, a result is presented within QS Services. All process data including all tolerance limits are displayed to the user as a graphic. In addition, conspicuous values are highlighted. In order to support the quality manager in his work, it will be possible to receive recommendations for quality improvement in the 'Measures' tab. This gives them the opportunity to solve the problem promptly.

Since there is a general requirement for the recommended measures to have a generally valid character, it will be necessary, despite all recommendations, that the quality engineer of the future be critically scrutinized in cooperation with experts and working together on the elimination of the quality problem. QS Services support them in the process of identifying causes and solutions. The results of this are finally documented in QS Services, so similar problems can be detected more quickly in the future on the basis of this data and prevented. This means that on the basis of the specified product quality problems, a very precise evaluation can be presented and specific measures can be recommended accordingly.

Thus, the development of QS Services offers the quality engineer an assistance system that comprehensively considers all aspects in the area of cause identification with regard to product quality problems. Based on the presented process, it becomes clear how far quality managers and inspectors as well as experts can be supported with this tool to improve the product quality. Due to the possibility of being able to intervene in the process at an early stage, organizations can save high costs due to rework, complaints or long fault clearance processes.

In the long term, the aim is to proactively predict quality problems, especially in the area of QS Services. The assistance system describes a first step towards predictive quality. Based on the results of the current research work, it will, therefore, be possible to implement the acquired knowledge and further develop it within the framework of further projects. In order to achieve this, it is essential for reliable predictions of product quality to apply the new findings in the field of data science in the production environment and to continuously improve the related technologies.

In particular, the approach of a holistic and universally valid view can only be carried out on the basis of a comprehensively networked, digitized world. In the future, this will give quality engineers the opportunity to intervene in the control of parameters already during process processing. The quality of the product can thus be positively influenced already during machining.

All in all, the QS Services offer SMEs the opportunity to record and analyse process data efficiently and cost-effectively and to derive measures from this data. This added value will, therefore, have a significant impact on competitiveness.

5 Conclusion

Both globally and regionally operating companies face new challenges due to the fourth industrial revolution. In addition to flexible production, organizations must be able to react quickly and flexibly to market requirements. The use of new and often expensive technologies offers an opportunity to further promote competitiveness. Projects such as Jump 4.0 have shown that especially SMEs do not have the financial means to implement new machines or industry 4.0-compatible methods.

With the help of the Industry 4.0 out of the box, SMEs will be able to integrate methods and technologies of the fourth industrial revolution into existing processes quickly and cost-effectively. All necessary hard- and software is inside the box without dependencies to external systems. This keeps the pause in production for the installation to 'Industry 4.0' as low as possible. The installation of sensors is carried out under instruction on existing machines so that the training costs remain low. In addition to displaying the production chain in the 'Industry Cockpit', it is possible to use the QS service to automatically identify causes and measures as soon as defects in the product have been identified. With the project 'Industry 4.0 out of the box', an integrated solution will be created that enables both financially strong and financially weak organizations to secure and expand their competitiveness in an increasingly digitized world in the long term and promotes the advantages of digitalization.

Acknowledgements This research and development project is funded by the European regional development fund. The author is responsible for the contents of this publication.

References

1. Plattform I4.0 Homepage. (2017). http://www.plattform-i40.de/I40/Navigation/DE/Industrie40/WasIndustrie40/was-ist-industrie-40.html. Last Accessed 12 Nov 2017.
2. IT-Innovation-Readiness-Index. (2015). *Pierre Audoin Consultants* (p. 13). München.
3. Kiem, R. (2016). *Qualität 4.0: QM, MES und CAQ in digitalen Geschäftsprozessen der Industrie 4.0*, p. 6. München: Carl Hanser Verlag GmbH & Co. KG.
4. Scheer, A. -W. (2013). *Industrie 4.0—Wie sehen Produktionsprozesse im Jahr 2020 aus?*, p. 15. Saarbrücken: Imc Information Multimedia Communication Ag.
5. Bajric, A., Mertins K., Rabe, M., & Jaekel, F. -W. (2010). A success story: Manufacturing execution system implementation. In *The Proceeding of the 6th International Conference on Interoperability for Enterprise Software and Applications (IESA 2010)*, Springer, Coventry, UK.
6. INTEROP NOE. (2007). INTEROP, Enterprise Interoperability. Framework and knowledge corpus Final, FP6–Contact n 508011, Deliverable DI.3.
7. Naudet, Y., Guédria, W., & Chen, D.: Systems science for enterprise. In *I-ESA'09 workshops, 5th International Conference*.
8. Weinert, N., Plank, M., & Ullrich, A. (2017). *Metamorphose zur intelligenten und vernetzten Fabrik: Ergebnisse des Verbundforschungsprojekts MetamoFAB*. Berlin: Springer.
9. JUMP 40. (2017). Homepage. http://www.jump40.de. Last Accessed 12 Jan 2017.
10. Spur, G., Mertins, K., & Jochem, R. (1993). *Integrierte Unternehmensmodellierung*. Berlin: Beuth.
11. ISO 19440. (2007). *Enterprise integration—constructs for enterprise modelling*. Genève: ISO.
12. Industrial Internet Consortium. (2017). https://www.iiconsortium.org. Last Accessed 23 Nov 2017.
13. Plattform Industrie 4.0. (2017). http://www.plattform-i40.de. Last Accessed 23 Nov 2017.
14. OpenIoTFog Homepage. (2017). https://openiotfog.org. Last Accessed 23 Nov 2017.

Collaborative Information System Toward a Robotic Process Automation Solution

Rami Torkhani, Jannik Laval, Hedi Malek and Nejib Moalla

Abstract Nowadays information systems need to be open and communicative across their environment. However, the processes that hold the enterprise applications often have a heterogeneity and a technological complexity which makes the task of re-engineering difficult and costly. In this paper, we adopted a robotic process automation approach for deploying a software automated interoperability framework according to an event-driven architecture. The robot is an agent able to analyze a graphical interface of any application in order to retrieve meaningful information and produce an adequate action. Once the action reaches the interoperability framework, the event will be interpreted according to the capability of the involved robots. The study is useful in fraud detection as well as risk assessment.

Keywords Interoperability · Business process automation · Standardization · Software engineering

1 Introduction

Nowadays software publishers are actively seeking solutions to evolve their application platform while keeping in mind the interoperability aspect.

Interoperability was always perceived as the faculty for two heterogeneous computer systems to function jointly and to give access to their resources in a reciprocal way [1]. However, several constraints may restrict the collaboration inter-information system. Among these constraints, we will find legacy applications. These solutions have inherited several problems: lack of documentation, extensive maintenance costs, and above all require skilled labor. Despite these limitations, companies cannot eliminate or replace these solutions since they represent a significant economic potential. Breaking with these solutions could have a considerable economic impact. Thus,

R. Torkhani (✉) · J. Laval · N. Moalla
University Lumière Lyon 2, Lyon, France
e-mail: r.torkhani@groupeonepoint.com

R. Torkhani · H. Malek
Groupe OnePoint, Paris, France

K. Popplewell et al. (eds.), *Enterprise Interoperability VIII*, Proceedings of the I-ESA Conferences 9, https://doi.org/10.1007/978-3-030-13693-2_5

legacy applications present a dilemma: They are vitally important to the business, and however, maintaining them incurs unjustifiable expenses [2].

To be able to inter-operate two solutions, we need to analyze the business process of the two involved systems. This analysis must include the qualification of the processes so that we can determine the functional gap between the two information systems [3].

We focused on our study on the information systems of financial companies. The subject of interoperability is ubiquitous in their activity. Recently, the robotic process automation **RPA** has appeared at the heart of the human–machine interface. This solution has played a role in the qualification and automation of business processes. RPA provides financial services stakeholders with a virtual workforce governed by rules and connected to corporate systems such as users.

In this study, we implement a framework that will use robotic process automation to ensure automated interoperability. Our goal is to design a tool that will be able to execute recurring tasks such as the management of new contracts as well as the diffusion of the updates on all the applications of the enterprise. This automated interoperability can be used also for risk detection purposes. Once we succeed in synchronizing all the applications in a financial enterprise, the detection of irrelevance and fraud becomes more recognizable.

Automated interoperability could have several advantages: Financially speaking, the results are very important. This approach reduces the cost of information systems re-engineering as well as labor costs. The quality of service is also affected by these improvements as the processing time of the files will be reduced exponentially.

2 Theoretical Background

In this section, we will discuss some theoretical concepts related to interoperability and process automation to initiate the comprehension of our contribution which will be illustrated in the next chapter.

2.1 Enterprise Interoperability

The objective of the project is to characterize the context of enterprise interoperability and contribute to the identification and structuration of the knowledge in the field of information systems. If today enterprises find several difficulties to inter-operate their solutions it is because there are barriers to interoperability.

These barriers have been defined by [4] as follows:

- Conceptual barriers: They focus on the syntactic and semantic heterogeneity of the data conveyed between two systems. These difficulties are involved in the modeling phase at the highest level of abstraction (IE: the enterprise models of a company).

- Technological barriers: These difficulties are caused by the incompatibility of the technologies of the various applications (information system architecture, deployment platforms, etc.). These problems are involved by the heterogeneity of the presentation standards, the different storage formats, as well as the communication protocols between the information systems.
- Organizational barriers: This category focuses on issues of liability and authority (who is responsible/who is authorized to do what?). It also deals with problems related to the incompatibility of organizational structure.

Interoperability can be interpreted at various enterprise levels. The technical architecture of the framework ATHENA [5] proposes the following allocation:

- The interoperability of data: It goes beyond harmonizing several data models and several querying languages together. It covers all the aspects of searching and sharing data between various information systems with multiple incompatible data sources and which may additionally present itself as a distributed system with different operating systems.
- The interoperability of services: It covers all problematics related to the identification and composition of different features (designed and developed in an independent context). Servicization in this context is not just about the computer-based features. It also refers to the business function of the company.
- The interoperability of processes: The objective is to synchronize a set of business process in order to achieve a well-defined goal. A process could be defined by a sequence of features that meet a specific requirement of a company. In the context of an enterprise network, the common process of interoperability must be generated from the internal processes of the companies involved.

2.2 *Standardization*

As we examine from the literature review [6–8], several interoperability standardization works have been launched. We have chosen in this study to introduce the ISO 16100 Manufacturing software capability profiling for interoperability. We opt for this standard to qualify the robots capabilities. The ISO 16100 series provides application integration by providing a common interface specification. This interface facilitates the understanding of processes. It offers a vision of the capabilities of the applications.

The capability profile is mainly divided into four parts shown in Fig. 1:

- Actions: It describes the set of methods that can be executed.
- Resources: The resources needed for the successful completion of the action executed.

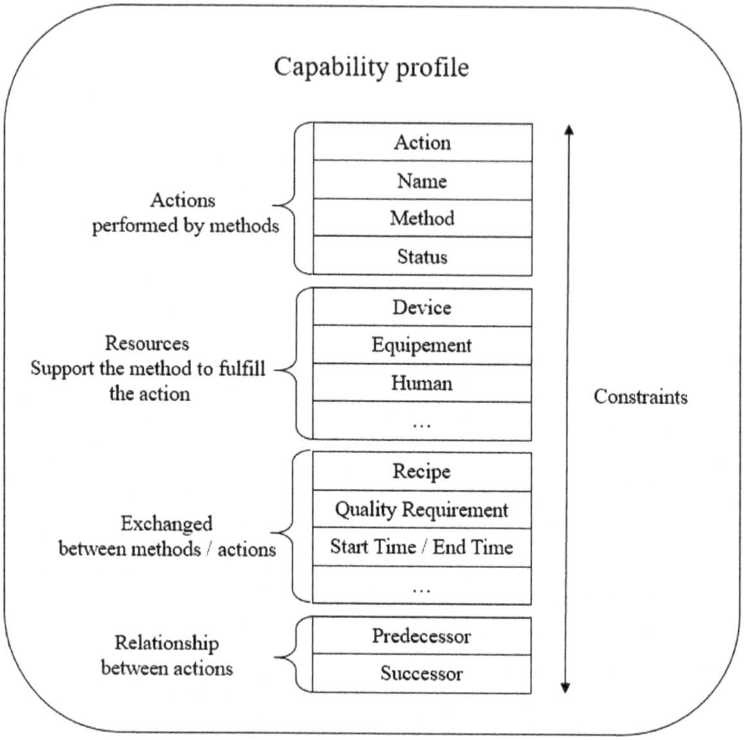

Fig. 1 Components of a capacity profile [10]

- Exchange: This part details the elements necessary to full fill the inter-connection between the involved applications.
- Relationship: It describes the relationships between the applications involved in the connection.

2.3 *Business Process Automation*

Business process automation is an activity that aims to control costs and turnaround times of business processes by integrating a maximum of automated devices that replace manual tasks.

This approach reduces the effort spent on manual tasks while improving the quality of goods and services produced (Fig. 2).

Three main techniques to automate business processes were recommended by [9]:

- Identification of processes eligible for automation: The only constraint is that the process must be digitized and rely on digital data.

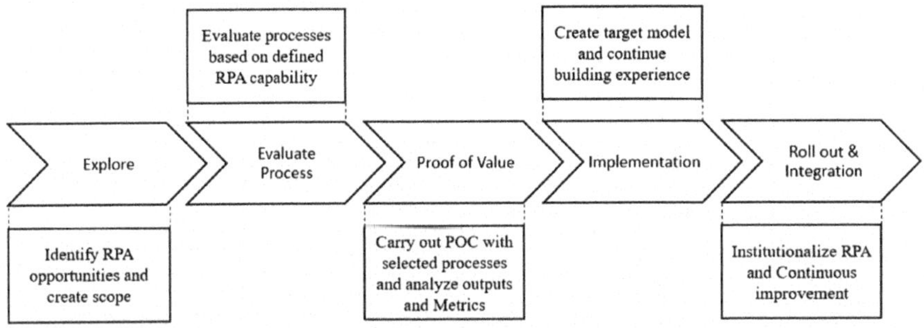

Fig. 2 RPA roadmap

- Validation of the feasibility of the project, the existing process must be stable and work properly.
- Qualification of the task volume is executed by the selected process to determine the interest of automation versus cost.

The organization benefits from various advantages are shown in Fig. 3 by integrating automation systems into its business processes: (i) restraint of information produced by the processes, (ii) supervision of business processes and various stakeholders, and (iii) the distribution of knowledge.

These benefits improve the overall productivity of the company. However, this approach requires a significant investment in the development and implementation of new automation systems [9].

3 Automated Interoperability Framework

Our contribution consists of an automated interoperability framework based on robotic process automation. In this study, we have tried to automate the collaboration between information system (SI) without impacting the stability of business services.

3.1 *The Robot*

The robot is an agent that can be deployed as a service in the operating system. It allows interacting with different interfaces of applications. Figure 4 depicts the robot components.

(a) UI element manager: To automate specific actions in the user interface, you are required to interact with various windows, buttons, drop-down lists, and many others. Most applications do this by relying on the screen position of UI

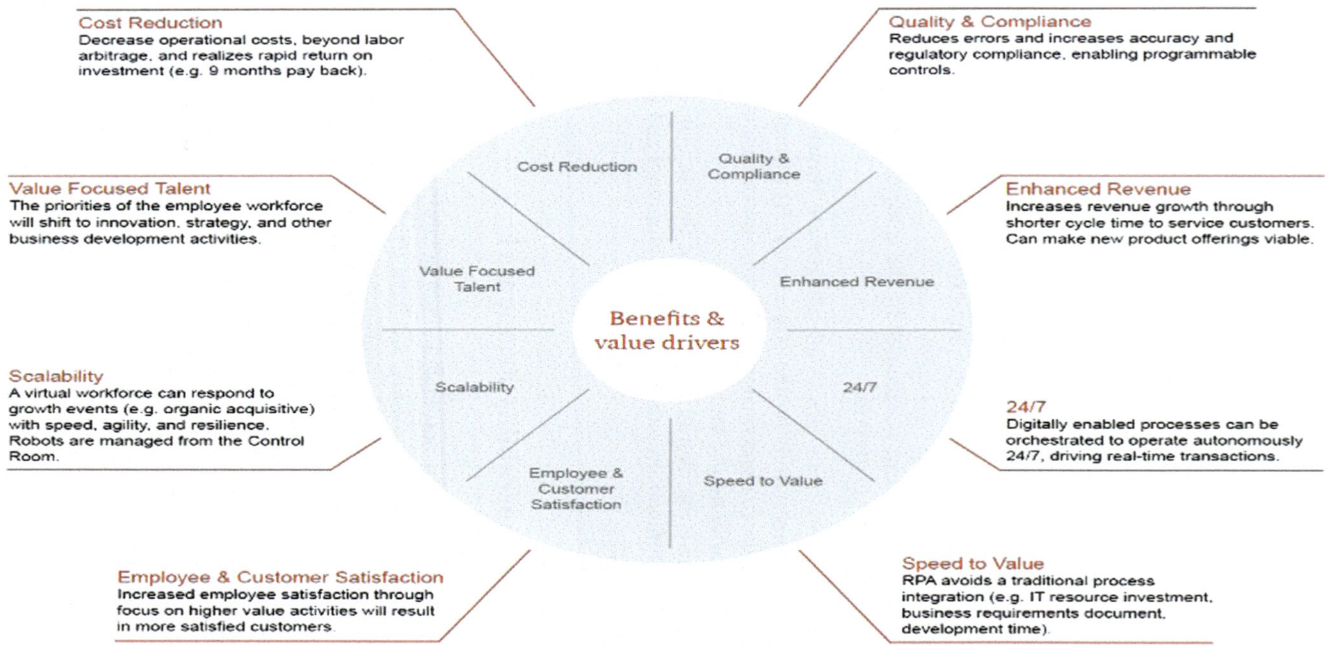

Fig. 3 RPA benefits [11]

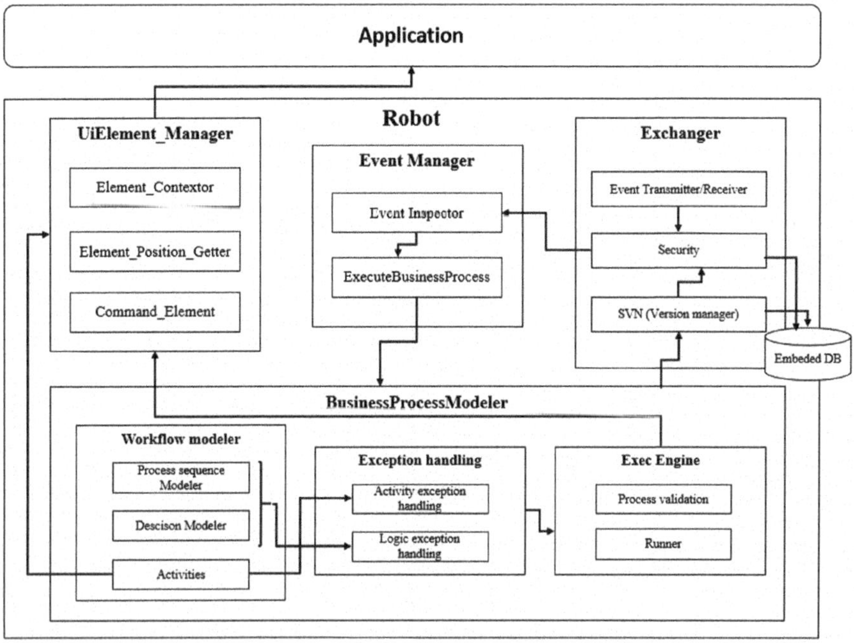

Fig. 4 Robot architecture

elements, a method that is not at all dependable. We tried to enrich this idea by the contextualization of element, and this latter becomes identifiable by name of the main window, index, and application.

(b) Business process modeler: To ensure proper automation, we need to model the business process. The business process modeler is composed of three entities:

- Workflow modeler: The entity is divided into three layers. (1) Process sequence modeler's role is to formalize the scheduling of processes while respecting their integrities. (2) The purpose of the decision modeler is to make controls on business processes. (3) Activities entity groups a set of functionalities that make it possible to build business logic or to control the interface of the application.
- Exception handler: We have designed two types of exception handler. (1) Logic exception handling aims to ensure consistency of treatment. (2) Activity exception handling ensures activities which are running properly.
- Execution engine: This entity has two main roles, the first is the validation of the modeling processes, and the second is the execution of the processes.

(c) Event manager: Event inspector analyzes events received from the exchanger entities. He tries to find the event in a database. If the event exists, the executor initiates the business process. Otherwise, the robot notifies the Interop Framework.

(d) Exchanger: Manages three important components:

- Event transmitter/receiver: It is a stream listener able to receive and emit events to the orchestrator.
- Security: It is a layer that verifies the origin of the events and ensures the relevance of the events compared to the security rules.
- Version manager (SVN): This entity manages revisions on the automation processes. As soon as a process modification is published, the version manager publishes the modification to the orchestrator in order to update the robot profile.

3.2 The Interoperability Framework

The interoperability framework is the core of the solution it is a mediator who ensures collaboration between robots. The framework is able to identify the capabilities of each robot. By means of Broker cloud hosted, the solution perceives and analyzes all the actions emitted by the robots. It adapts the event so that it can adequately be understood by the target robot. Figure 1 depicts the elements of the framework (Fig. 5).

(a) **Event manager**: Is the main component that manages and process all events. Upon receipt of an event, the interpreter must analyze the contents of the event in order to identify the target robot and the source robot. Subsequently, the profile manager tries to retrieve the capacity profiles of each robot.
In this study, we have taken over the capacity profile structure of ISO 16100 [10]. As it is described in Fig. 1 The profile is composed of three layers: (1) action performed by the robot which represents a capacity; (2) resources which support the robots to fulfill the action; (3) constraints that must be respected by actions emitted by robots; and (4) exchanger that defines the relationships between robots and actions.

(b) **Analysis entity**: The analysis step is based on three stages:

- *Involved robot analysis*: During this step, the profiles of the robots will be analyzed to identify the required resources ensuring the smooth running of the robot.
- *Resource analysis*: The resource analyzer compares the inputs and outputs of the involved robots. The analysis is proceeded by analyzing the structure of each parameter. Technically, this part is based on the Apache AVRO [11] solution which is a data serialization system. Once the data resources are analyzed, a comparison is made between the target robot capabilities and the action to be executed.

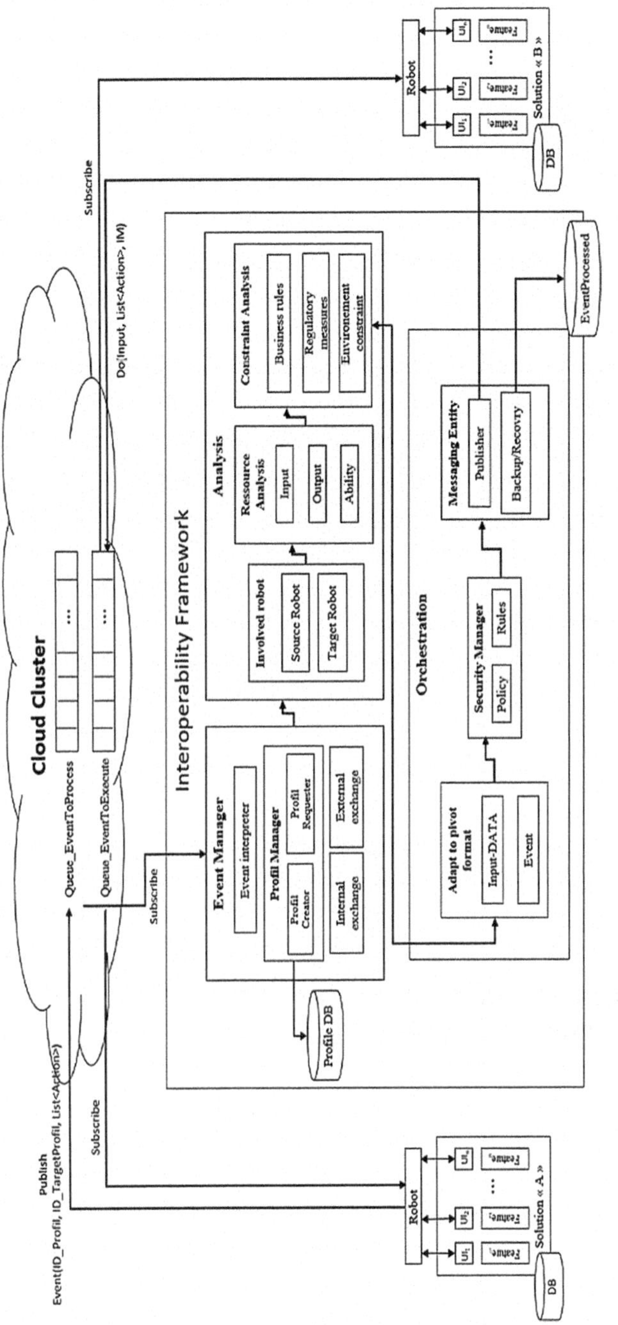

Fig. 5 Overview of the automated interoperability framework

- *Constraint analysis*: The constraint analyzer allows a consistency inspection of the actions regarding three types of constraints: (1) Business rules are requirements related to the logic of the business process. (2) Regulatory restriction verifies that the rules used by the government or the business communities are respected. (3) Environment constraints encompass the requirements related to the application environments that meet the demand of the robots.

(c) **The orchestration entity**:

- *Adapt to pivot format*: This component translates the data to a pivot format that can be understandable by any robot. The pivot format must be extensible so that the solution could manage the scalability of the solutions.
- *Security manager*: This entity manages the security of the exchanges. Its role is to consolidate a set of security rules by policy. In other words, it ensures applications credentials management. The second task is the assurance of the coherence of the actions in relation to the privileges granted.
- *Messaging entity*: This entity executes the event publishing task in the clouded MoM (Message-oriented Middleware). It also ensures another important role which consists of ensuring continuity of services. The module backup/recovery is actively pushing events into the database, in case of failure the framework can reset itself by recovering events that have not been processed.

4 Use Case

Our use case is inspired by the field of insurance. Indeed, in these establishments, several applications can be used. These applications are heterogeneous and do not necessarily share the same data structure. However, the business logic remains the links between these applications. Application governance also poses another problem, as software is managed by several business teams, making collaboration between systems complex and painful.

We have described the use case in Fig. 6 and below we detail the operating procedure:

1. The robot broadcasts a new contract in the internal cloud.
2. The interoperability platform takes responsibility for establishing the necessary links.
3. Retrieve or create the robot capability profile.
4. Adaptation of the initial message to a standardized format for the purpose of issuing understandable information.
5. The orchestrator selects the robots involved.
6. The action is shared on the internal cloud so that the concerned robot can retrieve and process it.

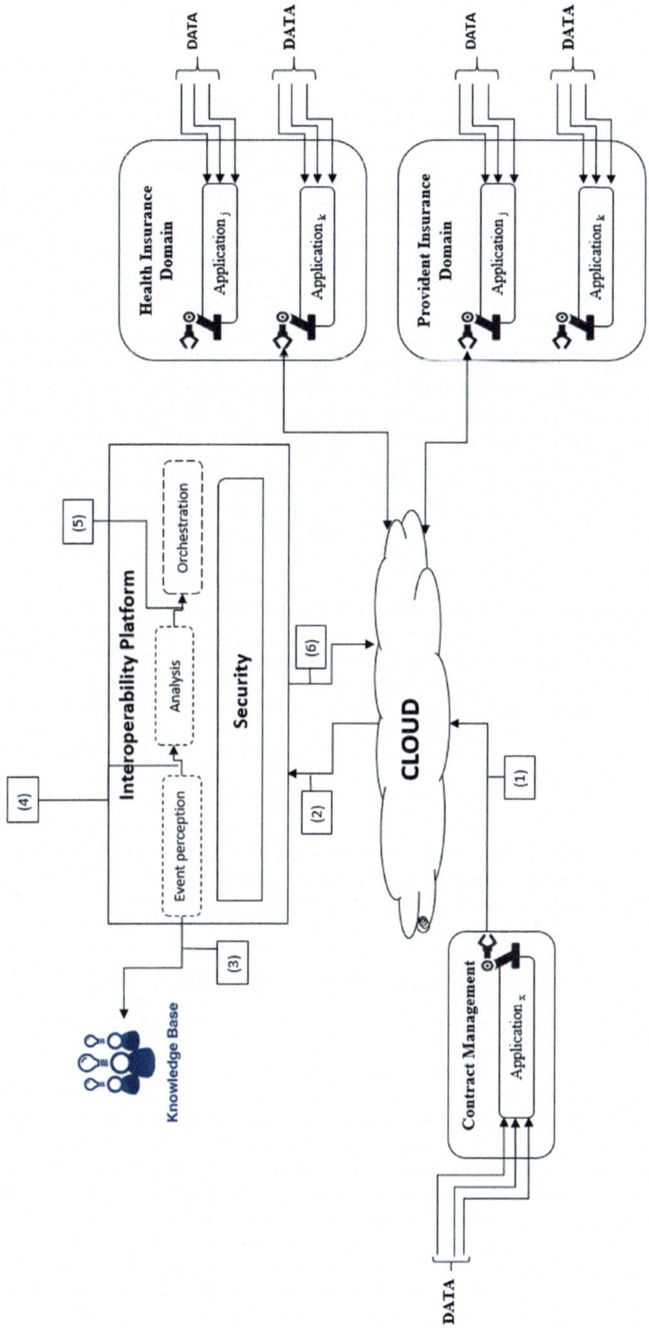

Fig. 6 Use case

5 Conclusion

In this article, we have proposed an automated interoperability framework. The framework is a multi-agent system based on the robotic process automation RPA approach. The added value proposed by this study against the work that was already published concerning interoperability of information system is demonstrated by the fact that we no longer need to tackle the structure of the information system.

Indeed, the robotic process automation is a solution that does not affect the architecture of the information system and it intervenes as an additional layer that operates with existing interfaces. This reduces the cost of legacy system migrating and ensures service continuity.

Moreover, this solution must face a security test. Soon we will integrate a Blockchain approach to cover the security aspect. The idea is to build a network of robots that can operate between them safely. The robots will have the opportunity with this device of exchange between them to check the relevance of the data or to make sure of the identity of the issuer of the actions.

References

1. Chen, D., Doumeingts, G., & Vernadat, F. (2008). Architectures for enterprise integration and interoperability: Past, present and future. *Computers in Industry, 59*(7), 647–659.
2. Bisbal, J., Lawless, D., Wu, B., & Grimson, J. (1999). Legacy information systems: Issues and directions. *IEEE Software, 16*(5), 103–111 (ISO 690).
3. Ostadzadeh, S. S., Shams, F., & Badie, K. (2015). An architectural model framework to improve digital ecosystems interoperability. In *New trends in networking, computing, e-learning, systems sciences, and engineering* (pp. 513–520). Cham: Springer.
4. Benali, H., Saoud, N. B. B., & Ahmed, M. B. (2014). Towards an ontology of SoS interoperability: Proposition of a SoS interoperability framework and a SoS conceptual interoperability model. In *Information system development* (pp. 85–97). Cham: Springer.
5. Yang, C., & Gao, S. S. (2015). A solution of collaboration and interoperability for networked enterprises. In *International Conference on Cooperative Design, Visualization and Engineering* (pp. 243–249). Cham: Springer.
6. Guédria, W., Naudet, Y., & Chen, D. (2015). Maturity model for enterprise interoperability. *Enterprise Information Systems, 9*(1), 1–28.
7. Rezaei, R., Chiew, T. K., Lee, S. P., & Aliee, Z. S. (2014). Interoperability evaluation models: A systematic review. *Computers in Industry, 65*(1), 1–23.
8. Charalabidis, Y., Lampathaki, F., Kavalaki, A., & Askounis, D. (2010). A review of electronic government interoperability frameworks: Patterns and challenges. *International Journal of Electronic Governance, 3*(2), 189–221.
9. Briol, P. (2014). *Bpms-L'automatisation des processus métiers*. Lulu. com.
10. Matsuda, M. (2012). Manufacturing software interoperability services which ISO 16100 brings about. In: *International IFIP Working Conference on Enterprise Interoperability* (pp. 60–70). Berlin: Springer.
11. Merten, T., Mager, B., Bürsner, S., & Paech, B. (2014) Classifying unstructured data into natural language text and technical information. In: *Proceedings of the 11th Working Conference on Mining Software Repositories, ACM* (pp. 300–303).

Maintenance 4.0 World of Integrated Information

Erkki Jantunen, Giovanni Di Orio, Csaba Hegedűs, Pal Varga, István Moldován, Felix Larrinaga, Martin Becker, Michele Albano and Pedro Maló

Abstract The condition-based maintenance (CBM) strategy has got new, **powerful toolset recently: the concepts of the Internet of Things (IoT) and Cyber-Physical Systems** (CPS). These can provide flexible but powerful data collection and analysis methods for proactive and predictive maintenance. In the landscape of new digitalization and interconnection of products, services, processes, enterprises and people, IoT/CPS-based platforms are increasing in their size and target applications in a steady manner. Beside the fundamental research challenges regarding the reference architecture, interoperability, performance, quality and deployment issues, the challenges regarding system maintenance are also burning. There are various issues that are specific to the maintenance domain: interoperability and data flow management, data representation models, and data processing models and tools. The paper describes a maintenance reference architecture and platform, which aims to tackle all these challenges. The architecture suggested by the MANTIS project covers edge and cloud level interoperability, data flow management and data processing issues. Furthermore, it provides domain-specific methods for root cause analysis, failure prediction and models for predicting remaining useful life. The architecture

E. Jantunen (✉)
VTT Technical Research Centre of Finland Ltd, P.O. Box 1000, 02044 Espoo, Finland
e-mail: erkki.jantunen@vtt.fi

G. Di Orio · P. Maló
Departamento de Engenharia Electrotécnica, FCT-UNL, UNINOVA-CTS, Lisbon, Portugal

C. Hegedűs
AITIA International Inc., Czetz Janos str. 48-50, Budapest 1039, Hungary

P. Varga · I. Moldován
Budapest University of Technology and Economics, Magyar Tudosok krt. 2, Budapest 1117, Hungary

F. Larrinaga
Mondragon University, Loramendi 4, 20500 Mondragon, Spain

M. Becker
Fraunhofer IESE, Fraunhofer Platz 1, 67663 Kaiserslautern, Germany

M. Albano
Department of Computer Science, Aalborg University, 9220 Aalborg, Denmark

K. Popplewell et al. (eds.), *Enterprise Interoperability VIII*, Proceedings of the I-ESA Conferences 9, https://doi.org/10.1007/978-3-030-13693-2_6

is strengthened by the concept of MIMOSA, a data model definition that allows data representation models that are easy to fit into relational object and information management models required by CBM. The MANTIS platform utilizes the Arrowhead framework for tackling interoperability and integrability issues.

Keywords Condition-based maintenance (CBM) · Internet of Things (IoT) · Cyber-Physical Systems (CPS)

1 Introduction

It is widely accepted that CBM is the most efficient maintenance strategy. The basic reasons for this are very simple. Running production without any condition monitoring or preventive maintenance actions can be extremely costly as any small fault can suddenly stop the production for a time that depends on how quickly the machine can be repaired, and this in turn might sometimes take a very long time causing lost production, lost labour and inefficient use of the whole investment. On the other hand, maintenance that is carried out in order to guarantee that no stoppages take place without measuring the need for maintenance can also be costly. A lot of maintenance might be done in vane possibly in addition to losing time and money in doing the maintenance but also possibly causing faults in the machinery, which in turn might be the cause of maintenance actions.

1.1 Internet of Things

With the advent of Internet of Things (IoT), novel strategies became feasible in industrial applications. The main difference between other concepts regarding embedded technologies (CPS, wireless sensor networks, machine to machine communication, body area networks, etc.) and the Internet of Things is that the latter considers all embedded systems as connected to the Internet. As far as functional requirements are concerned, this characteristic allows for ubiquitous access to the embedded systems. Moreover, since they make use of mature protocols and well-accepted software libraries, applications on top of IoT devices are faster to implement and easier to maintain, leading to lower time to market and cheaper maintenance.

1.2 Cyber-Physical Systems

Nowadays, conventional systems and processes are evolving into CPS in the most disparate contexts of applications (e.g. manufacturing, healthcare, automotive, white goods, logistics, etc.) and of different nature (e.g. mechanical, electrical and chemical). As stated in [1], the term "Cyber-Physical Systems" has been coined in 2006. Today, several definitions of the term CPS can be found in the literature. According to [2], CPS can be defined as transformative technologies that allow to manage interconnected systems between its physical assets and computational capabilities. The definitions in [3, 4] highlight the concept of collaboration and service provisioning. As a matter of fact, CPSs are defined as systems of collaborating computational entities that are strictly connected to the surrounding physical assets providing and using services to/from the internet. A working definition for CPS has been offered in [5], where a CPS is defined as a system consisting of computational, communication and control components combined with physical processes. Regardless to the specific definition of CPS, it is possible to identify their core elements and/or characteristics, extended from [6, 7]: (1) enhancement of physical entities with cyber capabilities; (2) networked at multiple and extreme scale; (3) dynamic behaviour (plug and unplug during operation); (4) high degrees of automation, the control loops are typically closed; (5) high degree of autonomy and collaboration to achieve a higher goal and (6) tight integration between devices, processes, machines, humans and other software applications. As deeply explained in [8], the CPS intrinsic characteristics are naturally pointing to ecosystems of interacting and connected CPS also called Cyber-Physical Systems of Systems (CPSoS) or Cyber-Physical Production Systems (CPPS) in the industrial domain. CPSoS and CPPS are promoting the design and development of advanced monitoring and control infrastructures that rely on a common virtualized space for collecting, processing, provisioning, analysing and visualizing large quantity of data [9]. This data can be potentially used for fast evaluation of the industrial assets performances to adapt and optimize (through reconfiguration) the overall behaviour of the production system while enabling the efficient and effective implementation of maintenance policies, e.g. CBM.

1.3 Challenges

The wider dissemination of CPS—and their aggregation into CPSoS—and IoT is creating new market opportunities and business models for all kind of European Industries. As a matter of fact, the new digitization and interconnection wave of products, services, processes, enterprises and people are expected to generate significant benefits for all the involved actors, assuming that the risks and challenges are properly addressed [10]. In this landscape, IoT/CPS-based platforms are increasing in their size and target applications in a steady manner. However, even if progresses are made every day supported by continuous technological advancements IoT/CPS design, development and deployment is still challenging. The great potential and enormous expectations around the IoT and CPS solutions are resulting in real challenges that the research community is asked to address to boost the progress and deployment of these solutions in real application context. The research challenges here summarized are extracted from [11–13] and clustered according to [2]:

a. Science and engineering foundations: a reference architecture for interoperable and interconnected CPS-populated systems in cross-sector applications. Enable seamless human-IoT/CPS interaction;
b. System performance, quality and acceptance: to create large, adaptive and resilient networked systems that are capable to operate in the specific environments where the physical entities are installed while delivering the required functionality in a reliable way. To develop science-based metrics for measuring system adaptability, flexibility, responsiveness, security safety and/or more in general method to predict the behaviour of highly dynamical systems and
c. Applied development and deployment: to provide mechanisms for representing highly distributed and heterogeneous systems. To provide methodologies for virtualization of physical entities and integration of heterogeneous systems. To deliver technology foundation for building interconnected and interoperable IoT/CPS-populated systems.

The platform that accommodates CBM data needs to tackle the following issues:

a. Provide interoperability at system level allowing the transmission of data from the CPS. How to transmit these data from the physical system and to where?
b. Use data representation models that enable the collection of CBM information (events, root cause analysis, fault prediction and remaining useful life results) related to CPS. How to create interoperable data representation and semantics?
c. Provide the mechanisms to process CBM data in real-time or in batch processes. How can we still maintain real-time restrictions and abide by communicational constraints?
d. What can be the back end that processes these inbound streams in a scalable manner?

1.4 Paper Structure

Section 1 of this paper introduces CBM, IoT, CPS and describes some of the challenges related to the adaption of these technologies. In Sect. 2, Predictive and Proactive Maintenance Platform and the MANTIS approach are discussed. In addition, the role of MIMOSA is presented in Sect. 2. In Sect. 3, MANTIS Reference Architecture and Implementation the MANTIS architecture and its implementation are discussed. Section 4 summarizes the paper in the format of a short conclusion.

2 Predictive and Proactive Maintenance Platform

The overall aim of the MANTIS project [14] is to develop platform for interoperable and interconnected CPS-based systems for proactive collaborative maintenance ecosystems, i.e. for facilitating the implementation of predictive and proactive maintenance strategies. The objective of predictive maintenance is to predict when maintenance should be performed and with better planning stops and avoiding unexpected interruptions (increase plant availability) [15]. Proactive maintenance seeks to detect and eradicate failure root causes [16]. Proactive maintenance depends on rigorous machine inspection and condition monitoring. Many technologies are employed to evaluate equipment condition (infrared, acoustics, vibration analysis, electrical motor power analysis, etc.). Site measurements are often supported by wireless sensor networks, and data analysis is essential.

2.1 MANTIS Approach

The MANTIS reference architecture and platform provides a practical mean for implementing collaborative maintenance by taking advantage of:

a. The omnipresence of intelligent devices—that combine physical entities with computational and communication capabilities—in modern processes, machines and other distinct application domains.
b. The maturity level reached by cloud-based infrastructure and the huge amount of computational and storage resources available and usable "on-demand".

Intelligent devices are the ones directly connected and/or installed to the physical resources and assets. They can potentially optimize and improve current maintenance activities and their related management systems by providing (often live) data—gathered during operation—that can be analysed (low-level data analysis) to understand the behaviour of the related physical resources and assets. Furthermore, the data gathered from physical resources and assets can be also combined and analysed globally (high-level data analysis) by using computational resources and complex algorithms running over the cloud (high level) to understand the collective behaviour of group of resources and assets. Therefore, within the MANTIS approach data extraction, transforming, loading and pattern analysis will take place at different levels, namely (see Fig. 1):

a. Low level: extraction, transforming, loading and analysis of simple signals to model and understand the behaviour of selected physical resources and assets.
b. High level: extraction, transforming, loading and analysis of complex data results of the low level to model and understand the global behaviour of assets.

Since data sources are typically characterized by distribution, heterogeneity and a high degree of dynamicity (e.g. data sources like sensors can be plugged and unplugged any time), the design of the MANTIS architecture has been driven by the following main requirements:

a. The provision of data structures that enable the collection of maintenance information (events, root cause analysis, fault prediction and remaining useful life results, etc.) related to systems and assets.

Fig. 1 MANTIS overall concept idea and data processing levels [14]

b. The provision of data structures that enable large volume of data to be processed in real-time or in batch processes.
c. Integration of complex and heterogeneous large-scale distributed systems from different application domain.
d. The design of CPS-populated systems to enable collaborative proactive maintenance strategies.

It is easy to understand that the design of interoperable and interconnected CPS-based maintenance ecosystems becomes a key element of the MANTIS implementations to allow to dynamical and on-demand addition or removal of data sources in/from the MANTIS platform to gather most of the maintenance relevant information automatically from the environment.

2.2 *Role of MIMOSA*

The use and benefit of CBM strategy are based on information and the knowledge that is gained from that information. From this it follows that a lot of information has to be managed, i.e. condition monitoring measurement data, information of the use of the machine in question, data about previous maintenance actions and exact information about the components of the machinery. Naturally, all the information has to be integrated in order to make it meaningful in defining the need for maintenance and the timing of these actions. What makes the situation challenging is the fact that there are many sources and representations for this information and that these systems do not usually interoperate. In most cases, the production machinery has been designed with some computer-aided design (CAD) software, which then holds information about all the components as designed. In the next phase, some computer-aided manufacturing (CAM) system has been used in the manufacturing phase. This CAM system then holds information about the components as manufactured. Usually, a manufacturing company uses some product lifecycle manufacturing (PLM) system to follow the whole lifecycle of a product. When production machineries are in use, the whole production process is managed with an enterprise resource planning (ERP) system that holds information about all the assets of the company and also about the personnel managing these machines. The machines are normally driven using some proprietary automation system. Many companies use computerized maintenance management system (CMMS) to handle the maintenance of their production machinery. Usually, the condition monitoring (CM) systems are separate from the CMMS. Since when following CBM strategy information from all the above-named systems is needed, it is clear that interoperability and integration of information are of the highest importance. Experience from industrial practice reveals that the heterogeneity and lacking integration of the information is a considerable roadblock towards Maintenance 4.0.

There exists an organization called Operations and Maintenance Information Open System Alliance (MIMOSA). MIMOSA sees their role as an integrator between various systems and says [17] that they provide a series of interrelated information standards. According to MIMOSA, the Common Conceptual Object Model (CCOM) provides a foundation for all MIMOSA standards, while the Common Relational Information Schema (CRIS) provides a means to store enterprise O&M information. MIMOSA also manages and publishes the Open O&M Web Service Information Service Bus Model (ws-ISBM) and Common Interoperability Register (CIR) specifications, while MIMOSA has aligned with POSC Caesar Association in the development of Reference Data Libraries and with Open Applications Group Integration Specification (OAGIS) in the use of its Business Object Document (BOD) architecture to support information exchange.

MIMOSA also states that they maintain strong industry ties with other formal standardization groups [17]. For example, MIMOSA is compliant with and forms the informative reference to the published ISO 13374-1 standard for machinery diagnostic systems. MIMOSA is hosting the Oil and Gas Interoperability (OGI) Pilot which is managed by the joint MIMOSA/PCA O&M SIG and working as part of the US TAG in ISO TC 184/WG 6, which is developing the ISO OGI Technical Specification.

3 MANTIS Reference Architecture and Implementation

The purpose of the MANTIS reference architecture is to support companies interested in predictive and proactive maintenance to set up an adequate system architecture, especially with issues that are new, risky and costly to change. Since large amounts of data can be collected from industrial devices, machines or vehicles, there is a sense in trying to utilize them. Obviously, the architecture of any Maintenance 4.0 solution has to be capable of handling data in a scalable manner, and even coming from different sources, stakeholders—in different formats. Furthermore, the data to be processed and managed will change over the lifecycle of the system.

One of the first drawbacks experienced, of course, was that although data has been collected for several years now, this data collection, aggregation and storage systems were not designed to be actually used later on in a Maintenance 4.0 solution. A further issue is that often meta-data for the already collected data is lacking. This aggravates the design of data analyses considerably.

Both issues can be addressed with the usage of the OSA-CBM domain model [17]. Other interoperability and data source-related issues are also handled in the project.

At the heart of any Maintenance 4.0 solution, there are innovative analysis, prediction and planning functions that operate on the different data sets. Within MANTIS, the main analytic services fall into three categories:

a. Remaining useful life (RUL) of components: continuous tracking of telemetry (usage) data and estimating how much time the given device or component has left before needs to be replaced
b. Fault prediction (FP): the system shall predict based on diagnostic data an inbound failure mode (different to wear-out to be detected by RUL),
c. Root cause analysis (RCA): when an unpredicted, complex failure occurs, the system shall deduct the actual module (cause) that caused the break.

The MANTIS reference architecture platform itself is designed to facilitate these services in a flexible and scalable manner. In here, we are relying on the following architectural decisions and patterns:

a. The edge computing paradigm can be used to reduce the data sent to the platform and to enable on-site maintenance operations with low latency. Therefore, a respective platform within the edge needs to be provided for storage, limited analytics, HMI and lifecycle management support. In here, CPSs are involved and are executing their general production operations.
b. The overall data flow architecture shall follow the IoT reference architecture model proposed by the Industrial Internet Consortium [18].
c. The MANTIS platform shall integrate into the existing enterprise infrastructure in a service-oriented manner [19] to ease integration and maintainability.
d. To enable multi-stakeholder interactions, a dynamical attachment procedure must be implemented (between edge and platform instances) so that many interested

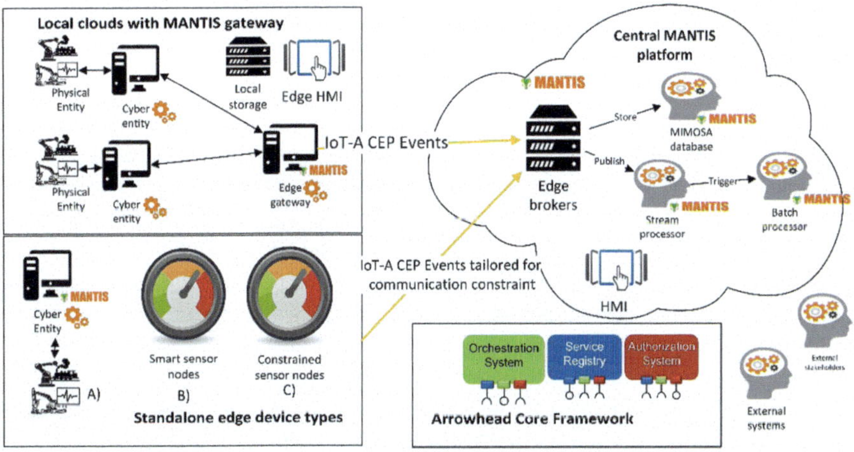

Fig. 2 Overview of the MANTIS reference architecture

partners can access a single edge establishment and acquire the selected data intended for them.

To this end, the MANTIS reference architecture extends the IoT reference model with two other aspects, as shown in Fig. 2.

Firstly, for data processing in the platform, MANTIS invokes a kappa or lambda architecture model [20] that is fitting the current trends in industrial big data analytic processes. According to the generalized lambda architecture pattern [19] defined by industry experts, data can be processed either as soon as it reaches the platform (stream processing), or later on, on demand fetched from storage (batch processing). In here, we are utilizing both stream- and batch-processing technologies described by lambda, to enable the three major maintenance related tasks listed above. This pipelined reference architecture model is implemented for 11 use cases within MANTIS.

The platform tier can be cloud-based, either in a public or local cloud running on corporate servers. The platform tier receives, processes and forwards control commands from the enterprise personnel or systems in general. It consolidates all the above-described processes and analyses data flows from the edge level. It provides management functions for devices and assets. It also offers non-domain-specific services such as data query and analytics. In order to provide the MANTIS partners with a concrete implementation of the MANTIS reference architecture, a reference implementation as shown in Fig. 2 has been developed.

The MANTIS reference implementation relies on the Arrowhead Framework [21] to enable an adequate edge-cloud interoperability. Besides the edge computing interoperability and connectivity issues, especially if real-time control loops are kept, Arrowhead is also used to resolve requirements #c and #d (Sect. 2.1), regarding multi-stakeholder applications. Multiple cloud platforms from various vendors are enabled to access one single production site or edge device in order to get the necessary information for their business purposes.

Additionally to provide access to the platform, the edge broker incorporates translation functionality to the solution. This enables the conversion of heterogeneous data formats and protocols to the requirements of the platform reinforcing the interoperability of the system with different CPS (requirement #c in Sect. 2.1). Interoperability at platform level is achieved by including the functionality of an enterprise service bus (ESB). Among the main features addressed by the ESB in MANTIS it is worth to mention its capability of mediation between communication protocols, data formats, and messaging standards coming from the CPS and storage repositories at platform level. The edge broker also addresses the translation or mapping of data formats and protocols to the reference architecture provided at platform level. The ESB enables the monitoring of those transactions, assures scalability, provides fault tolerance mechanisms or allows the dynamic provisioning of resources.

MIMOSA has been integrated into the MANTIS reference architecture as the information model that provides the data structures that enable the collection of maintenance information (requirement #a in Sect. 2.1). Distributed File System (DFS) storage resources are also available at platform level fulfilling requirement #b in Sect. 2.1.

4 Conclusion

After describing the current environment, expectations and challenges for the domain of CBM, this paper summarizes the MANTIS reference architecture and its reference implementation to give answer to many of the aforementioned challenges. The MANTIS reference architecture provides means for implementing collaborative maintenance by taking advantage of the omnipresence of intelligent devices and the maturity level reached by cloud-based infrastructure. The platform covers data collection from sensors, data pre-processing at the edge level, data flow management, batch- and stream-processing of data, as well as data presentation to application-specific human–machine interfaces. Maintenance-specific analysis is covered through cloud-based data processing, including methods and algorithms for estimating remaining useful life, predicting failure and providing root cause analysis results. The challenges of data representation and object modelling are tackled by the MIMOSA concept, with its Common Conceptual Object Model, and its Common Interoperability Register, and especially the Common Relational Information Schema that provides a means to store enterprise O&M information. Interoperability issues of all involved parties are covered by the support of the Arrowhead Framework. The edge broker enables access to the platform, translate and mediation services and data flow capabilities. The whole platform assures scalability, dynamic resource provision, monitoring and real-time and batch processing mechanisms and tools for intelligent management of industrial operations such as CBM.

Acknowledgements This work has been developed with the support of funds made available provided by the European Commission in the scope of ECSEL/H2020 MANTIS Research and Innovation Action (Project ID: 662189), by the Portuguese Fundação para a Ciência e a Tecnologia (FCT, I.P.) in the framework of project UID/EEA/00066/2013 PEST (Strategic Plan for Science and Technology) for the Centre of Technology and Systems (CTS), by the Finnish Funding Agency for Innovation Tekes and by Ministerio de Industria, Energía y Turismo (Spain).

References

1. Baheti, R., & Gill, H. (2011). Cyber-physical systems. *The Impact of Control Technology, 12*(1), 161–166.
2. Steering Committee. (2013). *Strategic R&D opportunities for 21st century cyber-physical systems, connecting computer and information systems with physical world*. NIST report.
3. Acatech. (2011). *Cyber-physical systems—Driving force for innovations in mobility*. Germany: Acatech.
4. Cengarle, M., Bensalem, S., McDermid, J., Passerone, R., Sangiovanni-Vincetelli, A., & Torngren, M. (2013). Characteristics, capabilities, potential applications of cyber-physical systems: A preliminary analysis, D2.1.
5. Huang, B. X. (2008). *Cyber physical systems: A survey*. Presentation report.
6. Sanislav, T., & Miclea, L. (2012). Cyber-physical systems—Concept, challenges and research areas. *Journal of Control Engineering Application Information, 14*(2), 28–33.
7. Omer, A. I., & Taleb, M. M. (2014). Architecture of industrial automation systems. *European Scientific Journal, 10*(3).
8. Jantunen, E., Zurutuza, U., Albano, M., di Orio, G., Maló, P., & Hegedus, C. (2017). *The way cyber physical systems will revolutionise maintenance*. Presented at the 30th conference on Condition Monitoring and Diagnostic Engineering Management.
9. Soldatos, J., Gusmeroli, S., Malo, P., & Di Orio, G. (2016). Internet of things applications in future manufacturing. In *Digitising industry—Internet of things connecting the physical, digital and virtual worlds*. Denmark: River Publishers.
10. Colombo, A. W., Karnouskos, S., Kaynak, O., Shi, Y., & Yin, S. (2017). Industrial cyberphysical systems: A backbone of the fourth industrial revolution. *IEEE Industrial Electronics Magazine, 11*(1), 6–16.
11. Jovane, F., Westkämper, E., & Williams, D. J. (2009). *The ManuFuture road: Towards competitive and sustainable high-adding-value manufacturing*. Berlin: Springer.
12. Candido, G. (2013). *Service-oriented architecture for device lifecycle support in industrial automation*. FCT-UNL.
13. Monostori, L. (2014). Cyber-physical production systems: Roots, expectations and R&D challenges. *Procedia CIRP, 17*, 9–13.
14. Jantunen, E., Zurutuza, U., Ferreira, L. L., & Varga, P. (2016). Optimising maintenance: What are the expectations for cyber physical systems. In *3rd International Workshop on Emerging Ideas and Trends in Engineering of Cyber-Physical Systems (EITEC)* (pp. 53–58).
15. Hunt, T. M. (2006). *The concise encyclopaedia of condition monitoring*. UK: Coxmoor Publishing Company. ISBN 1 901892 19 0.
16. Fitch, E. G. (1992) *Proactive maintenance for mechanical systems*. Amsterdam: Elsevier Science Publishers. ISBN 9781483292595.
17. Mimosa homepage. http://www.mimosa.org/what-mimosa. Last accessed 2017/11/20.

18. Industrial internet consortium homepage. http://www.iiconsortium.org/IIRA.htm. Last accessed 2017/11/20.
19. Erl, T. (2005). Service-oriented architecture (SOA) concepts, technology and design.
20. Hausenblas, M., & Bijnens, N. The lambda architecture website. http://lambda-architecture.net/. Last accessed 2017/11/20.
21. The arrowhead consortia, the arrowhead framework wiki. https://forge.soa4d.org/plugins/mediawiki/wiki/arrowhead-f/index.php/Arrowhead_Framework_Wiki. Last accessed 2017/11/20.

Part III
Platforms and Infrastructures for Enterprise Interoperability

Towards Interoperability Across Digital Manufacturing Platforms

Usman Wajid and Gash Bhullar

Abstract The ICT infrastructure of many manufacturing companies usually is composed of mixed or what is sometimes referred to as heterogeneous systems or platforms. This trend continues with the development of digital manufacturing platforms that are built around vendor-specific technologies or closed standards. This paper introduces a platform interoperability framework that aims to promote transparency across different levels of digital platforms. The framework supports the development of interoperable solutions and creation of an ecosystem where applications and services that can be deployed across multiple platforms. The data gathered from the study highlight the interoperability potential among different levels of the digital platforms. The usefulness of the framework is evident in not only promoting transparency across domain-specific platforms but also identifying potential for collaborations and ecosystem creations.

Keywords Interoperability · Industry 4.0 · Middleware · Manufacturing platform

1 Interoperability Across Digital Manufacturing Platforms

Industry 4.0 is now being deployed on shop-floors across Europe as part of smart factory solutions comprising of Internet of Things, Cyber-physical systems and cloud-based services. The plethora of technology-driven changes in traditional manufacturing activities brings its own challenges in the number of digital manufacturing platforms that have been developed to integrate and offer smart factory solutions. Each solution offers its own thematically focused solution such as condition monitoring, predictive maintenance or multi-sided marketplaces of smart factory

U. Wajid (✉)
Information Catalyst for Enterprise, 37, Crewe CW1 5QR, UK
e-mail: usman.wajid@informationcatalyst.com

G. Bhullar
Control 2K, Waterton Technology Centre, Bridgend CF31 3WT, UK
e-mail: gbhullar@control2k.co.uk

K. Popplewell et al. (eds.), *Enterprise Interoperability VIII*, Proceedings of the I-ESA Conferences 9, https://doi.org/10.1007/978-3-030-13693-2_7

solutions. To operate on such platforms, smart applications are either built on heterogeneous standards or follow specific protocols for data handling and processing.

As each vendor or provider offers their closed platforms (often for commercial and business reasons), the consumer is faced with an ever growing 'collection' of solutions that may each address a specific part of the business needs. These interoperability issues concerning cross-platform connectivity and utilisation of software applications and data integrity have only been recently acknowledged by industrial platforms like Siemens, Rockwell, etc. [1, 2]. The use of proprietary standards or even lack of transparency about standards, protocols and technologies used by the various platforms also raises the entry barriers or bars the integration of innovative solutions in existing platforms. However, all this make commercial sense to the platform providers or vendors, who want to capitalise on controlling the access and usage of their solutions. As a corner stone of Industry 4.0 solutions, Internet of Things (IoT) offers prospects of connecting things and heterogeneous systems. With the advancements in IoT, although many 'things' can interact with each other today, the reality remains that interoperability remains clunky, proprietary and largely controlled by technology vendors.

The challenge of interoperability across digital manufacturing platforms remains a key concern for not only manufacturers (as users and consumers) and software developers of smart factory solutions, but also for policy makers such as European Commission (EC) who have vested interest for businesses to take full advantage of interconnected systems in the manufacturing sector. Based on this interest, in 2016 EC funded 10 research projects under the factories of future public–private partnership to develop reference implementations of multi-sided market ecosystems of smart factory solutions.

With this background, this paper introduces a platform interoperability framework that is designed to gather insight from different digital manufacturing platforms to connect these systems in a structured way. The gathered information is analysed with aim to provide interoperability opportunities for the selected use-case scenarios and pathways for the reference implementation of connected systems and indeed connected factories. The interoperability framework is introduced in Sect. 2. Section 3 provides an overview of a specific use-case adopting the interoperability framework. Finally, Sect. 4 describes the related and future work in the area of platform interoperability to correlate and position the proposed framework with existing efforts.

2 The Platform Interoperability Framework

The proposed platform interoperability framework takes into account the need for vertical digital manufacturing platforms to be communicable and interoperable.

The framework, although in its infancy, is domain agnostic in nature; further development of the framework will focus on the granularity of information it can capture and represent as well as the support it can provide to platform interoperability initiatives, e.g., in the form of recommendations or standardisation. In essence, the

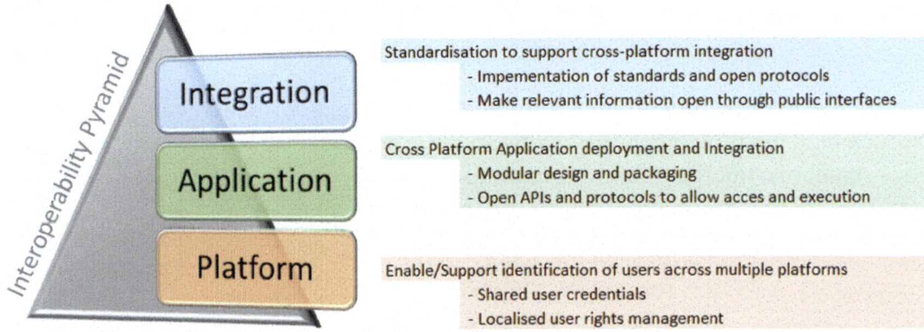

Fig. 1 Platform interoperability framework

interoperability framework broadly promotes transparency along different levels of the platform. Particularly, it breaks down the interoperability issues faced in the digital manufacturing environment into a three-tier hierarchy (i) platform, (ii) application and (iii) integration—as shown in Fig. 1.

2.1 Platform

The first or bottom tier of the proposed platform interoperability framework is concerned with *Platform Management*. It focuses on the separation of 'identification' from 'services' to allow shared access across different platforms. Steps taken and solutions implemented at this level should allow users to navigate through multiple platforms and identify points of interest. Approaches for single sign-on, policy-based access and user right management can all contribute towards interoperability at this level.

2.2 Application

The second or middle hierarchal level of interoperability framework is *Application Management*. Once users have access to the platform, they know what is on offer, so the platform developers and providers should look at ways to allow users to use the application and services on offer, either locally or through remote access. This can be realised by supporting the uptake of modular applications development and packaging approaches that make it easier to port application across multiple platforms. With that, the idea is to support development of interoperable and reusable functionalities that can be ported and offered via different platforms.

2.3 Integration

The top level of interoperability framework is *Integration*, which can prove more complex than the previous levels owing to the complexity of dealing with heterogeneous standards, interfaces and communication protocols. Here, the use of standards at all stages of the information/data flow can allow the applications, tools and services to be interoperable in an ecosystem environment.

3 Analysis of the Interoperability Framework and Use-Cases

The digital manufacturing environment and Industry 4.0 model allows platforms to become more transparent and support wider participation, e.g. the development and uptake of interoperable solutions and services.

With this context and to evaluate the platform interoperability framework, a qualitative study was conducted. During the study, the interoperability framework was introduced to a number of EU-funded research and innovation (factories of future) projects—all focusing on the development of smart factory platforms with multi-sided marketplaces. The projects were asked to provide two inputs:

a. High-level mapping of the project or the platform with the three tiers of the proposed framework as an introduction to the platform—as shown in Fig. 2.
b. Detailed mapping of project activities with the view to define use-cases or reference implementations of interoperable solutions. Some projects from previous step did not participate here—as shown in Figs. 3, 4 and 5.

Note: The names of the projects that participated in the study have been concealed in the following figures pending authorisation from their representatives. Nonetheless, the paper focuses on the usefulness of the proposed interoperability framework in highlighting the opportunities and scope of developing interoperable solutions. The specific collaboration opportunities exposed by different projects is not the focus of this paper.

The high-level mapping of the projects (that participated in the study) with the interoperability framework (as shown in Fig. 2) highlights the complementarities and cross-cutting themes across different projects. These rather generic themes were further investigated in the second step, where the information provided by the projects contributed towards setting up use-case scenarios that can help realise an ecosystem of interoperable smart factory solutions. The information gathered through the second step was organised in Figs. 3, 4 and 5.

The analysis of the information gathered through the Interoperability framework highlighted several similarities in the use of technologies and standards among different projects. The analysis also helped identifying the use-case scenarios that can be developed as reference implementations of interoperable solutions. The reference

FoF Project	Platform	Application	Integration
Project 1	Multi-sided collaborations with end-to-end security and identity management	Finding suppliers and providers of services; executing transactions with 3rd party services	Semantic annotation of products, processes and life-cycle data
Project 2	Distributed automation platform based on IEC-61499 standard	SDK to design distributed control solutions and HW/SW wrapping of legacy systems	IEC61499 based automation
Project 3	Multi-sided collaborations with standardised multi-aspect security and governance mechanisms	Tool-store for smart factory applications	OPC-UA based middleware with gateways to support integration of external applications
Project 4	Decentralised platform for automation, analytics and simulation, via Edge Computing and Ledger Technology	Ecosystem of applications developers, system integrators and OEMs	Extensive support for field automation standards (OPC UA, MQTT, IEC61499); Open API for cloud-based applications.
Project 5	Dynamic reconfiguration of production systems based on Cloud services	Security and privacy based on policy enforcement and DRM	Reconfiguration of production systems
Project 6	Multi-sided collaborations within Virtual Factory environment	SDK and Marketplace of smart factory applications	Data ingestion from shop-floor to smart applications
Project 7	Intra & Inter-Factory Collaboration & Optimisation Framework through Secure log-oriented ecosystem	Open Marketplace for supply & logistics, Open Ontology to support matchmaking services, Integrated Digital Factory Model, Secure Data Management in Manufacturing	Interoperability Among Different Data Collection Systems in Manufacturing, Adaptation Layer for Intra-Factory interoperability

Fig. 2 High-level mapping of platforms using the proposed framework

implementations pave the way for the creation of an ecosystem of solutions with cross-platform applicability. Some examples of the interoperability scenarios that emerged from the information gathered through the interoperability framework are discussed in the following sections.

3.1 Platform-Level Interoperability Scenarios

The analysis of the gathered information highlighted that the use of open standards for security and user authorisation such as OpenID Connect, OAuth 2.0 and SAML 2.0 and their implementation in open-source toolkits such as Keycloak [3] is common across multiple projects—as shown in Fig. 3. The use of common standards and open-source tools provides the platform developers or owners the opportunity to create federations or single sign-on solutions that contribute towards attracting more users to their platform. The challenge is to enable and promote interoperability while

Separation of 'Identification and Services			
FoF Project	**Authorisation and Authentication**	**User Access and Rights Management**	**Motivating Scenario / Use-case**
Project 1	**KeyCloak** - based on Standards (OpenID Connect, OAuth 2.0 and SAML 2.0)	**KeyCloak** based Identity and Access Management of Platform	SMEs can search for new supply chain partners and establish secure business collaborations via the platform
Project 2	Not foreseen for IEC-61499 but possible if performance (over real-time) overhead is acceptable. It needs industrial requirements to be justified.	Developing a solution specific for IEC-61499. Open to standardised methods to interface to existing SoA approaches in higher levels of automation.	Within IEC-61499, justified only if industrial requirements exists; the standard accepts the possibility to be extended, interfacing to other platforms, with open gateways.
Project 3	**KeyCloak** - based on Standards (OpenID Connect, OAuth 2.0 and SAML 2.0)	**KeyCloak** based Identity and Access Management of platform of the Project 3 Platform	Users from partner platform(s) can access the Project 3 Marketplace and view advertised tenders
Project 4	**OpenID Connect** is the candidate technology for securing the Open API)	New Generation Access Control (**NGAC**) framework for the specification and enforcement of access policies	
Project 5	Next Generation Access Control (**NGAC**) based authorisation and authentication	**NGAC** based access control management in the Project 5 solution	
Project 6	Centralised token based security framework for REST APIs	REST APIs for role based access management and security policy	Flexible security framework using open standards where Project 6 applications can use APIs to configure security at runtime
Project 7	**KeyCloak** (Authentication, Message Broker)	Role-based identity management through **KeyCloak** (Clients, Roles, Users)	Open & Secure (Log-oriented) Marketplace for bringing together Supply & Logistics value chain

Fig. 3 Platform-level mapping between different projects

preserving IPR, technology preferences, niche functionalities and business interests. Thus, a balancing act is required to support business and exploitation interests while establishing a horizontal interoperability layer across vertical platforms.

3.2 Application-Level Interoperability Scenarios

The interoperability at application layer can be supported by the uptake of modular applications development and packaging approaches (e.g. Docker [4]) that make it easier to port application across multiple platforms. With that the use or adoption of such technologies can support the development of reusable functionalities that

Cross Platform Application Deployment and Integration			
FoF Project	**Modular Design and Deployment Approaches**	**Open APIs and Communication Protocols**	**Motivating Scenario / Use-case**
Project 1	(1) **Docker**-based (2) Bluemix/Kubernetes-based cloud solution	Project Task 2.3 OpenAPI for the Platform	To enable an ecosystem that allows 3rd parties to connect their solutions to the services portfolio
Project 2	IEC-61499 is modular and OO in its approach to real-time automation. Moreover, it allows to "encapsulate" interfaces to other platform coherently with this OO approach.	IEC-61499 is open and interoperable by design. From within IEC-61499, interfaces to other platforms can be developed in C++ to generate specific APIs. Other platforms can interact over an IEC-61499 communication bus without changing internal programming.	Integrating "digital" applications running in other platform into the real-time automation engineering environment is a very interesting use-case
Project 3	**Docker** based application packaging and deployment is adopted in Project 3	Communication with middleware is supported by open data model to be developed	Project 3 applications are developed as standalone plug-and-play tools with dedicated interfaces. External applications (e.g. production planning, scheduling etc) can also be deployed on the tool store
Project 4	Platform composed of self-contained modules. Business logic for automation and analytics is packaged **Docker**.	The Platform defines its own Open API	
Project 5	**Docker** based application packaging and deployment will be used in the Project 5	Publish / Subscribe message system Apache Kafka will be used for communication in the platform	Each service will be developed as own **docker** container, to allow easy distribution and scalability of whole solution in the cloud
Project 6	**Dockerised** components to adapt a Platform instance to the requirements of a specific tenant	Middleware providing messaging and Pub/Sub Services to all components. VF Devices Drivers and Open APIs, a set of modules that provide interfaces to physical assets (eg, Sensors) and virtual assets (eg, ERP systems and data)	vApps (Applications) will be offered to manufacturing companies at the Manufacturing Applications Store at the Virtual Factory Platform (vf-P) to spread throughout the manufacturing domain
Project 7	**Docker**-Based application packaging for standalone & integrated framework	Integrated Digital Factory Model for Intra-Factory & MarketPlace Open Reference Data Model for Inter-Factory Support (bids, offers and transactions, etc.)	Log-Oriented Architecture through Blockchain Implementation for providing Audit Trail for Manufacturing & Supply Chain data, Matchmaking services (incl. semantic queries) as part of Marketplace

Fig. 4 Application-level mapping between different projects

Standardisation to Support Cross Platform Integration			
FoF Project	**Standards for Data Exchange**	**APIs and Integration Protocols**	**Motivating Scenario / Use-case**
Project 1	UBL for business processes, eClass for products, domain specific ontologies aligned via light-weight upper ontology	Mainly driven by UBL documents connected to defined business processes; OpenAPI	Finding new supply chain partners, establishing new supply chains and executing them via the platform
Project 2	IEC-61499 established both a programming language and the rules for open communication protocols for real-time control over distributed apps	A cross-platform acceptance of IEC-61499 for the real-time control domain for distributed applications guarantee interoperability	Interaction between non real-time and real-time IIoT could be guaranteed through the standard.
Project 3	**OPC-UA** based middleware and standardised authorisation protocols to support cross-platform data exchange and integration	Toolstore can host external/standalone applications	External applications can communicate with **OPC-UA** based middleware (subject to authentication etc) to ingest data e.g. from factories
Project 4	AutomationML, SenseML, **OPC UA**		
Project 5	IISF based access control framework to allow integration of additional external components into SAFIRE. **OPC-UA** plug-in for the data ingestion	Data Ingestion module will contain a plug-in architecture to allow integration of various data-source (e.g. **OPC-UA**)	External systems can ingest data into the Project 5 solution by either using pre-developed plug-ins (e.g. **OPC-UA**) or by developing own plug-in modules
Project 6	The pub-sub middleware compatible with major industry standards (i.e. JBI, SCA, BPEL or WSDL). IO Toolkit to implement OPC UA and MQTT. Other standards as ANSI/ISA-95 will be followed.	The IO Toolkit implements clients to the REST based kernel and middleware services facilitating the integration of virtually any technology	Software Developers can develop and upload to the marketplace device drivers and API Connectors for legacy platforms and services
Project 7	1) CyberSecurity For Factories, 2) Intra-Factory Business Process Modelling, 3) Inter & Intra-Factory Message Broker, 4) Data Persistence for IoT Sensor Data, 5) Manufacturing Shopfloor Modelling thro	1) Security Information & Event Management API, 2) BPMN standard as part of the Integrated Digital Factory Model, 3) RabbitMQ implementation, 4) OGC sensor things compliant API through Integrated Digital Factory Metadata Model, 5) Part of the Integrated Digital Factory Model	1) Traffic Monitoring in IoT-enabled shopfloors, 2) Intra-factory Optimisation & Decision support (i.e. forecasting services, trend analysis, predictive maintenance), 3, 4, 5) Robust module communication & message routing for Intra & Inter-scenarios

Fig. 5 Integration-level mapping between different projects

can be ported and offered via different platforms. Again, the information gathered through the interoperability framework showed that a number of ongoing projects are deploying their applications in containers or cloud-based services—as shown in Fig. 4. Thus, a number of use-case scenarios can be drawn from the cross-platform applicability of applications and services developed in different projects. The smart factory applications can be made available to a wider community and also protect the IPR and exploitation interests of individual projects.

3.3 *Integration-Level Interoperability Scenarios*

At the top level of interoperability framework, Integration can prove more complex than the other levels owning to the need for greater transparency both at platform and application ends. However, the integration of cross-platform applications can be facilitated through the use of open standards at both ends. For example, the information gathered through the interoperability framework shows (in Fig. 5) that many ongoing EC projects are investigating middleware solutions to link data sources with applications. Here, the use of standardised approaches (e.g. OPC-UA [5]) can allow developers to make their applications, tools and connectors to interconnect and communicate with the standard middleware. Also at the platform level, open APIs or publicly available application programming interfaces can provides developers with programmatic access to the underlying middleware that can expose information to the applications.

4 Related and Future Work on Platform Interoperability

Research on interoperability issues tends to focus more on organisational or enterprise level where interoperability is seen as a bottleneck in supply chains. There the approaches like Zachman Framework, ARIS, TOGAF, DoDAF and many more can be considered in the architecture of collaborative enterprises [6]. Similarly, the European initiatives such as ETHENA [7] and INTEROP's MDI framework [8] provide reference interoperability models defined at different levels of abstraction.

Platform interoperability is also relevant in the organisational context, e.g., European Interoperability Framework (EIF) [9] is a set of recommendations that specify how public-sector administrations communicate with businesses and citizens or the other way around. EIF supports the free movement of data or data portability and promotes seamless services and data flows between partner organisations in the public sector. In terms of interoperability at application level, Apache Ignite framework [10] allows different software development platforms, such as.NET, Java and C++, to interoperate with each other in a way that classes defined on different platforms could be converted to each other.

At integration level, the BIG IoT project [11] proposed a common API for enabling IoT ecosystem. The key components of the API are identity management, discovery of resources, access to metadata and data, tasking, vocabulary management, security management and charging which provides 'Cross-Platform Access', 'Cross Application Domain Access', 'Platform Independence', 'Platform-Scale Independence', and 'Higher-level Service Facades' Pattern.

These scenarios presented in [11] can be mapped to the three layers of the interoperability framework proposed in the paper. The analysis in Sect. 3 provides more concrete interoperability scenarios by highlighting the common features between different platforms. In this respect, the proposed interoperability framework can provide the foundation for the development of a common API (as in [11]) based on the mapping of similarities at different levels of the platform.

In addition, the coordination or clustering on IoT and smart factory solutions are also contributing towards harmonising various developments to shape up interoperable solutions. For example, the European Research Cluster on the Internet of Things (IERC) [12] is addressing the large potential for IoT-based capabilities in Europe and coordinating the convergence of ongoing IoT platform development activities. Similarly, the European Factories of Future Research Association (EFFRA) is coordinating a coordination and support action called connected factories to provide a structured overview of available and upcoming technological approaches and best practices in the smart factory and Industry 4.0 area.

This paper describes the initial outcomes of our contributions towards establishing an ecosystem of smart factory and Industry 4.0 solutions. The idea is not to achieve short-term objectives (e.g. working examples) but to promote more transparency and adoption of open tools and standards that guarantee interoperability in a sustainable way and not as a one-off target. Thus, interoperability agreements should be detailed enough to achieve their aim while leaving each stakeholder in the collaboration/value chain the maximum feasible internal autonomy.

References

1. https://www.rtaautomation.com/connecting-rockwell-siemens-automation-systems/.
2. http://news.autodesk.com/2016-03-01-Autodesk-and-Siemens-Sign-Agreement-to-Increase-Software-Interoperability.
3. Keycloak: Open source identity and access management. http://www.keycloak.org/.
4. Docker software containerisation platform. https://www.docker.com/.
5. OPC unified architecture. https://opcfoundation.org/about/opc-technologies/opc-ua/.
6. Chen, D., Doumeingts, G., & Vernadat, F. (2008). Architecture for enterprise integration and interoperability: Past, present and future. *Computers in Industry, 59*(7), 647–659.
7. Elvesæter, B., Hahn, A., Berre, A. J., & Neple, T. (2006). Towards an interoperability framework for model-driven development of software systems. In *Interoperability of enterprise software and applications* (pp. 409–420). London: Springer.
8. INTEROP NoE deliverables. http://interop-vlab.eu/ei_public_deliverables/interop-noe-deliverables.
9. The new European interoperability framework. https://ec.europa.eu/isa2/eif_en.

10. Apache ignite—In memory computing platform. https://ignite.apache.org/.
11. Bröring, A., Schmid, S., Schindhelm, C. K., Khelil, A., Kabisch, S., Kramer, D., et al. (2017). Enabling IoT ecosystems through platform interoperability. *IEEE Software, 34*(1), 54–61.
12. European research cluster on the IoT. http://www.internet-of-things-research.eu/.

Interoperable Algorithms for Its Implementation in a Cloud Collaborative Manufacturing Platform

Beatriz Andres, Raúl Poler, Raquel Sanchis, Josefa Mula and Manuel Díaz-Madroñero

Abstract Cloud collaborative manufacturing networks (C2NET), European H2020 funded, project has three main areas of research, including data collection, optimisation and collaboration tools. The three modules are integrated in the C2NET cloud platform. This paper focuses on the C2NET optimiser module that gathers a collection of algorithms to address enterprise' plans, which range from replenishment to delivery through manufacturing. The algorithms are designed considering a standardised data model, making the optimisation planning an interoperable process between the enterprises and the C2NET cloud platform. The first and second tiers of an automotive supply chain take part in C2NET project, and the specificities of automotive industries are identified. In this regard, the need of computing the injection moulding-sequencing plan. In order to fulfil the requirements of the automotive pilot, an optimisation algorithm is designed and presented as a proof of concept about the interoperable characteristics of the designed algorithms.

Keywords Cloud collaborative manufacturing platform · Smart service · Optimisation · Algorithm · Interoperability · Automotive sector

B. Andres (✉) · R. Poler · R. Sanchis · J. Mula · M. Díaz-Madroñero
Research Centre on Production Management and Engineering (CIGIP), Universitat Politècnica de València (UPV), Calle Alarcón, Alcoy, Spain
e-mail: bandres@cigip.upv.es

R. Poler
e-mail: rpoler@cigip.upv.es

R. Sanchis
e-mail: rsanchis@cigip.upv.es

J. Mula
e-mail: fmula@cigip.upv.es

M. Díaz-Madroñero
e-mail: fcofcodiama@cigip.upv.es

K. Popplewell et al. (eds.), *Enterprise Interoperability VIII*, Proceedings of the I-ESA Conferences 9, https://doi.org/10.1007/978-3-030-13693-2_8

1 Introduction

The European industry needs advanced methods and tools to support complex decision-making, namely the management of co-evolution of products-services and the related production and planning systems, the evaluation of alternative configurations of the network of actors involved in the global and collaborative supply chain or integration of new interoperable technologies in the factory. Particularly, small and medium enterprises (SMEs) have fewer possibilities to acquire and implement novel interoperable technologies and algorithms to deal with globalised, highly competitive and agile markets [1]. Despite the SMEs' willingness to face the troubles derived from the participation in collaborative relationships [2] and the operation in complex turbulent markets, there is still a gap to fill with regards providing more affordable, simple ubiquitous tools to support them. The research contribution goes further when these tools have also to provide interoperable functionalities. Besides designing interoperable tools, to aid enterprises agility and collaboration, there is a need to provide advanced algorithms to achieve holistic global and local optimisation of manufacturing assets and to respond faster and more efficiently to unforeseen changes. These algorithms should also work in an interoperable environment, so that they can provide planning results, regardless the nature of the input data needed to run the algorithms.

Cloud collaborative manufacturing networks (C2NET) European H20202 project [3, 4] has its main objective in overcoming the mentioned challenges, in the scope of enterprises planning, interoperability and collaboration. In the light of this, the main aim of the C2NET is to support enterprises in the automatic calculation of plans, the exchange of information in an interoperable way and the establishment of collaborative processes, by providing novel tools and methods embedded in the C2NET cloud platform in order to: (i) master complexity and data security within the supply network; (ii) store and share product, process and logistic data, in an interoperable way; (iii) optimise the manufacturing assets by the collaborative computation of production plans; (iv) optimise the logistics assets through efficient delivery plans; and (v) render the complete set of supply-chain management information on the any digital mobile device (PC, tablets and smartphones) of decision makers enabling them to monitor, visualise, control, share and collaborate. To this extent, C2NET cloud platform will incorporate interoperability functionalities, to facilitate the integration of enterprise legacy systems, in order to work in a harmonised way at all the levels of the enterprise network. C2NET modules will be designed and implemented following the model-driven interoperability paradigm, assuring the current and future interoperability with enterprise applications.

Finally, C2NET project has the purpose to connect the industrial, the research and the development perspectives. In this regard, this paper focuses on one of the project pilots, the automotive one, which groups the first- and second-tier suppliers. The study centres its attention on the needs that automotive enterprises have regarding their production plans [4], concluding that there is a need to compute the injection moulding-sequencing plan. In order to fulfil the requirements of the automotive pilot, an optimisation algorithm is presented as a proof of concept about the interoperable characteristics of the designed algorithms.

In order to show the innovative results obtained in C2NET project, this paper is organised as follows: Sect. 2 proposes an overview of the different modules of C2NET project, emphasising on the C2NET optimiser (C2NET OPT). Section 3 puts its complete attention in the C2NET OPT module, including (i) an overview and definition of the main concepts that made possible the creation of C2NET OPT, and a description of the main components involved in the optimisation; and (ii) a summary of the interoperable algorithms developed considering the needs of the automotive industrial pilot. As an application example, Sect. 4 shows as a proof of concept an interoperable optimisation algorithm designed to deal with the injection moulding-sequencing plan, fulfilling the specificities identified in the automotive pilot. Finally, Sect. 5 discusses the work performed in the paper and defines the future research lines.

2 C2NET Architecture

C2NET project has been developed with the aim of filling the gaps encountered in terms of SMEs' collaboration, data exchange and interoperability, and plans optimisation; making all its developments in a cloud platform. In this regard, C2NET project has designed an architecture composed by the cloud platform (C2NET CPL), the data collection framework (C2NET DCF), the repository of optimisation and interoperable algorithms (C2NET OPT), and the collaboration tools (C2NET COT) [5]. Very briefly, C2NET DCF ensures interoperability in a continuous data collection process from different supply network resources [6]; C2NET OPT gathers a set of interoperable algorithms to automatically deal with the enterprises' planning process [7]; C2NET COT provides support to the collaborative processes of the supply network [3, 4]; and C2NET CPL integrates the data module, the optimisers and the collaborative tools in the C2NET cloud platform [4].

This section centres its attention in C2NET OPT, developed for the optimisation of manufacturing and logistics assets of the supply network by the collaborative computation of production, replenishment and delivery plans, in an interoperable context, in which the algorithms are standardised enough using a standardised data model that allows enterprises to obtain planning results, regardless the nature of the input data needed to run the algorithms. The interoperable algorithms have been classified using the supply-chain operations reference (SCOR) model views, regarding the sourcing, production and delivery requirements [8]. Algorithms to solve source (S), make (M) and deliver (D) plans are designed, in the scope of C2NET OPT. Moreover, in order to address combinations of S, M or D plans, between two or more enterprises of the network, the following plans have been also considered: SM, MD, SD and SMD [9].

The identification of literature plans and algorithms, classified according to the type of Plans (S, M, D, SM, MD, SMD), has been performed and supported by the analysis of C2NET industrial pilots. From the analysis of literature plans and the industrial pilot plans, a set of input data, objectives and output data have been identified [7, 9–12].

In order to deal with the input data acquisition, needed to feed the designed algorithms from different sources, a proposal of a standardised data model has been created, enabling the input data homogenisation, and therefore, providing interoperability functionalities within the C2NET cloud platform modules. The standardised data model will allow collecting in C2NET cloud platform data from different enterprises using different legacy systems. For the homogenisation, of input data, a collection of fields are gathered in standardised tables (STables), allowing unifying all the input data in a consistent way, and managing big data collected from different industries [13]. The standardised data model, composed by the STables enables, C2NET cloud platform to be interoperable. Accordingly, C2NET cloud platform behaves as a modular architecture for interoperability of intra-plant and extra-plant processes (calculation of individual and collaborative plans) in conjunction with mobile, collaborative tools for data sharing, data analytics and knowledge-based systems, at factory and at supply network level. Achieving interoperability between legacy systems of the supply network partners and C2NET cloud platform embedded tools.

3 C2NET OPT Module

3.1 C2NET OPT Characterisation

This section introduces an overview and a definition of the main concepts in which C2NET OPT is based on. C2NET OPT will work with a set of entities with its own taxonomy and different levels of detail [9]:

- Units: are the set of nodes in a supply chain for which plans will be calculated, e.g. a unit can be one plant of an enterprise, several plants of the same enterprise or several plants of several enterprises;
- Plans: are the output information from C2NET OPT; it can be understood as the solution to the problems to be solved by the optimisation algorithms, e.g. a plan is calculated for a unit. The plans have been classified according to the SCOR model, including S, M, D, SM, SD, MD and SMD;
- Needs and resources: are the standardised input data; it can be understood as the problems to be solved by the optimisation algorithms in order to obtain the solutions (plans), e.g. needs and resources can pertain to the unit which calculates the plan or can be external to the unit; and
- Algorithms: are the means to solve the problems, from an interoperable perspective, answering the planning needs using the resources and standardised input data, coming from different enterprises' legacy systems, for carrying out an automated planning process.

C2NET OPT module is composed of four main components. The way how the components are designed allow identifying the best alternatives to compute S, M and D plans. A brief description of each component proposed next [5]:

- Optimisation algorithms (OA): hosts a set of 48 interoperable optimisation and heuristics algorithms, modelled using JuMP (Julia for mathematical optimisation), an algebraic modelling language embedded in Julia [14]. These interoperable algorithms are classified according to the type of problem to be solved (S, M or D).
- Solver manager (SM): It is the component incharge of managing algorithms. As such, it allows creating, editing and categorising algorithms and objective functions. SM automatically identifies the most appropriate interoperable algorithm to be used for solving the S, M and D plans. The automated selection is performed according to some criteria, such as the gap (accuracy of the results provided by the interoperable algorithm) and solving time.
- Optimisation problem configurator (OPC): It is incharge of creating, defining and configuring the different S, M and D plans.
- POMA manager: It is incharge of computing and calculating the optimisation of a specific optimisation problem.
- Standardised data model (STables): It is a meta-structure organised to store, in a standardised and interoperable way, the data coming from the different enterprises' legacy systems. The data model provides a common structured terminology to offer a shared understanding of all the different input data used to compute the interoperable algorithms and support the definition and calculation of replenishment (S), manufacturing (M) and delivery (D) plans [13].
- Plan data model (PTables): Following the same principles as STables, the PTables offer a standardised meta-structure to hold the optimisation results derived from the implementation of the interoperable algorithms.

3.2 *Interoperable Algorithms in the Scope of the Automotive Industry*

A set of interoperable algorithms have been developed, in the scope of the automotive industry pilot, considering the first- and second-tier specificities and their willingness to include automatic calculation of plans using real-time information, standardised and updated in the STables. Some examples are briefly described next (see Table 1). The algorithms designed by the authors of Table 1 are written in a mathematical notation. Nevertheless, in the algorithms repository located in C2NET OPT module, which is embedded in C2NET cloud platform, the algorithms listed in Table 1 are written considering the standardised data model provided in [13]. In order to have a better and extended insight, we refer readers to review the references.

The algorithms proposed in Table 1 are considered relevant by the enterprises belonging to the automotive pilot, which were more focused in the calculation of intra-enterprise materials requirement plan (S) [15], injection moulding plans (M)

Table 1 Algorithms designed for the automotive pilot

Author	Algorithm description	Plan type
[15]	A holistic algorithm for materials requirement planning in collaborative networks	S
[16]	Collaborative calculation of the materials requirement planning in the automotive industry	S
[17]	A MILP for multi-machine injection moulding sequencing in the scope of C2NET project	M
[18]	A MILP for mono-machine injection moulding sequencing	M
[19]	Capacitated lot-sizing and scheduling problem for second-tier suppliers in the automotive sector	M
[20]	A single-machine capacitated lot-sizing and scheduling model for the bi-part injection moulding problem	M
[21]	An optimisation approach for procurement transport operational planning in an automobile supply chain	D
[22]	A tabu search approach for production and sustainable routing planning decisions for inbound logistics in an automotive supply chain	D
[23]	Decentralised replenishment-production planning optimisation using negotiation rules in a collaborative network	SM
[24]	A mathematical programming model for integrating production and procurement transport decisions	SMD

[17–20] and procurement transport [21]. Collaboratively speaking, at the interenterprise level, the enterprises of the automotive pilot were interested in the calculation and negotiation of collaborative materials requirement plans (S collaborative) [16], the collaborative replenishment-injection plans (SM collaborative) [23] and the integrated replenishment, production and transport plans [24].

In order to show the standardised notation, established in the STables and PTables, and defined for the interoperability of C2NET cloud platform, an interoperable algorithm for mono-machine injection moulding sequencing based on [18] is presented. The interoperable algorithm proposed takes part of the repository of algorithms embedded in the C2NET cloud platform.

4 An Interoperable Algorithm for Mono-Machine Injection Moulding Sequencing

The proposed interoperable algorithm, for mono-machine injection moulding sequencing, is presented as a proof of concept about the interoperable characteristics of the algorithms designed in the scope of C2NET OPT, embedded in the C2NET cloud platform. The interoperable algorithm uses the fields of the STables and PTables defined in the standardised data model [13]. This algorithm is based on

Table 2 Nomenclature: STables and PTables

STable.Field	Definition
Tool.ToolID	C2NET unic identifier for moulds
Tool.SetupCost	Cost of changing a mould
Part.PartID	C2NET unic identifier of products
Part.AvailabilityAmount	Number of products in the inventory
Part.AvailabilityCost	Cost of storing a product in the inventory
Part.AvailabilityMaximumAmount	Maximum amount of products allowed in the inventory
Part.AvailabilityMinimumAmount	Minimum amount of products required in the inventory
Part.DelayCost	Cost product delay
Period.PeriodID	C2NET unic identifier for time periods
Period.NormalOperationTime	Amount of production time available in a period
Part_Period.RequirementAmount	Demand of product (PartID) in the time period (Period ID)
Part_Tool.NormalOperationAmount	Amount of products (PartID) produced in the mould (ToolID) in a specific period
S_PSE_A.ToolID	C2NET unic identifier for moulds
S_PSE_A.PeriodID	C2NET unic identifier for time periods
S_PSE_A.Sequence	1 if the tool becomes part of the sequence; 0 otherwise
S_PSE_A.SetupAmount	Number of set-ups of the tools in the period
S_PSE_B.PartID	C2NET unic identifier for products
S_PSE_B.PeriodID	C2NET unic identifier for time periods
S_PSE_B.AvailabilityAmount	Amount of parts available in the inventory in the period
S_PSE_B.DelayAmount	Amount of parts delayed in the period
S_PSE_B.NormalOperationAmount	Amount of the parts produced in the period

the MILP proposed by [18]; see Table 2 for the STables and PTables used in the algorithm and its definition. PTables are defined considering that the algorithm solves a make plan (M), specifically a production sequencing plan (PSE); the different PTables are ordered using consecutive letters (A, B, C, etc.). Finally, the algorithm is designed in Julia language notation [14] (see Table 3).

Briefly, the objective function minimises the inventory, backorders and set-up costs. Constraint C1 limits the number times to set up a mould in a time period. C2 indicates the production of products based on the production rate. C3 and C4 are related to the sequence and setting of the moulds. C5 and C6 represent the inventory balance equations. Finally, C7 and C8 limit the inventory levels for each product according to the available space for inventory holding. The detailed description of the standardised algorithm, described in Table 3, is presented in [18] using a nor-

Table 3 Interoperable algorithm Julia language notation

Indexes
oo = importdb(con,"Tool", "ToolID", "ToolID")
pp = importdb(con,"Part", "PartID", "PartID")
tt = importdb(con,"Period", "PeriodID", "PeriodID")
Data
SetupCost = importdb(con,"Tool", "SetupCost", "ToolID")
NormalOperationAmount = importdb(con,"Part_Tool","NormalOperationAmount", "ToolID, PartID", d(oo), d(pp))
AvailabilityMaximumAmount = importdb(con,"Part", "AvailabilityMaximumAmount", "PartID")
AvailabilityMinimumAmount = importdb(con,"Part", "AvailabilityMinimumAmount", "PartID")
AvailabilityAmount = importdb(con,"Part", "AvailabilityAmount", "PartID")
AvailabilityCost = importdb(con,"Part", "AvailabilityCost", "PartID")
DelayCost = importdb(con,"Part", "DelayCost", "PartID")
RequirementAmount = importdb(con,"Part_Period", "RequirementAmount", "PartID, PeriodID", d(pp), d(tt))
NormalOperationTime = importdb(con,"Period", "NormalOperationTime", "PeriodID")
Variables
@variable(m, 0<=vSequence[1:d(oo),1:d(tt)]<=1, Int)
@variable(m, vSetupamount[1:d(oo),1:d(tt)] >= 0, Int)
@variable(m, vNormalOperationAmount[1:d(pp),1:d(tt)] >= 0, Int)
@variable(m, vAvailabilityAmount[1:d(pp),1:d(tt)] >= 0, Int)
@variable(m, vDelayAmount[1:d(pp),1:d(tt)] >= 0, Int)
Expressions
@expression(m, TSetupCost, sum{SetupCost[o]*vSetupamount[o,t], o=1:d(oo), t=1:d(tt)})
@expression(m, TAvailabilityCost, sum{AvailabilityCost[p]*vAvailabilityAmount[p,t],p=1:d(pp),t=1:d(tt)})
@expression(m, TDelayCost, sum{DelayCost[p]*vDelayAmount[p,t],p=1:d(pp),t=1:d(tt)})
Objective
@objective(m, Min, TAvailabilityCost + TDelayCost + TSetupCost)
Constraints
@constraint(m, C1[t=1:d(tt)], sum{vSequence[o,t],o=1:d(oo)}<=1)
@constraint(m, C2[p=1:d(pp),t=1:d(tt)], vNormalOperationAmount[p,t]== sum{NormalOperationAmount[o,p]*NormalOperationTime[t]*vSequence[o,t],o=1:d(oo)})
@constraint(m, C3[o=1:d(oo),t=1], vSetupamount[o,t] == vSequence[o,t])
@constraint(m, C4[o=1:d(oo),t=2:d(tt)], vSetupamount[o,t] >= vSequence[o,t]-vSequence[o,t-1])

(continued)

Table 3 (continued)

@constraint(m, C5[p=1:d(pp),t=1], vAvailabilityAmount[p,t] == AvailabilityAmount[p]+vNormalOperationAmount[p,t]-RequirementAmount[p,t]+vDelayAmount[p,t])
@constraint(m, C6[p=1:d(pp),t=2:d(tt)], vAvailabilityAmount[p,t] == vAvailabilityAmount[p, t-1]+vNormalOperationAmount[p,t]-RequirementAmount[p,t]+vDelayAmount[p,t]-vDelayAmount[p,t-1])
@constraint(m, C7[p=1:d(pp),t=1:d(tt)], vAvailabilityAmount[p,t] >= AvailabilityMinimumAmount[p])
@constraint(m, C8[p=1:d(pp),t=1:d(tt)], vAvailabilityAmount[p,t] <= AvailabilityMaximumAmount[p])

malised mathematical notation in which the indexes, parameters, objective function and constraints are provided.

5 Conclusions and Future Research Lines

C2NET project has provided, especially to SMEs, affordable interoperable tools to exchange real-time information and automatically compute individual and collaborative plans. A literature review of algorithms to solve plans in the scope of the automotive pilot is carried out. The reviewed algorithms have been included in C2NET OPT interoperable algorithms repository. Amongst all the reviewed algorithms the automotive industrial partners have identified as relevant the injection moulding-sequencing plan. In order to fulfil the requirements of the automotive pilot, an interoperable optimisation algorithm is designed and presented as a proof of concept about the interoperable characteristics of the designed algorithms. Julia language is used for implementing the algorithm. The interoperability associated to the algorithm allows using the same algorithm to compute plans in other enterprises, belonging to other industrial sectors, whose input data is defined differs from one enterprise to another. The standardised data model provides a set of STables, and fields associated to these STables.

The algorithms are designed considering a standardised and interoperable data model, making the optimisation planning an interoperable process between the enterprises and the C2NET cloud platform. The validation of the proposed interoperable algorithm has been carried out in the C2NET cloud services. Future work leads to identify more enterprises plans and needs, for the proposal of novel interoperable algorithms to automatically compute collaborative plans, such as MD, SD or SMD. In this regard, the main objective of the novel interoperable algorithms is to automate the calculation of plans regardless the nature of the input data used to run the algorithm, with the aim of reducing costs and increase flexibility in the decision-making; considering both individual and collaborative perspectives. Future research lines will be devoted to continue developing interoperable algorithms (using STables and PTables) for enterprises, of different industrial sectors, considering specific needs, in order to complete the repository of algorithms in C2NET OPT, embedded in C2NET cloud service.

Acknowledgements The research that led to these results forms part of the 'Cloud Collaborative Manufacturing Networks' (C2NET) Project, which has received funding from the EU Horizon 2020 Research and Innovation Programme with grant agreement No. 63690.

References

1. European Commission. (2012). *Reaping the benefits of globalization*. European competitiveness report. Commission staff working document. SWD(2012)299 final.
2. Andres, B., & Poler, R. (2015). Models, guidelines and tools for the integration of collaborative processes in non-hierarchical manufacturing networks: A review. *International Journal of Computer Integrated Manufacturing, 29*(2), 166–201.
3. Lauras, M., Lamothe, J., Benaben, F., Andres, B., & Poler, R. (2015). Towards an agile and collaborative platform for managing supply chain uncertainties. In *International IFIP Working Conference on Enterprise Interoperability* (pp. 64–72). Berlin: Springer.
4. Andres, B., Sanchis, R., & Poler, R. (2016). A cloud platform to support collaboration in supply networks. *International Journal of Production Management and Engineering, 4*(1), 5–13.
5. Sanchis, R., Andres, B., Poler, R., Mula, J., & Diaz-Madroñero, M. (2017). The C2NET optimisation solution. *Dirección y Organización* (in press).
6. Mohamed, W. M., Ramis, B., Sanchis, R., Andres, B., & Agostinho, C. (2017). A multi-agent approach for processing industrial enterprise data. In *23rd ICE/IEEE International Technology Management Conference*.
7. Andres, B., Sanchis, R., Lamothe, J., Saari, L., & Hauser, F. (2017). Integrated production-distribution planning optimization models: A review in collaborative networks context. *International Journal of Production Management and Engineering, 5*(1), 31–38.
8. APICS. (2017). SCOR framework. *Supply chain operations reference model (SCOR)*. [Online]. Available: http://www.apics.org/apics-for-business/products-and-services/apics-scc-frameworks/scor.
9. Andres, B., Saari, L., Lauras, M., & Eizaguirre, F. (2016). Optimization algorithms for collaborative manufacturing and logistics processes. In M. Zelm, G. Doumeingts, & J. P. Mendonça (Eds.), *Enterprise interoperability in the digitized and networked factory of the future* (pp. 167–173). iSTE.
10. Andres, B., Sanchis, R., Lamothe, J., Saari L., & Hauser, F. (2016). Combined models for production and distribution planning in a supply chain. In *Building bridges between researchers and practitioners. Book of Abstracts of the International Joint Conference CIO-ICIEOM-IISE-AIM (IJC2016)* (p. 71).
11. Orbegozo, A., Andres, B., Mula, J., Lauras, M., Monteiro, C., & Malheiro, M. (2018). An overview of optimization models for integrated replenishment and production planning decisions. In *Closing the gap between practice and research in industrial engineering. Lecture Notes in Management and Industrial Engineering* (pp. 239–247).
12. Andres, B., Poler, R., Saari, L., Arana, J., Benaches, J. V., & Salazar, J. (2018). Optimization models to support decision-making in collaborative networks : A review. In *Closing the gap between practice and research in industrial engineering. Lecture Notes in Management and Industrial Engineering* (pp. 249–258).
13. Andres, B., Sanchis, R., Poler, R., & Saari, L. (2017). A proposal of standardised data model for cloud manufacturing collaborative networks. In *IFIP Advances in Information and Communication Technology* (Vol. 1, pp. 77–85).

14. Julia. (2017). [Online]. Available: https://julialang.org/.
15. Andres, B., Poler, R., & Sanchis, R. (2017). A holistic algorithm for materials requirement planning in collaborative networks. In *IFIP Advances in Information and Communication Technology* (Vol. 506, pp. 41–50).
16. Andres, B., Sanchis, R., Poler, R., & Saari, L. (2017). Collaborative calculation of the materials requirement planning in the automotive industry. In *23rd ICE/IEEE International Technology Management Conference*.
17. Andres, B., Sanchis, R., Poler, R., Mula, J., & Díaz-Madroñero, M. (2017). A MILP for multi-machine injection moulding sequencing in the scope of C2NET project. *International Journal of Production Management and Engineering, 6*(1), 29–36.
18. Sanchis, R., Andres, B., Poler, R., Diaz-Madroñero, M., & Mula, J. (2017). A MILP for mono-machine injection moulding sequencing. In R. Poler, J. Mula, M. Díaz-Madroñero, & R. Sanchis (Eds.), *New global perspectives on industrial engineering and management. The 3rd International Joint Conference ICIEOM-ADINGOR-IISE-AIM-ASEM (IJC2017) Proceedings*. ISBN: 978-84-697-7199-0.
19. Diaz-Madroñero, M., Mula, J., Andres, B., Poler, R., & Sanchis, S. (2018). Capacitated lot-sizing and scheduling problem for second-tier suppliers in the automotive sector. In *Closing the gap between practice and research in industrial engineering* (pp. 121–129). Switzerland: Springer International Publishing.
20. Diaz-Madroñero, M., Mula, J., Andres, B., Poler, R., & Sanchis, R. (2018). A single-machine capacitated lot-sizing and scheduling model for the bi-part injection moulding problem. *Central European Journal of Operations Research* (in Press).
21. Díaz-Madroñero, M., Peidro, D., & Mula, J. (2014). A fuzzy optimization approach for procurement transport operational planning in an automobile supply chain. *Applied Mathematical Modelling, 38*(23), 5705–5725.
22. Peidro, D., Díaz-Madroñero, M., Mula, J., & Navalón, A. (2015). A tabu search approach for production and sustainable routing planning decisions for inbound logistics in an automotive supply chain. In *Enhancing synergies in a collaborative environment* (pp. 61–68). Switzerland: Springer International Publishing.
23. Andres, B., Poler, R., Mula, J., Díaz-Madroñero, M., & Sanchis, R. (2016). Decentralised replenishment-production planning optimisation using negotiation rules in a collaborative network. In *IFIP Advances in Information and Communication Technology* (Vol. 480, pp. 228–240).
24. Díaz-Madroñero, M., Mula, J., & Peidro, D. (2017). A mathematical programming model for integrating production and procurement transport decisions. *Applied Mathematical Modelling, 52*, 527–543.

OpenPaaS::NG: A Cloud-Based Interoperable Enterprise Platform to Support Inter-organizational Collaborations

Sarah Zribi and Jean-Pierre Lorré

Abstract The rapid evolution of the means of communication has led to the multiplication of collaboration tools that are constantly evolving and are now indispensable and introduced as a daily practice of all employees. Enterprises have become aware of the need and the importance of these productivity tools. As such, OpenPaaS::NG is an open-source, interoperable, scalable, extensible enterprise collaboration platform that aims to remove internal collaboration barriers, promote digital openness, and to reduce the gap between the new communication trends of the daily life and the outdated usage of software in business. It also includes artificial intelligence-based recommendation system in order to provide advanced real-time contextual recommendations for meetings' participants. In this paper, we report the work resulting from our involvement in several research projects and we present the main features of the OpenPaaS::NG platform.

Keywords Collaboration · Cloud · Interoperability · Collaboration editors · Web conference · Real-time recommendations · Enterprise social network · Open-source

1 Introduction

Ubiquitous computing, with geolocation and new media, is characterized by the permanent connection, multiplication and synchronization of fixed and mobile devices, videoconferencing and new uses of video, new formats of real-time communication and the omnipresence of social networks. In a few years, the landscape of digital communication and work organization has changed dramatically.

The worlds of business and general public are intertwined in social applications as well as hardware [1] that are both used for business and individual use.

S. Zribi (✉) · J.-P. Lorré
Linagora, 75 route de Revel, 31400 Toulouse, France
e-mail: szribi@linagora.com

J.-P. Lorré
e-mail: jplorre@linagora.com

K. Popplewell et al. (eds.), *Enterprise Interoperability VIII*, Proceedings of the I-ESA Conferences 9, https://doi.org/10.1007/978-3-030-13693-2_9

This multiplication of the modes of interindividual exchanges is not without posing new difficulties:

- Professional and/or individual users are confronted with the infobesity, the fragmentation of their exchanges in silos, the difficulty of reorganizing these dispersed exchanges, threats to their private lives, and the emergence of economic models for which they do not always master the scope;
- Developers or providers of these different solutions are often unable to exceed their initial silos, especially because of their deep technological options. They are then unable to develop substantially tools and especially methods of communication;
- In general, it is difficult to understand the content of these exchanges, to formalize them, and to take advantage of them. Indeed, most of these interactions take place in the form of natural language exchanges, intended for human understanding.

The extended intranet space works more and more as an office from which a multitude of tasks can be performed. Proof of this is the steady growth of instant messaging, web conferencing, webmail tools or direct access to document databases.

According to the latest edition (2017) of its "e-transformation and intranet Observatory" study, the Arctus firm shows that companies are more and more numerous to equip themselves with information exchange platforms or to enrich their collaborative spaces. Indeed, the main intranet trends are:

- Collaborative tools: 75% of companies have collaborative workspaces. Project communities continue to rise to 80%, organization spaces also to 71%. Organizations that have put in place an advanced facility with collaborative functions are developing collaborative branches (collaborative workspaces, sharing functions, etc.).
- Social facilities: a strong growth is noted in the social interaction functions available within companies. The "simple" social functions are the most diffused: Publishing comments 87% and Like function 80%.
- Mobile access: Mobile access is involving. 75% of companies have developed a mobility policy.

Companies have nowadays become aware of the need and the importance to federate into a single interface all collaboration tools and channels of communication of the employee based on enterprise social network (ESN) concepts. In this context, this paper aims to propose a new interoperable enterprise platform-based cloud that is open source and supports both inter- and intra-organizational collaborations and provides real-time recommendations during a meeting.

The remainder of this paper is organized as follows. Section 2 presents a brief state-of-the-art on the existing enterprise collaborative platforms. We sketch in the third section an overview of the proposed architecture. Section 4 describes the main characteristics of the OpenPaaS::NG platform. Finally, the last section concludes and gives insights of future works.

2 Enterprise Collaborative Platforms

Collaboration and communication are becoming an essential business tool. Nowadays, several solutions are available in the related literature to facilitate team collaboration. Among these solutions, we find Office 365, Google Suite, Jive and Podio.

Office 365 is the collaborative solution of Microsoft that provides online services through a single management platform [2]. It covers several sub-categories of products; one of them is intended for individuals and the seven others for enterprises. The solution takes the form of a Web portal from which several services hosted in the cloud are accessible. Office 365 includes a large set of features: collaborative editors, the respect of confidentiality and privacy, data security, authentication, resilience and network autonomy, video conference (Skype), recommendations (Delve), and survey forms and meeting minutes (only world document). Office 365 is an expensive solution and all the features may not be used by most small businesses, but for those that need the capabilities it is the best option. In addition, business process modeling is not covered by the solution.

Google Apps (also called G suite or Google Apps) is a business solution offered by Google, a subsidiary of Alphabet [3]. It comes in the form of a suite of office automation, collaboration, and productivity applications. The solution provides access to a centralized management interface that allows users to navigate between the different applications in their subscription. Each service is an independent application whose access depends on the type of subscription and the choice of the deputy head. Several features are proposed by the Google Apps tool such as: collaboration editors, the respect of confidentiality and privacy, data security, authentication, resilience and network autonomy, video conference (Hangouts), social network (Google+), and survey form. Google Apps provide businesses with a cheaper alternative to Office 365 and is sufficient for users who only need to do basic work with the software. Indeed, Google Apps tool does not provide meeting minutes at the end of the meeting and is not open source. In addition, Google Apps does not offer an app store. All Google Apps are all accessible and launched via the same panel. Moreover, it does not offer a business process modeling editor or recommendation feature beyond suggestions for contacts when reading an e-mail for example.

Jive-n and Jive-x are the two main solutions implemented by Jive company [4]. The first one, Jive-n, is a social intranet with social networking features. It is based on collaboration and knowledge management. Its main features include: creation and participation in online communities, micro-blogging social networking, discussion forums, blogs, wikis, instant messaging, RSS feed management, and integration of messaging services (Google Apps and Office 365). The second one, Jive-x, is a BtoC social network that is to say it allows the user to manage communities of customers, partners, etc. Several features are available within Jive such as the respect of confidentiality and privacy, data security, authentication, resilience and network autonomy, app store, chat and video conference (WebEx), and social network. Nevertheless, collaboration editors, recommendation, and meeting minutes are not provided by Jive.

Podio, a solution owned by Citrix, is a SaaS-based application that focuses on workspaces [5]. Each of these spaces, created by the users or by the administrator, is an empty shell on which will be grafted an environment of applications. Each user will also be able to customize some of their spaces to build the interfaces that correspond to them. Podio is centered on simple tooling of the company's business processes, represented through its forms modeler and app store. Podio offers several features such as: the respect of confidentiality and privacy, authentication, app store, chat and video conference, and business process modeling (Globiflow). Although there are synchronous services, Podio does not offer collaborative publishing tools, meeting minutes, does not have recommendation intelligence, and has some applications to exchange around content or files, but it remains light. In addition, Citrix does not provide any warranty regarding the availability of the service.

However, in addition to the weaknesses of each of these aforementioned solutions, all are proprietary and our principal focus is open source. Indeed, there are no real open-source platforms that combine collaboration, productivity, and organization. The market still lacks maturity.

3 OpenPaaS::NG: Overall Architecture

OpenPaaS::NG is an open-source, interoperable, scalable, extensible enterprise collaboration platform. We illustrate its overall architecture in Fig. 1.

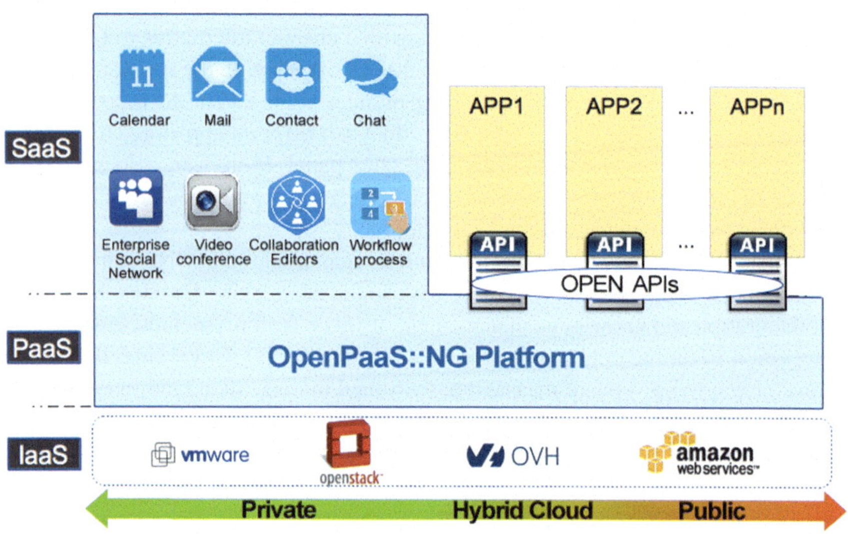

Fig. 1 OpenPaaS::NG architecture

Our approach consists on a Platform as a Service (PaaS) technology dedicated to enterprise collaborative applications deployed on multi-cloud infrastructures (Private, hybrid, and public). Our main goal is to build a platform that encourages collaboration and teamwork inside an easy user experience. Indeed, OpenPaaS::NG platform rely on collaboration and PaaS approaches in order to set up a cloud framework providing innovation real-time Software as a Services (SaaS) collaboration services (mail, ESN, shared calendar, advanced video conference, document edition, business process modeling, etc.) on top of different Infrastructures as a Service (IaaS) such as OpenStack, Amazon EC2, OVH, and VMware. Based on ESN and PaaS concepts, it allows also defining organization, communities, users' profiles and roles. In addition, OpenPaaS::NG provides facilities for users to develop and deploy new applications leveraging hosted collaboration services available in a store. This allows the user to reinforce his platform with customized modules and adapt it to his specific professional needs.

The platform comes bundled with collaboration and communication open APIs that the developers can use to speed up their development flows. The API covers a vast amount of collaboration business flow, like users, address books, calendars, instant messaging, files storage and sharing, collaboration editors, video conferencing, e-mails, etc. Moreover, the OpenPaaS::NG user interface is a pluggable Web application. Developers can add modules inside this application. The module may have an entry in the application grid; it may provide its own set of pages, or enhance other existing modules. Besides, the platform supports multi-tenant configurations, and can be managed by an LDAP compatible user directory.

4 OpenPaaS::NG Collaborative Platform

The OpenPaaS::NG platform offers a next-generation cloud enabled virtual desktop based on ESN concepts to provide advanced collaborative and recommendation services: (i) a real-time edition of documents, spreadsheets, presentations, and business processes and (ii) a context-based recommendation and automatic summarization for virtual meetings. In addition, as a PaaS cloud technology, OpenPaaS::NG allows to host new third-party collaborative services thanks to an app store.

As illustrated by Fig. 2, OpenPaaS::NG is accessible from any device (smartphone, tablet, and laptop). It provides a single user-friendly interoperating environment aggregating multiple collaborative services (e-mails, shared files, contacts, calendar, WebRTC web conference, chat, ESN, collaborative editors, etc.), thus presenting a number of competitive features, even compared to Google or Microsoft.

OpenPaaS platform[1] aims to cater for enterprise needs and is playing the role of a centralized gatekeeper between the user data and the collaborative services. In the next sub-sections, we detail these collaborative services offered within the OpenPaaS::NG.

[1] Plateforme OpenPaaS: https://open-paas.org/.

Fig. 2 OpenPaaS::NG features

4.1 OpenPaaS::NG Communication Tools

OpenPaaS::NG platform provides several communication tools: an unified mail, a contact module, and a video conferencing solution.

Contact module provides standards features as CRUD (create, read, update, delete) contacts operations and more advanced ones as full text search; synchronization with external contacts lists, which allows to import contacts from social networks (Wikipedia and Google API contacts) in order to invite them to join; and finally collected contacts. Indeed, OpenPaaS::NG provides a contact collector feature whose goal is to automatically create contacts from several sources without any user interaction. These contacts are then available in a specific user address book called "collected" and available in the contact module as other contacts.

Unified mail[2] is more than a simple e-mail service. It allows the user to bring together his social media accounts and synchronized communication channels. Thus, he can get notified, interact, and edit content in one interface.

[2]www.open-paas.org.

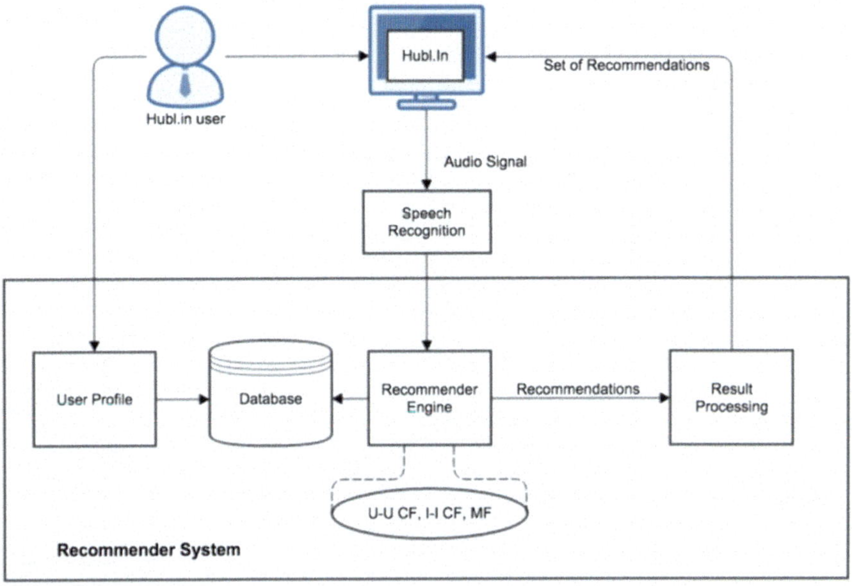

Fig. 3 OpenPaaS::NG: recommender system architecture

Video conferencing Solution called "Hubl.in" is based on peer-to-peer (P2P) and Web real-time communication (WebRTC) standard that provides browsers tools for real-time communications. It initiates a communication anywhere with any person without using additional plug-ins.

4.2 OpenPaaS::NG Recommendation Mechanisms

The recommendation mechanisms are one of the main innovative parts of the OpenPaaS::NG platform. We implemented an artificial intelligence-based recommendation system within the video conferencing tool "Hubl.in" (presented in the subsection above) in order to provide advanced real-time contextual recommendations for meetings' participants.

In Fig. 3, we present the global architecture of our real-time recommender system and its interaction with the video conferencing tool "Hubl.in"

Each block of the recommender system represents a step or process that will be completed in order to be able to recommend relevant items to users such as documents, meeting summaries, and colleague profiles. The first step of the recommender system is the *Automatic Speech Recognition* (ASR). Classical ASR systems are dedicated to the recognition of speech coming from one speaker and used to schedule simple actions [6]. Virtual meeting context is far more complex: multiple participants are speaking about different topics with potentially noise, hesitations, and people speaking simultaneously. Transcription of each meeting's participant speech-to-text (speech-to-text—STT) step is then a scientific and technical challenge not successfully addressed by current on the shelf technologies. In order to choose the most appropriate speech-to-text (STT) system to our OpenPaaS::NG platform, a comparative study is done and is presented in the table below.

<table>
<tr><th></th><th>HDecode</th><th>Julius</th><th>Sphinx-4</th><th>PocketSphinx</th><th>Kaldi</th></tr>
<tr><td>Programming language</td><td>C</td><td>C</td><td>Java</td><td>C</td><td>C++</td></tr>
<tr><td>Set up, prepare, run, time optimization</td><td colspan="2">Most</td><td colspan="2">Less</td><td>It has the best results in short time but has the highest computational cost because of the use of deep neural network (DNN) [7]</td></tr>
<tr><td>Decoding skills</td><td colspan="2">Most difficult toolkit. Setting up the system required the development of the training pipeline, which is time consuming and error prone</td><td colspan="2">The results obtained are similar to hidden Markov model toolkit, but the effort to get these is less</td><td>It outperforms all the other recognition</td></tr>
<tr><td>Building model difficulties</td><td colspan="2">Medium</td><td colspan="2">Easy</td><td>Difficult</td></tr>
<tr><td>Hot words</td><td colspan="2">No</td><td colspan="2">Yes</td><td>Yes</td></tr>
<tr><td>Real-time functions</td><td colspan="2">Likely no because spends much time to return results</td><td colspan="2">Yes</td><td>Yes</td></tr>
</table>

Kaldi represents the tool commonly used since it implements the latest phonetic modeling techniques such as neural networks for the construction of ASR systems [8, 9]. In order to build our ASR system for French spoken, we used French broadcast *Corpora* (ESTER 1 et ESTER 2, around 180 h manually transcripted). This allows us to train our model using Kaldi; indeed, our model consists on triphone ASR system (hidden Markov model—HMM [10]/Gaussian mixture models—GMM [11]) with speaker adaptive training method. We obtained 22% of word error rate for 10 h of broadcast evaluation, which is one of the best score comparing to the last evaluation campaign [12]. In Fig. 4, we illustrate our recommendation approach based on Kaldi for the ASR system for an advanced video conferencing, as well as the different interactions between Hub.in and OpenPaaS::NG.

During the video conference meeting using Hubl.in, the voice signal of each participant is extracted (1) then sent to the speech-to-text engine for decoding, which is

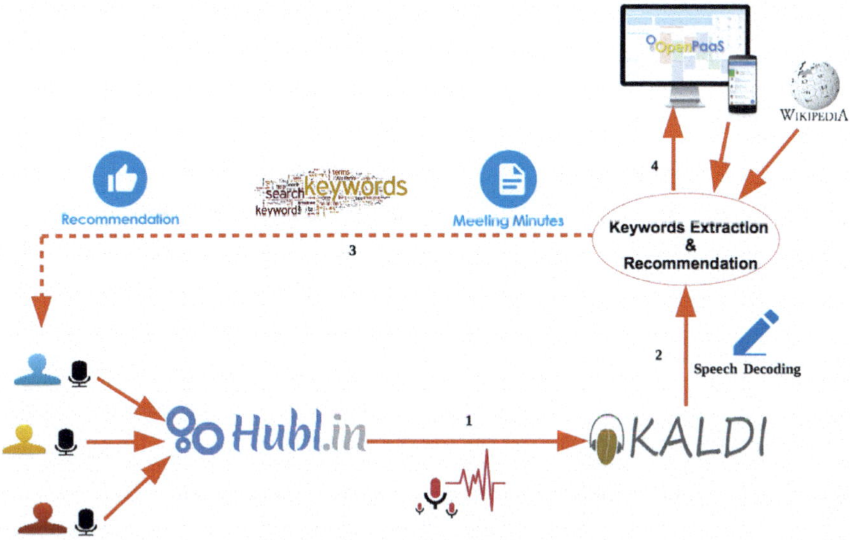

Fig. 4 OpenPaaS::NG: recommendation mechanisms approach

based on the Kaldi open-source toolbox for ASR 97 and our implemented acoustical and language models (2). Resulting transcription is then sent in real time to the keyword extraction component in charge of identifying the most important words from the text stream. The keywords are then used by the recommendation engine to propose to the participants' relevant documents from the OpenPaaS::NG knowledge base (e-mails, files, previous meetings minutes, etc.) or relevant topics from Wikipedia (3). Recommendations are based, on the one hand, on the understanding the system has about the topics of the current conversation taking place between participants and, on the other hand, on the available knowledge stored into the OpenPaaS::NG platform (e-mails, shared calendars and files, previous meetings' minutes, etc.). At the end of the meeting, in addition to the deduced recommendation and keywords, the minutes of meeting is automatically generated and sent to all participants. Meeting minutes, keywords, and recommendation are stored in the platform (4).

4.3 *OpenPaaS::NG Collaboration Editors*

In this sub-section, we present the document, spreadsheet, presentation, and business process modeling collaborative editors.

Document, *Spreadsheet,* and *Presentation* collaborative editors are available within OpenPaaS::NG. Indeed, CryptPad and Only Office are integrated within the platform and both of them allow to use word processor for all reports and summaries (document editor), to gather and analyze all data in one place (spreadsheet editor),

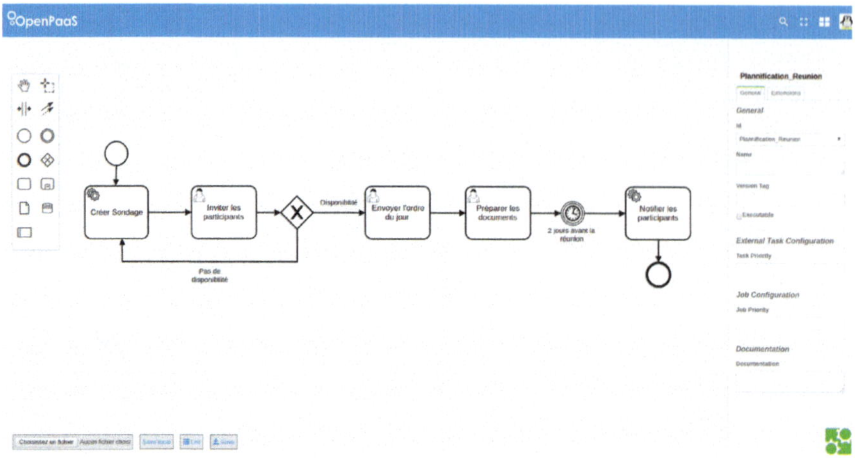

Fig. 5 OpenPaaS::NG: business process editor

and to present projects and work (presentation Editor). The first one, CryptPad, is a real-time collaborative rapid and light pad; the second one, OnlyOffice, is more complete. According to his needs, the user can choose the tool.

A graphical *Business Process Modeling* is implemented within the OpenPaaS::NG platform. It complies with the business process model and notation (BPMN) 2.0 [13]. The BPMN editor allows creating a new business process or editing an existing one. As illustrated by Fig. 5, it is divided into three main parts: (i) BPMN palette—in the left—containing all BPMN 2.0 elements allowing users to create business processes, (ii) modeling area—in the center— which allows users to draw their business processes by dragging and dropping the needed elements from the palette, and (iii) properties panel—in the right— which allows users to specify the properties for each element of the business process (id, name, functional description, etc.).

4.4 OpenPaaS::NG Teamwork Experience

In this sub-section, we introduce the community, calendar, and instant messaging [14].

In order to organize communication, sharing, and collaboration, OpenPaaS includes a module called *Community* where members can work together on collective projects in order to achieve a common goal and featuring other collaborative tools like surveys and events.

Community is a set of persons, has a name, an avatar, and a description. This module has an intrinsic rule regarding the visibility of information that is shared inside. It can be public, restricted, private, or confidential. The community's administrator manages these parameters.

Every user (member or not) can participate in a public community and join it unlike a private one where only members can access it (read and add content). A restricted community allows all users to read its content but only members can add and share information inside. A confidential community is invisible for all users that are not members. Finally, non-members of a private community can only see its name, avatar, and description [14].

Shared Calendar Supports two kinds of calendars: (i) delegated calendars represent a privileged access for a given calendar to a specific user. Precisely, when an owner delegates his own calendars to other users. For example, a manager can delegate his calendars to his secretary; (ii) public calendars when an owner can set his calendars to public. Other users, or external ones, can subscribe to such calendars and see/create events. Shared calendars come with rights that are applied to users who are looking at other users calendars. Four types of rights are considered (free/busy, read, write, and administration): in the free/busy, the details of the events are not available. User can only see that the calendar owner is free or busy in a time period; in the read mode, user has access to all the event information but cannot change anything; in the write mode, user has access to all the event information and can change anything; finally, in administration mode, user can do anything on the calendar (edit, delete, etc.) as well as on the calendar events. In addition, users can book physical resources from the calendar module such as meeting room, video projector, car, etc.

Instant Messaging is an activity stream where members can exchange, in real time, several kinds of messages create channels (public and private), attach files in real time.

5 Conclusion

In this paper, we illustrate our open-source solution of providing an innovative, flexible, and interoperable collaborative-based cloud platform for enterprises which originality relied on (i) its ability to offer to users to work together from any device, anywhere, and at any moment and (ii) its artificial intelligence-based recommendation system in order to provide advanced real-time contextual recommendations for meetings' participants. Indeed, OpenPaaS::NG provides a complete suite, including communication tools, advanced video conferencing tool offering real-time recommendations and providing an automatically generated meeting minutes at the end of the meeting that will be sent to all the participants, and collaboration editors and teamwork experience. More than facilitating inter and intra-organizational collaborations, is also provides in one platform all what a user need.

As future work, in order to provide more facilities to users, a professional conversational assistant will be interfaced with the OpenPaaS::NG platform. In addition, we are working for adding English language to our ASR system and to use more complex modelization of acoustic modeling such as sequence modeling LSTM (Long short-term memory.

Acknowledgements This work has been partially supported by both C2Net European Project (Horizon 2020), and OpenPaaS::NG French Project (PSPC—*Projets de recherché et développement Structurants pour la Compétitivité*). The work reflects only the author's views. The community is not liable for any use that may be made of the information contained therein.

References

1. Ballagas, R., Rohs, M., Sheridan, J. G., & Borchers, J. (2004, September). Byod: Bring your own device. In *Proceedings of the Workshop on Ubiquitous Display Environments*, Ubicomp (Vol. 2004).
2. Murray, K. (2011). *Microsoft Office 365: Connect and collaborate virtually anywhere, anytime*. Microsoft Press.
3. Conner, N. (2008). *Google apps: The missing manual*. O'Reilly Media, Inc.
4. Jive-n, Jive-x. https://www.jivesoftware.com/fr/
5. Podio. https://www.citrix.fr/products/podio.html
6. Rabiner, L. R., & Juang, B. H. (1993). *Fundamentals of speech recognition*.
7. Deng, L., Hinton, G., & Kingsbury, B. (2013, May). New types of deep neural network learning for speech recognition and related applications: An overview. In *2013 IEEE International Conference on Acoustics, Speech and Signal Processing (ICASSP)* (pp. 8599–8603). IEEE.
8. Price, M., Glass, J., & Chandrakasan, A. P. (2017, February). 14.4 A scalable speech recognizer with deep-neural-network acoustic models and voice-activated power gating. In *2017 IEEE International Solid-State Circuits Conference (ISSCC)* (pp. 244–245). IEEE.
9. D. Povey, Ghoshal, A., Boulianne, G., Burget, L., Glembek, O., Goel, N., & Vesely, K. (2012). *The Kaldi speech recognition toolkit*
10. Zucchini, W., MacDonald, I. L., & Langrock, R. (2016). *Hidden Markov models for time series: An introduction using R* (Vol. 150). CRC Press.
11. Reynolds, D. (2015). Gaussian mixture models. *Encyclopedia of Biometrics*, 827–832.
12. Gravier, G., Bonastre, J. F., Geoffrois, E., Galliano, S., McTait, K., & Choukri, K. (2004, May). The ESTER evaluation campaign for the rich transcription of French broadcast news. In *LREC*.
13. Allweyer, T. (2016). *BPMN 2.0: Introduction to the standard for business process modeling*. BoD–Books on Demand.
14. Zribi, S., Montarnal, A., Bénabén, F., Lauras, M., Lamothe, J., Bailly, M., & Lorré, J. P. (2015, October). Towards an enterprise social network to support inter-organizational collaborations. In *Working Conference on Virtual Enterprises* (pp. 421–428). Cham: Springer.

Part IV
Semantic Interoperability

Semantic Modeling of Cascading Risks in Interoperable Socio-technical Systems

Alex Coletti, Antonio De Nicola, Giordano Vicoli and Maria Luisa Villani

Abstract We present a method to create semantic representations of cascading risks of interoperable socio-technical systems. This is based on a structured domain ontology representing socio-technical systems, their interdependencies, environmental and anthropic hazards, and the related threats. The ontology is accompanied by a software application, i.e., the CREAtivity Machine that generates cascades of risks by means of semantic and computational creativity techniques. The presented running prototype refers to risk assessment of critical infrastructures; however, the same method can be applied to risks concerning other system types like businesses, ecosystems, and financial networks.

Keywords Interoperable socio-technical systems · Risk assessment · Ontology · Computational creativity

1 Introduction

Interoperable socio-technical systems are those systems consisting of both humans and technological elements that need to work together to perform their functions. Examples of socio-technical systems are critical infrastructures including both technological and organizational infrastructures and human resources that are necessary for their operation. These systems are highly complex also because of many interdependency links existing between them. Hence, a failure in a system could generate

A. Coletti
SMRC, Ashburn Virginia, USA
e-mail: acoletti@smrcusa.com

A. De Nicola (✉) · G. Vicoli · M. L. Villani
ENEA-CR Casaccia, Via Anguillarese 301, 00123 Rome, Italy
e-mail: antonio.denicola@enea.it

G. Vicoli
e-mail: giordano.vicoli@enea.it

M. L. Villani
e-mail: marialuisa.villani@enea.it

K. Popplewell et al. (eds.), *Enterprise Interoperability VIII*, Proceedings of the I-ESA Conferences 9, https://doi.org/10.1007/978-3-030-13693-2_10

a failure in another one. Figuring out cascades of failures and of the associated risk situations is not simple, but it is necessary to assess the impact of a damaging event or to create simulation models.

In this context, we propose a semantic modeling framework with two objectives. The former is to provide a formal specification of knowledge related to risks of socio-technical systems in the form of ontology [1]. The latter is to automatically infer new knowledge related to possible risk situations. In particular, we show how the method generates new risk cascades from descriptions of situations identifiable in interoperating systems.

The TERritorial Management and Infrastructures ontology for institutional and industrial USage (TERMINUS) is the ontology we have built for this purpose. The ontology is a follow-up of a knowledge elicitation experiment involving water system stakeholders in the United States [2].

The CREAtivity Machine (CREAM) is the software application we implemented to create cascades of risk mini-models, i.e., cascades of conceptual representations of risk situations. This application leverages on both semantic and computational creativity techniques. Indeed, it is a shared belief in this domain that creativity, other than knowledge and experience, is required to identify possible scenarios of system performance degradation, including (indirect) threat causes, failures, and consequences. Also, the limited interdisciplinary technical expertise from the individual system managers/analysts may lead to incomplete or inaccurate risk assessment.

This work is part of a wider research activity aimed at risk assessment of complex socio-technical systems. In fact, in [3], we put the foundations of the TERMINUS ontology by defining the vulnerability upper model (VUM) design pattern to be used as building block for our ontology engineering project. In [4], we presented CREAM to generate single and independent risk mini-models for a given system. Finally, in [5], we proposed a mobile app to elicit knowledge from domain experts, to enrich the ontology, and to create/validate risk mini-models.

The rest of the paper is organized as follows. Section 2 presents the related work in the area. The conceptual framework for risk assessment is described in Sect. 3. Then, Sect. 4 presents the semantic model of cascading risks and how CREAM generates them. Finally, Sect. 5 concludes the paper with some considerations on the work done until now.

2 Related Work

Risk of cascading failures is one of the most relevant issues to cope within the field of critical infrastructures protection [6]. In this context, Rinaldi et al. [7] present probably one of the seminal works dealing with the problem and contributing to define the related conceptual entities. In particular, Rinaldi et al. [7] define critical infrastructures interdependencies as the "connections among agents in different infrastructures in a general system of systems" that increase their complexity. They are classified across several dimensions including: type (i.e., physical, cyber, geographic, and logical), environment (e.g., health, safety, and technical), coupling and response behavior (e.g., adaptive and linear/complex), type of failure (i.e., common cause, cascading, and escalating), infrastructure characteristics (i.e., organizational, operational, temporal, and spatial), and state of operation (i.e., repair/restoration, normal, and stressed/disrupted). Then, Kotzanikolaou et al. [8] present a method to identify and assess multi-order interdependencies by means of a dependency risk table.

While these efforts uncovered and systematized the several aspects of the problem, Haase [9] demonstrated how communities can review, discover, and recognize vulnerability causes. In fact, Haase [9] presents a participatory study concerning flood risk, vulnerability, and adaptive capacity and Coletti [2] proposes a collaborative experiment aiming at assessing vulnerabilities of water systems. Then, Maiden [10] presents a system enhancing a collaborative platform for risk detection and resolutions in three manufacturing plants with creativity support. CIrcle [11] is a software application to collect data concerning cascading effects of the impact of a hazard on critical infrastructures. Our proposed method shares with these works the goal of gathering knowledge related to risks regarding socio-technical systems, but adds a conceptual model that makes experiences transferrable between communities. The set of logical axioms in the model organizes vulnerabilities of individual assessments in ways that similarities, differences, and contextual features can be stored and shared. The ontology model defines the axioms needed to describe failures that occur deep inside a system so they can be searched, recognized, and chained together for different systems with similar structures. By applying the same sets of axioms, users can benefit from the analysis of parallel experiences and prioritize them within their own social, geographic, and cultural context.

Although there are similarities between the mentioned approaches and ours, we focus on risk modeling and cascade of risks. The unique ability of the method to infer cascading risks from computational creativity procedures that access a shared ontology model can add a significant capability to decision support systems, vulnerability portals, and applications for cost analysis of resilience plans.

3 A Conceptual Framework for Risk Assessment

A conceptual framework, illustrated in Fig. 1, is proposed as a basis for risk assessment processes. Namely, conceptual representations of cascades of system risks are based on a domain-specific ontology and risk design patterns. A modeling approach has been defined to develop both the ontology and the design patterns from some upper-level models focusing on system risks, system characteristics, and critical infrastructure interdependencies. In what follows, we briefly present the domain ontology and the risk mini-model, according to our modeling approach.

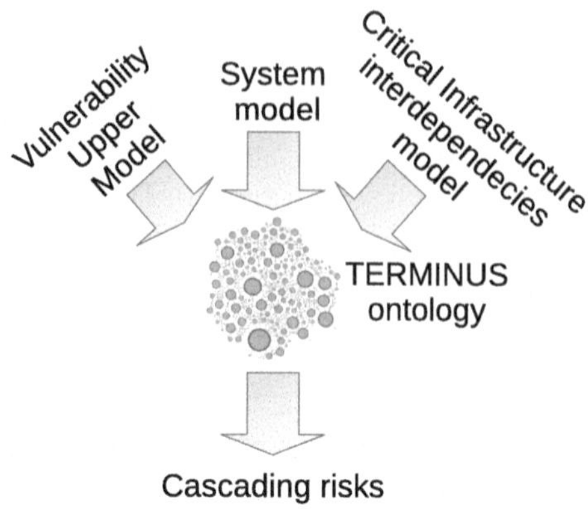

Fig. 1 Overview of the conceptual framework for risk assessment

3.1 TERMINUS Domain Ontology

The TERritorial Management and Infrastructures ontology for institutional and industrial USage has been built to represent knowledge concerning environment, critical infrastructures and related hazards, risks, and threats. TERMINUS is a structured ontology that has been developed by extending some ontology design patterns defined. Ontology design patterns [12] are reusable conceptual structures aimed at supporting the ontology engineering process. In [3], we presented the *vulnerability upper model* (VUM) design pattern to support modeling of concepts concerning risk assessment of systems. We refer to [7] to model interdependencies between critical infrastructures. Then, here, we present the *system aspect* design pattern, aiming at supporting modeling systems according to different perspectives.

The *system aspect* design pattern in Fig. 2 includes a *stakeholder* with interest on a *system* and its sub-systems. Examples of systems are socio-technical systems as the *water system*, the *energy system*, and the *transportation system*. A *system* can be viewed from the following different perspectives, i.e., *system aspect*s: *system service*, *system operation*, *asset*, *commons*, *infrastructure*, and *managed object*. *System service* models the output of a system provided to stakeholders. *System operation*s are the (internal) activities performed in system and that are required preconditions to deliver services. *Asset*s model the items of value owned by the system. *Commons* refer to the cultural and natural resources accessible to all members of a society, including natural materials such as air, water, and a habitable earth. Examples of commons are lake, water spring, river, and glacier. *Infrastructure*s model the physical, technological, and organizational structure a system. *Managed object*s model the entities that are handled by the system, as water in case of water system or fuel in case of oil system.

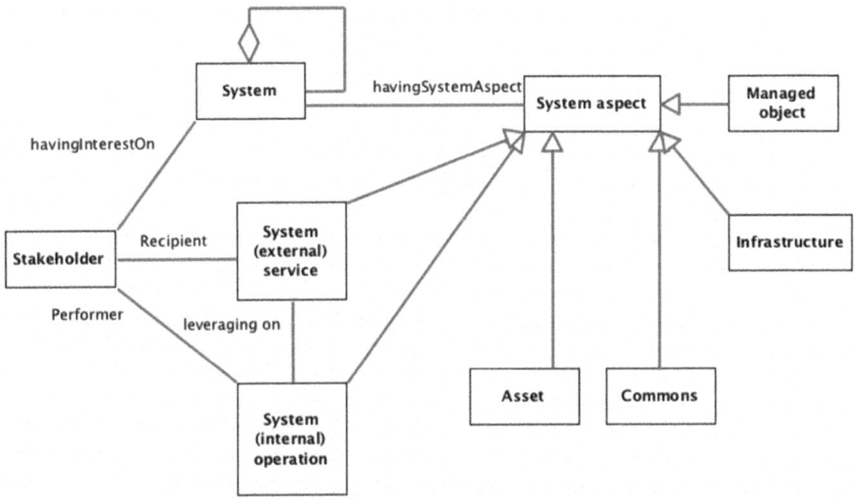

Fig. 2 *System aspect* design pattern

3.2 *Risk Mini-Models*

The atomic element at the base of our method for semantic modeling of risks is the *risk mini-model* [4] that is defined as a fragment of conceptual model representing a new idea related to a risk. A *risk mini-model* is generated by CREAM by querying and combining ontology concepts by means of SPARQL queries [4].

In the following, we present an example of risk mini-model representing an oil system risk that is used in the paper as the source risk of a cascade.

```
                 Oil system risk
Hazard: Hurricane
Threat: Flooding of gasoline stations
System aspect: Provision of fuel
Vulnerability: Availability of gasoline stations
Severity of risk: Closed gasoline stations
Stakeholder: Highway users
```

This can be interpreted as follows. "One of the vulnerabilities of the oil system is the availability of gasoline stations as it is the most relevant channel for provision of fuel. A hurricane causes flooding of gasoline stations. As mentioned, closed gasoline stations are a severe issue for the oil system as this impacts highway users and, hence, their mobility."

```
<owl:ObjectProperty  rdf:about="ns#physical_interdependency">
    <rdfs:subPropertyOf rdf:resource="ns#interdependency"/>
    <rdf:type rdf:resource="ns#SymmetricProperty"/>
        <rdfs:domain rdf:resource="ns#System_aspect"/>
        <rdfs:range rdf:resource="ns#System_aspect"/>
</owl:ObjectProperty>
```

Fig. 3 An excerpt of OWL code defining the "physical interdependency" object property

4 Cascading Risks in Interoperable Socio-technical Systems

4.1 Semantic Model of Cascading Risks

Interdependences could increase the number of failures in interoperable socio-technical systems [7]. We propose to semantically represent them in the ontology as symmetric object properties where both domain and range are system aspects. The interdependency property can be further specialized in the following sub-properties: physical interdependency, cyber interdependency, geographic interdependency, and logical interdependency. The fragment of owl code concerning physical interdependency is presented in Fig. 3.

Rinaldi et al. [7] classify the types of failures originating from interdependencies as cascading, common cause, and escalating. Here, we address cascading failures and, in particular, the associated cascading risk mini-models. We define a cascading risk mini-models (or cascade of risk mini-models) as a sequence of risk mini-models where the risk concerning a system aspect depends on the risk concerning a different system aspect. These system aspects could belong either to different systems or to the same one. In the following, we address cascading risk mini-models of the former case (i.e., different systems).

Figure 4 presents a cascade of risk mini-models involving the oil system, the transportation system, and the water system. Accordingly, the risk mini-model presented in Sect. 3.2 is the source risk situation that originates a transportation system risk situation of highway users unable to circulate due to petroleum dependence of cars and trucks and to closed gasoline stations. In turn, this second risk mini-model originates a new risk situation concerning the water system, that is, drinking water system users are without service as distribution of water by trucks is not possible due to the lack of fuel.

From an ontological point view, the cascades of risk mini-models are originated from object properties connecting two different system aspects that allow creating a link between two different risk mini-models. Figure 5 presents the excerpt of OWL code concerning the `physical interdependency` object property between the concepts `Cars_and_trucks_move` and `Provision_of_fuel`. This relationship is used, as explained in Sect. 4.2, to infer a cascade between different risk mini-models.

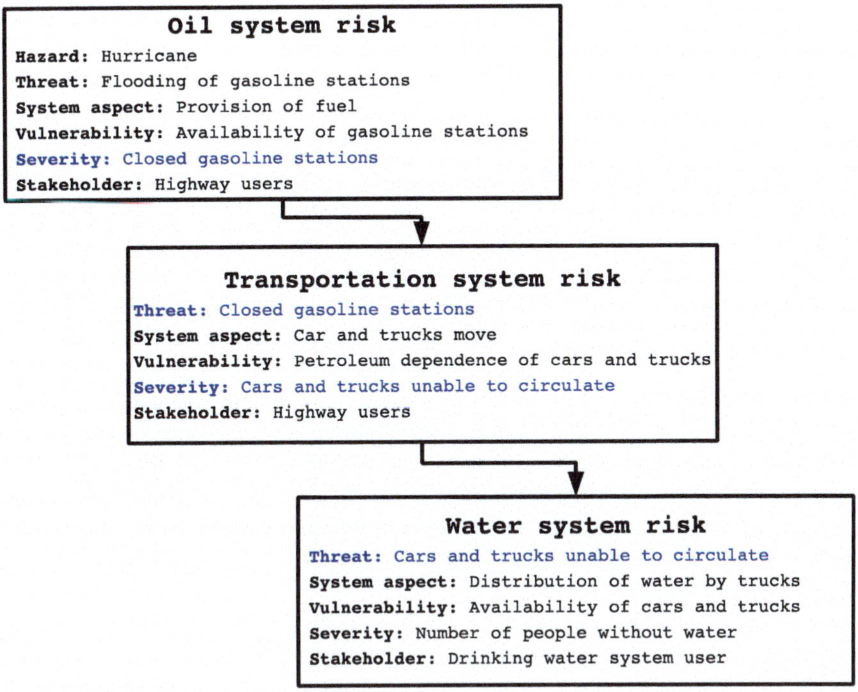

Fig. 4 Example of cascades of risk mini-models. Blue text depicts the "hooks" between different risk mini-models

```
<owl:Class rdf:about="ns#Cars_and_trucks_move">
    <rdfs:subClassOf rdf:resource="ns#Road_system_operation"/>
        <rdfs:subClassOf>
            <owl:Restriction>
                <owl:onProperty rdf:resource="ns#physical_interdependency"/>
                <owl:someValuesFrom rdf:resource="ns#Provision_of_fuel"/>
            </owl:Restriction>
        </rdfs:subClassOf>
</owl:Class>
```

Fig. 5 An excerpt of the OWL code representing the "`physical interdependency`" object property between the concepts "`Cars_and_trucks_move`" and "`Provision_of_fuel`"

4.2 *Generation of Cascading Risk Mini-Models*

Cascading system risks are based on interdependencies between different systems, and having such interdependencies modeled in the TERMINUS ontology allows us to experiment methods for automatic generation of new risk mini-models for those systems in a creative way.

Indeed, as described in [4], we developed a software tool for automatic suggestion of risk mini-models leveraging on computational creativity techniques, whose general aim is defining computational systems that create artifacts and ideas [13]. In our software system, we modeled the risk mini-model detection as a search process within a space consisting of the domain ontology, possibly constrained by contextual rules, which is updated and refined by accounting for the end users risk mini-model definitions, once they are validated by the risk analysts. In this work, we show how this method can be applied to automatically generate "chains" of candidate risk mini-models referring to different interdependent systems. After validation, such chains would lead to a set of cascading risk representations of the type presented in Fig. 4 of Sect. 4.1. Essentially, whenever updates on risk definitions of some system involve aspects that are source of interdependency links with aspects of other systems, risk mini-models for these interdependent systems are automatically generated by CREAM following the system interdependency relationships represented in the TERMINUS ontology (i.e., through sub-properties of the "`interdependency`" object property of Fig. 3). Chains of these risk mini-models, with each mini-model representing a risk situation of a certain system that may lead to a risk situation of another system, will be automatically suggested as cascading risk representations to risk experts for validation.

Generally, computational creativity methods address the problem of thinking something new, e.g., a risk situation, by varying and/or combining one or more aspects of what already exists. In particular, the *transformation* method consists of modifying the form of some particular features of an existing design; the *analogy* method of matching and transferring specific aspects of the conceptual structure of one domain with those of another domain; finally, the *combination* method consists of creating a new configuration from features of an existing design. In the case of generating cascading risks, we apply these methods as risk mini-models *variation* operations.

First the analogy method attempts to matching/transferring a severity-specific concept sev_A of a risk mini-model of some system A with/to a threat-specific concept th_B of another system B. This is done whenever an interdependency int_{AB}:= (asp_A, asp_B) exists from an aspect asp_A of system A to an aspect asp_B of system B. Then, the transformation method is used to modify some existing risk mini-models of system B containing asp_B by replacing in them every threat-specific concept with th_B. Finally, we interpret the combination method as the variation of an existing systems' cascading risk by means of the previous two methods. Generally, cascading risks $(Rm_{A1}, \ldots, Rm_{An})$ are identified from generated chains of risk mini-models where the termination criteria for the chain is either there or not exists A_{n+1} such that int_{AnAn+1}:= (a_{An}, a_{An+1}) or $A_{n+1} = A_1$. The automatic generation method is sketched in Table 1.

The method was implemented in Java and uses SPARQL queries and Apache Jena technology [14] to retrieve system interdependency chains and validated risk mini-models from the TERMINUS ontology.

Table 1 Method for automatic suggestion of cascading risks from a given risk mini-model

	generateCascadingRisks	
Input	$Rm_A = (th_A, vul_A, asp_A, sev_A, sk_A)$, *TERMINUS ontology*	*// Rm_A risk minimodel of System risk R_A of A*
Output	$\{(Rm_{A=Ai1},\ldots, Rm_{Aij})\}$, $1<j<=n$, $i=1,..,k$	*// k possible cascading risks of length at most n from Rm_A*
begin	*compute* $Int(A):= \{(int(asp_A, asp_{Ai2}),\ldots,int(asp_{Aij-1}, asp_{Aij})), j<=n\}$	*// chains of interdependencies from A of length <=n*
for each chain i do	$l:=2$, $A_{i1}:=A$	
for each interdependency	$int(asp_{Ail-1}, asp_{Ail})$, $1<l<=j$	
and	asp_{Ail-1} *is a component of* Rm_{Ail-1}	
do	$th_{Ail:} = sev_{Ail-1}$	*// set new threat for system A_{il}*
	$Rm_{Ail}: = generateMinimodel(th_{Ail}, asp_{Ail})$	*// choose an existing risk mini-model having system aspect asp_{Ail} and variate threat with th_{Ail}*
end for		
	build chain $(Rm_{A=Ai1},\ldots, Rm_{Ais})$, $s<=j$	
end for		
return	*union of chains*	

4.3 Experts Validation

The validation of a candidate cascading system risk, automatically generated by CREAM, requires a collaborative process with the participation of risk experts of the various socio-technical systems involved in its description. This process is supported by a mobile application we developed (the ICE tool [5]) whose general aim is to enable distributed information gathering finalized to the creation of new risk mini-models, avoiding organization of ad hoc meetings. The ICE tool uses a gamified collaborative process to engage risk knowledge contributors to share their experiences and specialized system knowledge.

In this approach, given the possibility of having both a cascading risk and risk contributors making entries from different systems, the individual risk mini-models composing it are automatically flagged by the ICE tool. The flags are then notified to the team members by the ICE Tool, as new possible risk mini-models for their system. To avoid undesired information leakage, the validation of each risk mini-model is performed internally by each organization by means of collaborative support functions of the ICE Tool described in [5] which may include modifications on the original version. Finally, all the validated risk mini-models are used to eventually rebuild the cascading system risk with the support of the team members.

We are designing a case study on risks for the oil system, the transportation system, and the water system and planning to invite some academic and practitioner experts to validate CREAM capability to generate cascading risk models for those critical infrastructures. Our primary objectives of this validation are: (a) to evaluate correctness and precision of the results; (b) to evaluate the recall of the generation algorithm by using lists of known cascading risks also "manually-defined" by the experts; (c) to evaluate the effectiveness of the computational creativity methods by measuring the number of new cascading risks the experts are able to recognize by using the ICE-CREAM tool.

5 Conclusion

The current open collaboration approaches to risk identification in complex socio-technological systems have been enhanced by a novel tailored risk management system, which, through a lightweight and engaging approach, is capable to collecting structured risk information that can be processed automatically through semantic reasoning techniques [3–5]. Such automatic support is even more required when eliciting cascading risk situations originated from interdependency relationships with other systems, due to the complexity of a complete system-of-systems representation.

In this paper, we presented an extension of our system risk generative approach [4], based on computational creativity techniques, toward elicitation of cascading risk situations (risk mini-models) involving different systems. This relies on a collaborative and distributed process of risk mini-models validation, performed by risk experts of the various systems, and enabled by the ICE mobile application.

Practitioners, i.e., analysts of critical infrastructures, would benefit from our proposal as conceiving new cascading risks requires multi-disciplinary competences that are extremely rare in the job market. With our approach, such multi-disciplinary knowledge is automatically provided by CREAM.

The presented approach opens some challenges for researchers. For instance, guaranteeing plausibility of generated risks requires coping with the problem of collecting, specifying, and exploiting common sense knowledge [15]. Finally, ensuring completeness of the ontology for these large interconnected domains requires improving automatic ontology building techniques to support the work of ontology engineers .

References

1. Gruber, T. R. (1993). A translation approach to portable ontology specifications. *Knowledge Acquisition, 5*(2), 199–220.
2. Coletti, A., Howe, P. D., Yarnal, B., & Wood, N. J. (2013). A support system for assessing local vulnerability to weather and climate. *Natural Hazards, 65*(1), 999–1008.
3. Coletti, A., De Nicola, A., & Villani, M. L. (2016). Building climate change into risk assessments. *Natural Hazards, 84*(2), 1307–1325.
4. Coletti, A., De Nicola, A., and Villani, M. L. (2017). Enhancing creativity in risk assessment of complex sociotechnical systems. In O. Gervasi et al. (Eds.), *Computational science and its applications—ICCSA 2017*. Lecture Notes in Computer Science, Vol. 10405. Cham: Springer.
5. Coletti, A., De Nicola, A., Vicoli, G., & Villani, M. L. (2017). A gamified approach to participatory modelling of water system risks. In *Proceedings of the 12th International Conference on Critical Information Infrastructures Security*.
6. Buldyrev, S. V., Parshani, R., Paul, G., Stanley, H. E., & Havlin, S. (2010). Catastrophic cascade of failures in interdependent networks. *Nature, 464*(7291), 1025–1028.
7. Rinaldi, S. M., Peerenboom, J. P., & Kelly, T. K. (2001). Identifying, understanding, and analyzing critical infrastructure interdependencies. *IEEE Control Systems, 21*(6), 11–25.
8. Kotzanikolaou P., Theoharidou M., Gritzalis D. (2013). Interdependencies between critical infrastructures: Analyzing the risk of cascading effects. In S. Bologna, B. Hämmerli, D. Gritzalis, S. Wolthusen (Eds.), *Critical Information Infrastructure Security. CRITIS 2011*. Lecture Notes in Computer Science, Vol. 6983. Berlin, Heidelberg: Springer.
9. Haase, D. (2013). Participatory modelling of vulnerability and adaptive capacity in flood risk management. *Natural Hazards, 67*(1), 77–97.
10. Maiden, N., Zachos, K., Lockerbie, J., Hoddy, S., & Camargo, K. (2017). Establishing digital creativity support in non-creative work environments. In *Proceedings of the 11th ACM Creativity and Cognition Conference*. ACM.
11. Deltares. CIrcle. Retrieved October 31, 2017 from www.deltares.nl/circle.
12. Gangemi, A., & Presutti, V. (2009). Ontology design patterns. In S. Staab & R. Studer (Eds.), *Handbook on ontologies. International handbooks on information systems*. Berlin, Heidelberg: Springer.
13. Colton, S., Wiggins, G. A. (2012). Computational creativity: The final frontier? In *Proceedings of the 20th European Conference on Artificial Intelligence* (pp. 21–26). IOS Press.
14. Apache Jena, version 2.11.1. Retrieved October 31, 2017 from http://jena.apache.org/.
15. Speer, R., Havasi, C., Lieberman, H. (2008). AnalogySpace: reducing the dimensionality of common sense knowledge. In *Proceedings of the 23rd National Conference on Artificial Intelligence, AAAI 2008* (Vol. 1, pp. 548–553). AAAI Press.

Toward the Use of Upper-Level Ontologies for Semantically Interoperable Systems: An Emergency Management Use Case

Linda Elmhadhbi, Mohamed-Hedi Karray and Bernard Archimède

Abstract In the context of globalization and knowledge management, information technologies require an ample need of unprecedented levels of data exchange and sharing to allow collaboration between heterogeneous systems. Yet, understanding the semantics of the exchanged data is one of the major challenges. Semantic interoperability can be ensured by capturing knowledge from diverse sources by using ontologies and align these latter by using upper-level ontologies to come up with a common shared vocabulary. In this paper, we aim in one hand to investigate the role of upper-level ontologies as a mean for enabling the formalization and integration of heterogeneous sources of information and how it may support interoperability of systems. On the other hand, we present several upper-level ontologies and how we chose and then used basic formal ontology (BFO) as an upper-level ontology and common core ontology (CCO) as a mid-level ontology to develop a modular ontology that defines emergency responders' knowledge starting from firefighters' module for a solution to the semantic interoperability problem in emergency management.

Keywords Semantic interoperability · Ontology · Upper-level ontology · BFO · Mid-level ontology · CCO · Emergency management

1 Introduction

Today, the more information systems are becoming connected, the more the word is getting smaller and smaller. To manage the integration and interaction of these linked complex systems and the evolution of the amount of data that should be exchanged and shared, interoperability is considered as the key feature. It refers to the "Ability for two (or more) systems or components to exchange information and to use the information that has been exchanged" [1]. From this definition, it is possible to decompose interoperability into two distinct components: "syntactic interoperability" is the ability to exchange information and "semantic interoperability" is the

L. Elmhadhbi (✉) · M.-H. Karray · B. Archimède
University of Toulouse, Toulouse 31013, France
e-mail: linda.elmhadhbi@enit.fr

K. Popplewell et al. (eds.), *Enterprise Interoperability VIII*, Proceedings of the I-ESA Conferences 9, https://doi.org/10.1007/978-3-030-13693-2_11

ability to use the information once it has been received [2]. That is to say, semantic interoperability ensures that these exchanges make sense—that the requester and the provider have a common understanding of the "meanings" of the requested services and data [3]. The semantic heterogeneity of data leads to very serious issues since there are several interpretations of one expression. Let us take the example of the term "tank". In an information system of armored vehicles, the term normally refers to a certain kind of specialized armored vehicle used by army, but in an information system that store zoological equipment, the term "tank" refers to a kind of container which can hold water. Now, suppose that a military basis uses the two information systems and that the two information systems are to interoperate within a base-wide facility management system. In this case, it is not evident how to interpret the expression "three tanks" [4].

To overcome semantic heterogeneity and to guarantee a consistent shared understanding of the meaning of information, the use of ontologies is crucial [5]. Ontologies are expressed in a logic-based language, so that accurate, consistent, and meaningful distinctions can be made among the classes, instances, properties, attributes, and relations to reveal the implicit and hidden knowledge in order to understand the meaning of the data. Thus, they offer the richest representations of machine-interpretable semantics for systems and databases [6]. They serve as both knowledge representation and as mediation to enable heterogeneous systems interoperability [7]. However, the question that arises is how to match these ontologies in order to provide semantic interoperability of multiple information systems. The key way for integrating heterogeneous knowledge across various ontologies is to make use of upper-level ontologies. It provides a common ontological foundation for domain ontologies which describe the most general domain independent categories of reality as: time and space, individuals, objects, events, process, instantiation, and so on [8].

Many upper-level ontologies have been developed over the years and used in different domains such as emergency management. This field is often challenging; it evolves the correlation of different actors and various pieces of information. Emergency management is the ability of an organization to quickly respond to an incident in order to reduce the negative impacts. It includes coordination of service efforts and strategic directions. In such domain, information interoperability is essential during an emergency to exchange data between the different stakeholders to successfully respond to day-to-day incidents and large-scale events.

This work aims in one hand to investigate the role of upper-level ontologies as a mean for enabling the formalization and integration of heterogeneous sources of information in the field of systems interoperability. On the other hand, we work toward defining the knowledge of emergency responders by developing a modular ontology starting with firefighter's module to solve the issue of semantic interoperability during emergencies. Hence, this paper is organized as follows: in the next section, we discuss the four levels of abstraction specifically the upper, mid-level, domain and sub-domain ontologies and we look into the advantages and the possibilities opened by the use of upper-level ontologies for semantic interoperability of systems and then we discuss several upper-level ontologies. In Sect. 3, we justify our choice for selecting the appropriate upper-level ontology. Section 4 goes into the details of how we used basic formal ontology (BFO) and common core ontology (CCO) to develop our ontology. At last, the conclusion and the future work are presented.

2 Background

2.1 Ontologies Levels of Abstraction

There are three levels of abstraction of ontologies specifically upper, mid-level, domain and sub-domain ontology as illustrated in Fig. 1; First, the upper-level ontology, as defined in [9], it "describes very general concepts that are the same across all domains and usually consist of a hierarchy of entities and rules (both theorems and regulations) that attempt to describe those general entities that do not belong to a specific problem domain." They provide a high-level domain independent conceptual model that describes abstract concepts such as object, process, events, and quality. Examples of upper-level ontologies include: basic formal ontology (BFO), descriptive ontology for cognitive and linguistic engineering (DOLCE), general formal ontology (GFO), suggested upper merged ontology (SUMO), common semantic model (COSMO), Cyc project, and so on. Second, middle-level ontology presents the bridge between the abstract concepts of upper-level ontologies and the rich details of domain ontologies by adding more specific modules like space and time. Domain ontologies or lower ontologies describe concepts of a domain of interest in a very specific way and it may also extend concepts from mid-level ontologies. Ontologies from different domains may be as well integrated by alignment to an upper-level ontology. Finally, the lowest level of abstraction is sub-domain ontologies. They describe concepts that depend on a specific task in a particular domain. These concepts often correspond to the roles played by the entities.

Reusing well-established ontologies in the development of a domain ontology allows one to take advantage of the semantic richness of the relevant concepts and logic already built into the reused ontology. In this way, ontologies may provide a web of meaning with semantic decomposition of concepts [10].

2.2 Upper-Level Ontologies for Interoperability of Systems

With the increasing amount of data coming from different sources, there is a strong need to determine the meaning of these information to be exchanged precisely enough that a software application can interpret them. So many applications of ontologies address the problem of semantic interoperability, in which we have different users using various software tools that need to cooperate by exchanging data with unambiguous, shared meaning. Interoperability could then be achieved by using ontologies

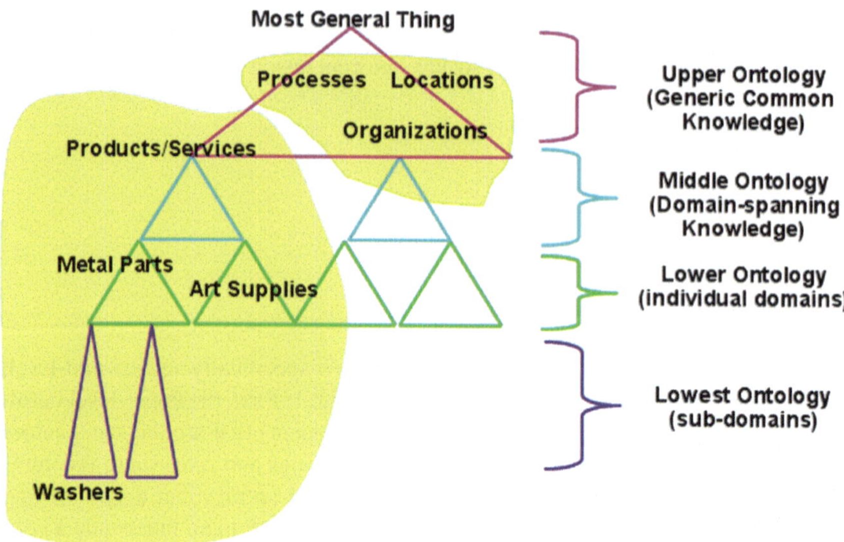

Fig. 1 Ontology's levels of abstraction [11]

that define concepts and their relationships and moreover deduce new knowledge from combing existing facts. Then, semantics searches can be performed basing on the meaning of each concept, for example, one could make the difference between horses and cars which both have the same label of "mustang" [10]. Furthermore, the use of upper-level ontologies facilitates the alignment between several domain ontologies. In other words, if the ontologies to be mapped are driven from a standard upper-level ontology, this will make the mapping task very easy. In addition, upper-level ontologies play the same role as libraries in software programming tasks. Once they are used, one could reuse the defined concepts and relationships and so as inherit the inferencing capabilities furnished by them. In this way, developing a domain ontology is an easier task that requires less time than usual. Moreover, the aim is to avoid having several incompatible domain ontologies. The usage of upper-level ontologies for integrating information and sharing knowledge among heterogeneous sources has been motivated in various related works [12]. Moreover, they have been used in various domains including situation awareness, pervasive systems [13], biomedical information systems, government and US military system [9], and especially emergency management [14].

Over the years, several upper-level ontologies have been already developed and well established, including BFO, SUMO, DOLCE, GFO, Cyc, and COSMO.

Cyc project was founded in 1984 by D. Leant as a lead project in the microelectronics and computer technology corporation (MCC). The aim of Cyc ontology is to enable the usage of knowledge across domains. The ontology includes a wide range of categories. The fundamental distinction of entities in the ontology is between collections and individuals. It is intended to capture concepts such as temporality, mathematics, and relationship types [15].

GFO (general formal ontology) project was launched in 1999 in the context of GOL project (general ontological language) at the University of Leipzig. It is an upper-level ontology presenting a multi-categorial approach that integrates universals, concepts, and symbol structures and their interrelations. It contains several novel ontological modules, in particular, a module for functions and a module for roles. It exposes a three-layered meta-ontological architecture consisting of an abstract top level, an abstract core level, and a basic level [16].

SUMO (suggested upper merged ontology) [17] is an upper-level ontology developed in 2000 by the Standard Upper Ontology Working Group, an IEEE-sanctioned working group composed of researchers from different fields such as engineering, philosophy, and information science. It proposes definitions for general purpose terms as a foundation that intend to be expanded for more specific domain ontologies. The idea of SUMO was the merging of several existing upper ontologies that did not have licensing restrictions, including John Sowa's upper-level ontology, Russell and Norvig's upper-level ontology, James Allen's temporal axioms, Casati and Varzi's formal theory of holes, Barry Smith's ontology of boundaries, Nicola Guarino's formal mereotopology, and various formal representations of plans and processes. Indeed, SUMO is a mixed upper ontology that contains both elements of realism as well as cognitively specific categories [18].

BFO project was initiated in 2002 under the auspices of the project Forms of Life sponsored by the Volkswagen Foundation. It is designed for use in supporting information retrieval, analysis, and integration in scientific and other domains. It does not contain specific terms such as physical, chemical, or biological terms. BFO is a realist, formal, and domain-neutral upper-level ontology; it is designed to represent at a very high level of generality the types of entities that exist in the world and the relations that hold between them. It is utilized as a starting point for the categorization of entities and relationships by more than 250 domain ontology [19, 20].

DOLCE (descriptive ontology for linguistic and cognitive engineering) [21] is the first module of a Foundational Ontology Library for the Semantic Web being developed within the WonderWeb project19 that started in 2002. It is not intended to be a universal or standard upper ontology, but instead, it serves as an ontology of instances. The most fundamental distinction between entities made in DOLCE is related about their behavior in time. On one hand, «*Perdurants*» are entities that unfold in time, on the other hand, «*Endurants*» are entities that are present "all-at-once" in time.

COSMO (common semantic model) project started in 2006, it arises from the efforts of the COSMO working group (COSMO-WG) and its parent group, the Ontology and Taxonomy Coordinating Working Group (ONTACWG). It is the result of merging some upper-level ontologies, COSMO integrates concepts from the Cyc project, SUMO ontologies, DOLCE and BFO [22].

3 Toward the Choice of "Basic Formal Ontology" and "Common Core Ontology"

To select the appropriate upper-level ontology among several ones, we first looked for a realist upper ontology that represents the world as is and not underlying natural language and human common sense. This criterion excludes DOLCE, SUMO, COSMO, and Cyc in view of the fact that they are more particular and descriptive than realist [23]. Then, to ensure that, the upper-level ontology can be extended to an emergency management ontology, it should be universal. Universal classes are often characterized as natural classes that abstract or generalize over similar particular things. Person, location, process, etc., are examples of universals [24]. So, this disqualifies GFO. Accordingly, in this work, we employed BFO as an upper-level ontology. The choice of a BFO-based approach lies in the fact that it focuses on the universals in reality,—we might say that the ontology encapsulates the knowledge of the world that is associated with the general terms used by scientists in the corresponding domain [25]. As a starting point, BFO uses the term «*entity*» as a common representation of anything that exists in the world from the point of view whether of philosophers or scientific researchers. Then, it incorporates two categories of entity «*Continuants*» and «*Occurrents*» in a single framework as a top-level distinction between entities. Continuants are entities that persist through time including three axes; objects (material entity) or and spatial regions (immaterial entities) as independent continuant, functions, and qualities as specifically independent continuant and finally, generically dependent continuant. Occurrents are entities that happen or develop in time such as process.

As a mid-level ontology, we decide that CCO meets most our requirements since it inherits from BFO as an upper-level ontology and defines a modular set of extensible classes and relations that can be connected to our domain ontology. The ten mid-level ontologies that compose the common core ontology are: The information entity ontology, the agent ontology, the quality ontology, the event ontology, the artifact ontology, the time ontology, the geospatial ontology, the units of measure ontology, the currency unit ontology, and the extended relation ontology. A simplified explanation of the diverse modules is presented in [26]: "In CCO, Agents (People and Organizations), use Artifacts to perform Actions that occur in both Time and Space, and are differentiated from other Agents and Artifacts via Attributes." The development of CCO started since 2010 in IARPA's knowledge, discovery and dissemination programs. The purpose of this core ontology is to provide a structured base vocabulary that serves as the unified semantics. Once extended, it represents the content of any data sources [27].

4 Firefighters Use Case

To develop our ontology, we adopted the following four steps [28]; First, we identified the purpose of defining this ontology; Basing on the feedbacks of emergency management experiences, there is a strong need to solve real issues that cause slower decision making in emergency situations such as heterogeneity of data, deficiency of interoperability in emergency management systems, and misunderstanding between stakeholders (firefighters, police, army, medical team, etc.). To solve these issues, there is an ample need to define the complex knowledge of the different stakeholders so as to come up with a common shared vocabulary.

For all we know, the modular ontology proposed in this paper is the first ontology based on the BFO and CCO that aims to define the emergency responders' knowledge starting with firefighter's module, the rest of the modules will be presented in future works.

In the second step, interviews were conducted with firefighters so as to capture their needs and to identify their technical vocabulary (Commandment hierarchy, means, types of intervention, roles, etc.). In the third step, we used Protégé, an open-source ontology editor, to create our modular ontology.

In order to ensure a better understanding of the created ontology, the architecture of the ontology development is shown in Fig. 2. The three levels are layered from top to bottom. As a starting, we integrate the basic formal ontology (BFO) as an upper-level ontology. It contains a total of 35 classes including one top class «*entity*» and all classes are connected by means of "is-a" relation. The most general categories in this level are «*Continuant*» and «*Occurrent*» as explained in the previous section.

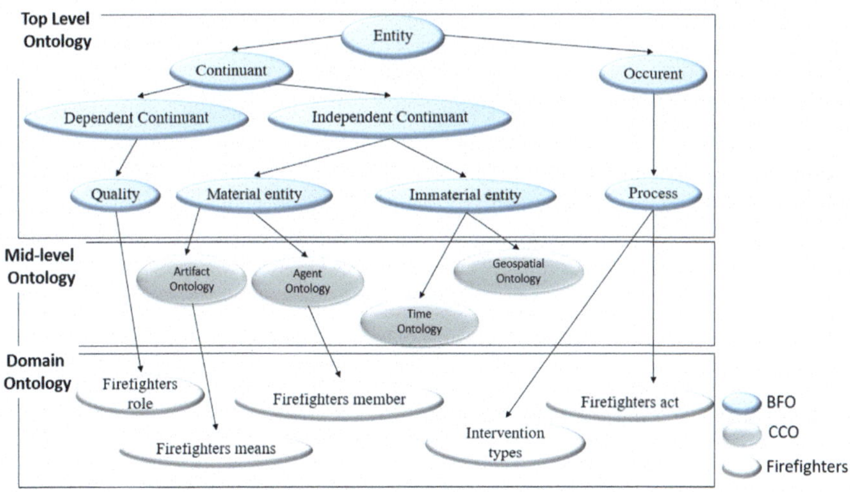

Fig. 2 Architecture of ontology development

Once the upper-level ontology is integrated, it is time to incorporate the mid-level ontology which is common core ontology. The ability to reuse modules in a flexible manner is a key feature of modular ontologies. In our work, we reused only four modules that will be extended according to the domain level needs which are (1) agent ontology, (2) artifact ontology, (3) time ontology, and (4) geospatial ontology [18].

(1) Agent ontology: The notion of agent includes both person as an individual agent and organization as a group of individuals. In addition, it contains agents' roles and agents' quality.
(2) Artifact ontology contains concepts representing general types of artifacts like communication artifact, facility, tool, vehicle, and weapon. Furthermore, the ontology enables a user to make assertions about which qualities or functions an artifact is designed to have.
(3) Time ontology provides the basic vocabulary for describing when events occur.
(4) Geospatial ontology offers the basic vocabulary for describing the locations of agents and occurrences of events.

In the firefighters' module, as regards to the continuant part, we extended the agent ontology to cover the different members of firefighter organization. Under agent quality, we incorporated the firefighter hierarchy of commandment and we attributed a grade for each member. In the class quality of dependent continuant, we affected the role of each member. In the artifact ontology, we classified firefighter different means by specifying their functions. Concerning occurrent entities, we added the different types of firefighters' interventions and its needs in terms of means and staff. Furthermore, Time ontology and geospatial ontology will be very helpful in emergency management context; it will determine when and where events occur. To summarize, the ontology we created, once it is complete and all the modules are integrated, can be used to be a common shared vocabulary for emergency management systems.

At the end of this stage, the firefighters' ontology had around 429 classes and 246 relations. The classes are labeled in English and in French. The final step consists on the evaluation of the proposed ontology by domain experts in term of inconsistency, incompleteness, and redundancy [29]. Once all the emergency management actors' modules are created, the ontology will be instantiated to test it by means of a concrete use case and it will be used in an emergency management system as a common shared vocabulary. Domain expert and users should then evaluate and validate the obtained results.

5 Conclusion and Future Work

In this paper, we presented how we employed BFO as an upper-level ontology and CCO as a mid-level ontology to propose a modular ontology that defines firefighter's knowledge (vocabulary, graphical charter, data representation, etc.). The use of upper ontologies improves data quality, reduces development time, and especially facilitate large-scale information integration by avoiding ambiguities or inconsistencies to guarantee semantic interoperability of systems. The suitable interoperability among emergency response systems can ensure the speed, efficiency, and appropriateness of emergency management. This work is an important step toward defining and formalizing emergency responder's knowledge.

As a future work, in emergency management situation, there are other stakeholders beyond firefighters, including police, medical team, army, etc. The idea is to formalize their knowledge to come up with a common shared vocabulary that will be used latter in an emergency management system to ensure a better coordination and cooperation between these stakeholders so as to guarantee the efficiency and appropriateness of emergency management.

Acknowledgements This research was conducted as part of the POLARISC (Plateforme OpérationnelLe d'Actualisation du Renseignement Interservices pour la Sécurité Civile) project. It was funded by the regional operational program FEDER/FSE «Midi-Pyrénées et Garonne 2014–2020» as part of the call for projects "Easynov2016."

References

1. IEEE. (1990). *IEEE standard computer dictionary: a compilation of IEEE standard computer glossaries*.
2. Komatsoulis, G. A., et al. (2008). caCORE version 3: Implementation of a model driven, service-oriented architecture for semantic interoperability. *Journal of Biomedical Informatics, 41*(1), 106–123.
3. Heiler, S. (1995). Semantic interoperability. *ACM Computing Surveys (CSUR), 27*(2), 271–273.
4. Bittner, T., Donnelly, M., & Winter, S. (2005). Ontology and semantic interoperability. In *Large-scale 3D data integration: Challenges and opportunities* (pp. 139–160).
5. Antunes, G., et al. (2013). Using ontologies to integrate multiple enterprise architecture domains. In *International Conference on Business Information Systems*. Springer, Berlin, Heidelberg.
6. Obrst, L. (2003). Ontologies for semantically interoperable systems. In *Proceedings of the Twelfth International Conference on Information and Knowledge Management*. ACM.
7. Song, F., Zacharewicz, G., & Chen, D. (2013). An ontology-driven framework towards building enterprise semantic information layer. *Advanced Engineering Informatics, 27*(1), 38–50.
8. Degen, W., et al. (2001). GOL: Toward an axiomatized upper-level ontology. In *Proceedings of the International Conference on Formal Ontology in Information Systems* (Vol. 2001) ACM.
9. Conesa, J., Storey, V. C., & Sugumaran, V. (2010). Usability of upper level ontologies: The case of ResearchCyc. *Data & Knowledge Engineering, 69*(4), 343–356.
10. Semy, S. K., Pulvermacher, M. K., & Obrst, L. J. (2004). *Toward the use of an upper ontology for US government and US military domains: An evaluation. No. MTR-04B0000063*. Bedford MA: MITRE Corp.
11. Leo Obrst: The Semantic Spectrum & Semantic Models, a Powerpoint presentation. http://ontolog.cim3.net/file/resource/presentation/LeoObrst_20060112/OntologySpectrumSemanticModels–LeoObrst_20060112.ppt.
12. Baumgartner, N., & Retschitzegger, W. (2006). A survey of upper ontologies for situation awareness. In *Proceedings of the 4th IASTED International Conference on Knowledge Sharing and Collaborative Engineering*, St. Thomas, US VI.
13. Stevenson, G., et al. (2009). Ontonym: A collection of upper ontologies for developing pervasive systems. In *Proceedings of the 1st Workshop on Context, Information and Ontologies*. ACM.

14. Liu, Y., Chen, S., & Wang, Y. (2014). SOFERS: Scenario ontology for emergency response system. *JNW, 9*(9).
15. Matuszek, C., et al. (2006). An introduction to the syntax and content of Cyc. In *AAAI Spring Symposium: Formalizing and Compiling Background Knowledge and Its Applications to Knowledge Representation and Question Answering*.
16. Herre, H. (2010). *General formal ontology (GFO): A foundational ontology for conceptual modelling: Theory and applications of ontology: Computer applications*. Netherlands: Springer.
17. Niles, I., & Pease, A. (2001). Towards a standard upper ontology. In *Proceedings of the International Conference on Formal Ontology in Information Systems* (Vol. 2001). ACM.
18. Farrar, S., & Bateman, J. (2004). *General ontology baseline, deliverable D1, I1-[OntoSpace]; workpackage 1*.
19. Smith, B., & Ceusters, W. (2010). Ontological realism: A methodology for coordinated evolution of scientific ontologies. *Applied ontology, 5*(3–4).
20. Seppälä, S., Smith, B., & Ceusters, W. (2014). Applying the realism-based ontology-versioning method for tracking changes in the basic formal ontology. In: *FOIS*.
21. Gangemi, A., et al. (2002). Sweetening ontologies with DOLCE. In *Knowledge engineering and knowledge management: Ontologies and the semantic Web*.
22. COSMO, Home Page. Retrieved January 15, 2018 from http://ontolog.cim3.net/wiki/COSMO.html.
23. Mascardi, V., & Cordì, V., & Rosso, P. (2007). *A comparison of upper ontologies*. WOA.
24. Semy, S. K., Pulvermacher, M. K., & Obrst, L. J. (2004). *Toward the use of an upper ontology for US government and US military domains: An evaluation*. Bedford, MA: MITRE Corp.
25. Arp, R., Smith, B., & Spear, A. D. (2015). *Building ontologies with basic formal ontology*. MIT Press.
26. Rudnicki, R. (2016). *An overview of the common core ontologies*. CUBRC, Inc.
27. Cubrc Homepage. Retrieved November 20, 2017 from http://www.cubrc.org/index.php/data-science-and-information-fusion/ontology.
28. Uschold, M., & Gruninger, M. (1996). Ontologies: Principles, methods and applications. *The Knowledge Engineering Review, 11*(2).
29. Lovrencic, S., & Cubrilo, M. (2008). Ontology evaluation-comprising verification and validation. In *CECIIS*.

Enabling Semantic Interoperability for Risk and Vulnerability Analysis of Public Buildings

Ling Shi, Bjørg E. Pettersen and Dumitru Roman

Abstract The risk and vulnerability analysis of buildings identifies buildings with high exposure to natural hazards. It is crucial to prevent or reduce damage and economic loss for the construction and property management industry. The government, as well as private landowners, benefit from risk and vulnerability assessment of their real estate portfolio. In this paper, we propose a semantic-enhanced method to simplify the risk and vulnerability assessment process and improve interoperability of real estate and natural hazards data. The method relies on the publication and integration of data from various sources as linked data. Visualization of the integrated data is also presented, together with application scenarios.

Keywords Building · Risk and vulnerability analysis · Real estate · Natural hazard · Semantic interoperability

1 Introduction

Real estate is the property consisting land and buildings on it. A building is usually a roofed and walled structure built for permanent use (as for a dwelling).[1] The building is built on a site, which is an owned or leased cadastral parcel, within an administrative unit such as a municipality. It is affected by the surrounding environment such as ground, terrain, air and climate. The physical building is continuously changing due to ageing, rehabilitation, extension and also damages caused by human or natural

[1] https://www.merriam-webster.com/dictionary/building.

L. Shi (✉) · B. E. Pettersen
Statsbygg, Pb. 8106 Dep, 0032 Oslo, Norway
e-mail: ling.shi@statsbygg.no

B. E. Pettersen
e-mail: bjorg.pettersen@statsbygg.no

D. Roman
SINTEF, Pb. 124 Blindern, 0314 Oslo, Norway
e-mail: dumitru.roman@sintef.no

K. Popplewell et al. (eds.), *Enterprise Interoperability VIII*, Proceedings of the I-ESA Conferences 9, https://doi.org/10.1007/978-3-030-13693-2_12

hazards. Over the last decades, climate change [1] has increased the intensity of storms,[2] deforestation has worsened flooding[3] and precipitation changes have led to increased flooding and flash floods. These changes have led to increased damages on real estate, especially buildings and thereby huge economic loss for the society.

Knowing which natural hazard poses the greatest risk to buildings is crucial for preventing damage [2]. Risk and vulnerability analysis of buildings identifies the natural hazards buildings are exposed to. It provides important information to owners or land-use planners for the maintenance and development of individual buildings as well as groups of buildings.

The risk and vulnerability analysis process requires access to datasets that are created and maintained by different organizations. For example, the buildings' ownership information, geographical location and the natural hazards data covering the corresponding area are provided by different sources. The location of a building can be geocoded by searching for the building's street address using Google Maps Geocoding API[4] or similar services. However, the addresses are often changed due to change of postal codes or street names. The national cadastral system is a more authoritative data source of buildings including the location information. Cadastral data are normally difficult to access, and it requires domain expertise to understand the data. To increase the availability and accessibility of cadastral data, there have been efforts to make cadastral data available as exemplified in [3, 4]. The natural hazards risk data is traditionally published as map files, and most of the natural hazards risk assessment studies use GIS[5] as a tool for spatial analysis. The analysis is also dependent on how the data are integrated, and the algorithms used for calculating risk exposure indices. Access to natural hazards data is a known barrier for users without GIS knowledge. It requires multiple domain knowledge to integrate risk data with real estate data before calculating the risk exposure indices. To simplify the risk assessment process and make both the data and the process more accessible to potential users, we propose in this paper a method for generating a risk and vulnerability analysis service for buildings, using semantic interoperability techniques. The method relies on the publication of data from various sources as Linked Data,[6] enriching data of buildings with risk exposure indices and generating service for risk and vulnerability analysis based on the resulting dataset.

The paper is organized as follows. Section 2 introduces the background and scope of the paper. Section 3 presents the data integration approach used in the risk and vulnerability analysis service. Data visualization and application scenarios are presented in Sect. 4. Section 5 summarizes the paper and outlines further work.

[2]https://earthobservatory.nasa.gov/Features/RisingCost/rising_cost5.php.
[3]https://earthobservatory.nasa.gov/Features/RisingCost/rising_cost3.php.
[4]https://developers.google.com/maps/documentation/geocoding/intro.
[5]https://en.wikipedia.org/wiki/Geographic_information_system.
[6]https://www.w3.org/standards/semanticweb/data.

2 Scope and Background

The government is one of the largest landowners and administrates a large portfolio of real estates which stands for a major part of state-owned assets. Risk and vulnerability analysis of state-owned buildings provides an indication of buildings with high risk exposure and it also identifies areas with high percentage of buildings exposed to risk. In this paper, the scope is defined to cover state-owned buildings in Norway mainly due to the accessibility of data and domain knowledge from property management and GIS experts at Statsbygg[7]—an important player in the construction and property management industry in Norway. The Norwegian State of Estate (SoE) dataset was previously created to include all the state-owned real estates in Norway [5]. Many types of natural hazards data are openly available or accessible from various sources, though only in GIS format.

A single risk for an individual building can be derived from the risk zone the building is located in. It is also important for real estate owners or land-use planners in a certain municipality to know the aggregated risks for their portfolio of buildings. Based on the statistics from the Norwegian Natural Perils Pool,[8] *storms*, *floods* and *landslides* cause most economical loss while other kinds of natural hazards such as earthquakes are not that representative in Norway. The relative share of insurance compensation can be used to weigh the indices for damage caused by the corresponding hazards. The common exposure index (EI) is constructed by adding weighted exposure indices as EI = storm (64%) + flood (28%) + landslides (8%).[9] In our case, we excluded the landslide hazard in the calculation because the landslide hazard map is not geographically complete, and also because it causes less economic loss compared to storm and flood in Norway. Storm and flood risk datasets are therefore necessary to assess risk exposure for individual buildings in Norway and to generate the common exposure index. Risk datasets need to be integrated with real estate data, such as the location, belonging municipality and owner, in order to develop the risk and vulnerability analysis service.

3 Risk Exposure Indices as Linked Data

This section presents the storm and flood datasets (Sect. 3.1), the overall approach for integration of real estate and risk datasets (Sect. 3.2), the actual data workflow for generating the risk and vulnerability analysis service (Sect. 3.3), and the resulting dataset (Sect. 3.4).

---

[7]http://www.statsbygg.no/Om-Statsbygg/About-Statsbygg/.

[8]http://www.naturskade.no/statistikk/.

[9]http://setebos.svt.ntnu.no/viewexposed/learn_about_vul/.

Table 1 Flood susceptibility zones dataset fields and examples

Field name	FID	Shape	objType	Opphav	digiDato
Description	Shape ID	Geometry type	It indicates if it is a potential flood risk area	Data owner	Data digitized date
Example 1	123	Polygon ZM	PotensiellFlomfareOmr	NVE	20091231
Example 2	124	Polygon ZM	PotensiellFlomfareOmr	NVE	20091231

Table 2 "MeanStormHours" dataset fields and example

Field name	FID	Shape	GridCode	MeanStormH
Field description	Shape ID	Geometry type	Grid identifier	Mean storm hours
Example 1	21	Polygon	4505536	4.16189
Example 2	2207	Polygon	954562	0.954562

3.1 *Source Datasets for Flood and Storm in Norway*

The original datasets are available in ESRI shapefile.[10] The flood susceptibility zones dataset [6, 7] is published by the Norwegian Water Resources and Energy Directorate (NVE)[11] and can be downloaded at NVE's GIS portal,[12] and the meta-description is available as PDF.[13] The dataset includes more than 12,000 flood zone polygons and the most relevant fields of the dataset are shown in Table 1 with two examples. The risk value is a implicit "yes" within a flood zone, but it does not show how vulnerable a flood zone is. If a building is located wholly or partly inside one or more flood susceptibility zones, it indicates flood risk.

The "MeanStormHours" dataset is provided by Kjeller Vindteknikk A/S[14] via Statsbygg. The wind speed is measured at 10 m above the terrain. MeanStormHours is the average hours of storm when the wind speed is above the minimum magnitude (20.8 m/s). The dataset contains float values that cover 1 $\times$ 1 km grid each. Table 2 lists up the relevant fields and their descriptions, and it also provides two examples from the MeanStormHours dataset. The value 4.16189 for MeanStormH means that the grid has 4.16189 h of storm in average per year, while a value of zero means no storm exposure. The value indicates how exposed a grid area is to storm.

[10]https://en.wikipedia.org/wiki/Shapefile.

[11]https://www.nve.no/english/.

[12]http://nedlasting.nve.no/gis/.

[13]http://gis3.nve.no/metadata/produktark/Produktark_FlomAktsomhet.pdf.

[14]http://www.vindteknikk.com/services/analyses/wind-engineering.

3.2 Integration of Real Estate and Risk Data

Understanding and using the original risk datasets are not easy for users without domain knowledge in natural hazards risk and GIS. Semantic technologies can be used to improve the interoperability of risk and real estate data. For example, the ontology model for risk data provides a standard for interpreting, processing, and transforming risk data. The risk data can be published and shared in a semantic web format, which enhances the accessibility of the data.

Integration of real estate data and risk data requires expertise in real estate and statistics, in addition to natural hazards risk and GIS. Semantic definitions of the attributes in the datasets help to define and evaluate the integration alternatives. The real estate dataset we used, in this case, was the Norwegian SoE dataset, a result of integrating cadastral data provided by the Norwegian mapping agency, business entity data and building accessibility data. The dataset includes cadastral parcels and buildings, and both have geospatial attributes as boundaries and centre points. Each real estate has normally an address point. The risk datasets also have geospatial attributes as zones or grids. Integration between real estate and risk datasets can, therefore, have a couple of alternatives depending on which kind of real estate geospatial attributes are utilized. We can choose between cadastral parcel boundary, building boundary, cadastral parcel centre point, building centre points or address points.

A cadastral parcel boundary defines a two-dimensional area of a real estate. The size of a cadastral parcel area can be large enough to cross several risk or non-risk zones, or a very small part of the area is inside a risk zone, making it rather difficult to design a proper algorithm to calculate the risk and vulnerability for such cases. The centre point of a large irregular formed cadastral parcel can be a point which may not be as representative as it should, for it to be used as the element based on which the integration is carried out. Normally, the buildings represent most of the values of real estates though exceptions may occur in some special cases. By focusing on the buildings, we reduced the complexity of integration though we may have a similar challenge with large buildings. Earlier risk assessment studies [2, 8] used address points because they focused on where people live, work or study. However, our risk assessment objects are real estates and buildings; a building may have several address points, which leads to uncertainty in the analysis. Buildings boundaries are not available in our dataset and therefore the building centre points are chosen as the integration keys because of their representativeness and availability, causing less complexity in the analysis.

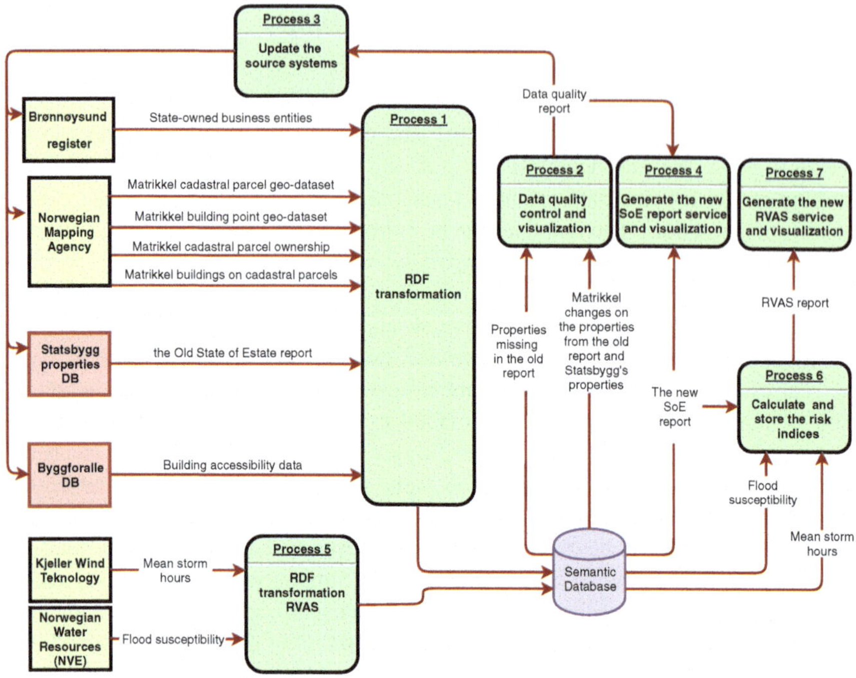

Fig. 1 Data workflow diagram for the risk and vulnerability analysis service (RVAS)

3.3 *Data Workflow for the Risk and Vulnerability Analysis Service*

The Norwegian state-owned buildings are published as a subset of the Norwegian SoE dataset. Figure 1 illustrates the data workflow which consists of seven processes. It starts with Process 1 which transforms data from four source systems to RDF and stores the result in the semantic database. Then it continues with Processes 2 and 3 in a loop to improve the data quality. When the quality is acceptable, it moves to Process 4 to generate the new SoE Report which is both dynamic and up to date compared to the old static SoE Report. Afterwards, the new SoE Report dataset is enriched by two natural hazards datasets on mean storm hours and flood susceptibility in order to generate the risk and vulnerability analysis service (RVAS) through Processes 5–7.

In Process 5, the two risk datasets are transformed into RDF, mapped to the GeoSPARQL vocabulary[15] and the proDataMarket vocabulary[16] [9] and stored in the semantic database.

Process 6 calculates the risk indices and stores the data in the semantic database. It includes several steps as follows:

[15]http://www.opengeospatial.org/standards/geosparql.

[16]http://vocabs.datagraft.net/.

- Calculate storm exposure and flood exposure for individual buildings
  - The storm exposure (ES) and flood exposure (EF) are calculated for each building and added as extra triples in the semantic database. The GeoSPARQL function[17] *sfWithin* is used to decide whether a building is inside a flood or storm zone;
- Calculate the aggregated exposure indices for groups of buildings.
  - Buildings are grouped by municipality or owner.
  - The storm exposure index (EIS) and flood exposure index (EIF) are the result of dividing the number of buildings exposed to risk with the total number of buildings in this group.
- Generate normalized exposure indices for groups of buildings.
  - In order to compare relative exposure, we must normalize the index values (x) to a common scale (X') by the minimum (min) and maximum (max) index values using the formula $X' = ((x - \min)/(\max - \min)) * 100$. The results are the normalized storm exposure index (EIS100) and the normalized flood exposure index (EIF100);
  - Total normalized exposure index storm and flood by municipality or organization are a weighted result of EI100 = (0.64 * EIS100) + (0.28 * EIF100).

In Process 7, the RVAS is generated and the result is visualized in maps and graphs (examples will be shown in Sect. 4).

3.4 *The Result Dataset for RVAS*

The result dataset for RVAS is an extension to the Norwegian SoE dataset. The flood susceptibility and MeanStormHours values are calculated and attached to each building. Table 3 gives an example of SPARQL query to extract flood risk for buildings owned by the central government in Norway, and some examples of results are shown in Table 4. A similar query can be run to extract storm risk values for buildings, and the top five buildings with high storm risk are presented in Table 5. The results include a building's cadastral building number (*nr*), coordinates (*Coords*), owner, and risk values. The cadastral building number can be used later as a key to integrate with real estate data to obtain additional information on buildings. The data from both Tables 4 and 5 are visualized in Sect. 4.

An example of aggregated risk exposure indices is presented in Table 6. It shows that 20.4% of buildings owned by Statsbygg in Oslo are exposed to flood while there is no building exposed to storm.

[17] http://www.opengis.net/def/function/geosparql/.

Table 3 SPARQL query to extract flood risk for buildings

PREFIX prodm-cad: <http://vocabs.datagraft.net/proDataMarket/0.1/Cadastre#>
PREFIX prodm-com: <http://vocabs.datagraft.net/proDataMarket/0.1/Common#> *PREFIX schema: <http://schema.org/>* *PREFIX gsp: <http://www.opengis.net/ont/geosparql#>* *PREFIX dul: <http://www.ontologydesignpatterns.org/ont/dul/DUL.owl#>* *PREFIX dbo: <http://dbpedia.org/ontology/>* *SELECT DISTINCT?floodRisk?nr?owner?coords* *WHERE {* *?bygg a prodm-cad:Building;* *prodm-cad:hasCadastralBuildingNumber?nr;* *gsp:hasGeometry [gsp:asWKT?coords;];* *prodm-cad:isBuiltOn?cp.* *?rr a prodm-cad:RealRights;* *dul:defines?cp;* *dul:defines?org.* *?org a prodm-cad:RightsHolderOrganization;* *schema:legalName?owner.* *OPTIONAL {?bygg prodm-com:hasIndicator [a prodm-com:Indicator;* *dbo:type "FloodRisk";* *rdf:value?flood;];} bind(if(bound(?flood), "1", "0") as?floodRisk)}*

Table 4 Examples of flood risk for buildings near the opera in Oslo

Owner	nr	Flood risk	Cords
Statsbygg	81622124	1	Point (10.7534402517774 59.9072013693557)
Statsbygg	81770859	1	Point (10.7531675849471 59.9074564779126)
Statsbygg	80466536	0	Point (10.7402902014331 59.9087857351063)
Statsbygg	81791627	0	Point (10.7398048706453 59.9087382526236)
Statsbygg	81826013	0	Point (10.7388619729803 59.9084902330054)

Table 5 Examples of five buildings with the highest storm risk

Owner	nr	Storm risk	Coords
Meteorologisk Institutt	300082441	27.1	Point (5.12766397110468 62.1871443440619)
Meteorologisk Institutt	178242741	27.1	Point (5.12769187185718 62.1871712798231)
Miljødirektoratet	179418584	22.3	Point (5.4914954370634 62.3211685144463)
Miljødirektoratet	179360063	22.3	Point (5.49161979646997 62.3212615128426)
Miljødirektoratet	179394219	19.4	Point (5.63211878525755 62.402970688029)

Table 6 Aggregated risk exposure indices dataset fields and examples

Field name	Field description	Example
Municipality	Name of the municipality	Oslo
Organization	Name of the organization	Statsbygg
Sum_ap	Total number of buildings	2051
EIF	ExposureIndexFlood	0.204
EIS	ExposureIndexStorm	0

4 Data Visualization and Application Scenarios

Assessing the natural hazards risk values for individual buildings helps a real estate manager to take proactive actions to prevent or reduce possible natural hazards damages. The visualization in Fig. 2 is based on data from Table 4. It shows flood risk of buildings owned by Statsbygg near the opera in Oslo. The buildings within the flood risk zone are coloured in red, and buildings without flood risk are coloured in green. This analysis and visualization on the map aim to give the real estate owners, government authorities and the public a good perspective of the flood risk in their targeted areas. The visualization in Fig. 3 is based on Table 5—the top five buildings with storm risk. The five buildings are located in the west coast and owned by two organizations: Norwegian Environment Agency and The Norwegian Meteorological Institute.

5 Summary and Outlook

Risk and vulnerability analysis of public buildings is an important service to prevent and reduce possible economic loss caused by natural hazards. RVAS is targeted at the construction and property management industry and land-use planners in municipality though it also covers risk exposure of individual buildings. This paper introduced the interoperability problem to be addressed for risk and vulnerability analysis, and presented a semantic-enhanced method to publish and integrate natural hazards data with buildings data. Visualization and application scenarios were presented for single risk for individual buildings and aggregated exposure indices for a municipality.

In terms of future work, the preparation of source datasets currently includes several manual steps which can be automatized to a certain level. Landslide can be added as the third biggest natural hazard to calculate the common exposure index when landslide data become more complete. The experience and process methods can further be tested on all buildings in Norway or similar datasets from other countries. We can also evaluate the effect of using building boundaries when they are available instead of building centre points for integration with risk data. Furthermore, the study can be extended with financial consequences of natural hazards on buildings.

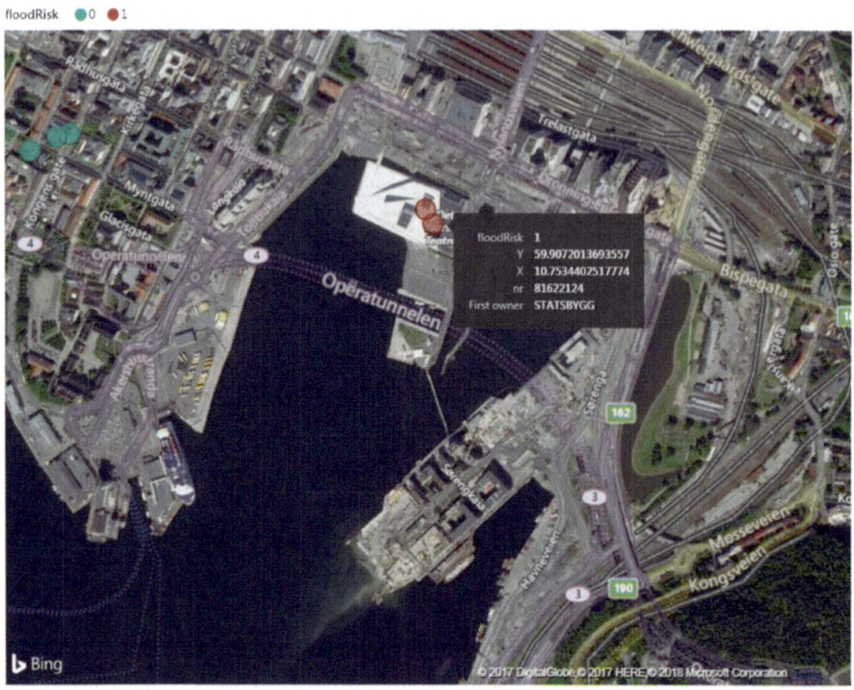

Fig. 2 Flood susceptibility dataset and building points

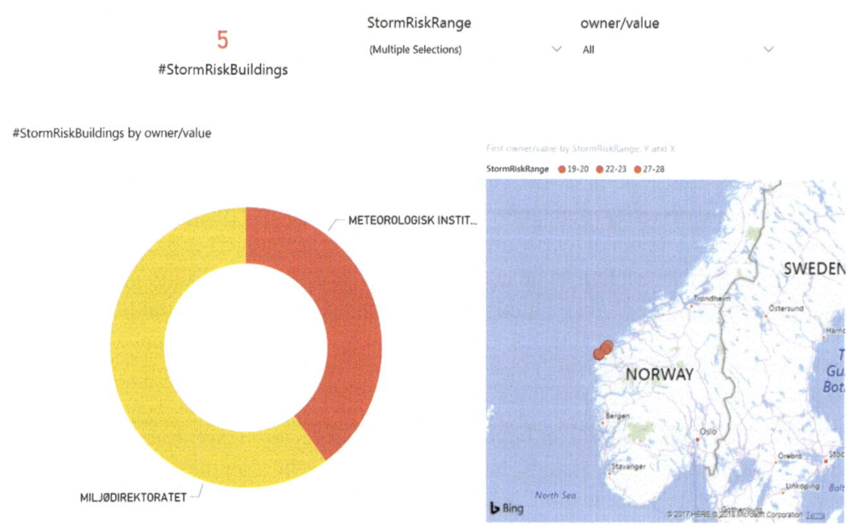

Fig. 3 Top five buildings with high storm risk

Acknowledgements The work has been funded by the European Commission under the projects proDataMarket (Grant number: 644497) and euBusinessGraph (Grant number: 732003). The authors would like to thank the proDataMarket consortium for additional contributions.

References

1. Field, C. B. (Ed.). (2012). *Managing the risks of extreme events and disasters to advance climate change adaptation: Special report of the intergovernmental panel on climate change*. Cambridge University Press.
2. Opach, T., & Rød, J. K. (2013). Cartographic visualization of vulnerability to natural hazards. *Cartographica: The International Journal for Geographic Information and Geovisualization, 48*(2), 113–125.
3. Vilches-Blázquez, L. M., Villazón-Terrazas, B., Corcho, O., & Gómez-Pérez, A. (2014). Integrating geographical information in the Linked Digital Earth. *International Journal of Digital Earth, 7*(7), 554–575.
4. Saavedra, J., Vilches-Blázquez, L. M., & Boada, A. (2014). *Cadastral data integration through Linked Data*.
5. Shi, L., Sukhobok, D., Nikolov, N., & Roman, D. (2017, September). Norwegian state of estate report as linked open data. In *OTM confederated international conferences "On the move to meaningful internet systems"* (pp. 445–462). Cham: Springer.
6. Peereboom, I. O. (2011). *Identification of areas exposed to flooding in Norway at a national level*.
7. Peereboom, I. O. (2013). Flood mapping for PFRA and NVE's spatial planning guidelines. In Nve (Ed.). NVE.
8. Rød, J. K., Opach, T., & Neset, T. S. (2015). Three core activities toward a relevant integrated vulnerability assessment: Validate, visualize, and negotiate. *Journal of Risk Research, 18*(7), 877–895.
9. Shi, L., Nikolov, N., Sukhobok, D., Tarasova, T., & Roman, D. (2017). The proDataMarket ontology for publishing and integrating cross-domain real property data. *Journal "Territorio Italia. Land Administration, Cadastre and Real Estate", 2*, 1. https://doi.org/10.14609/Ti_2_17_1e.

Part V
Interoperability Testing

IoT-Based Automatic Non-conformity Detection: A Metalworking SME Use Case

Maria Marques, Adriana Cunha, Wael M. Mohammed, Ricardo Jardim-Gonçalves and Carlos Agostinho

Abstract Industrial production's main goal is to achieve adaptability, resource efficiency, as well as, to integrate the complete value and supply chains, including customers, in business and value processes. To this end, manufacturing systems need to be as generic as possible in order to answer the different needs of a variety of industries. Industry4.0 paradigm stands as the baseline to answer these requirements, and data collection capabilities represent a major pillar in this strategy. Moreover, the way companies interact and communicate, being able of sharing information among themselves as well as to take full advantage of the data and knowledge being generated (even within the same company) demand huge attention to solving interoperability issues. The C2NET project (Cloud Collaborative Manufacturing Networks project), intends to implement the Industry 4.0 vision aiming to provide a cloud-based platform for managing the company interactions and promoting enterprise interoperability. This paper presents how the Data Collection Framework (DCF) developed within C2NET project can be used to collect data and support an automatic non-conformity detection case in a Portuguese metalworking SME. The developed components are briefly described as well as the implemented use case. The results obtained are also presented and discussed.

Keywords Enterprise interoperability · Data collection · Resource virtualization

1 Introduction

"Industry 4.0" has become the new paradigm in what regards the reality of industrial production in the years to come. The main goal of Industry 4.0 is to realize the

M. Marques (✉) · R. Jardim-Gonçalves · C. Agostinho
Centre of Technology and Systems, CTS, UNINOVA, 2829-516 Caparica, Portugal
e-mail: mcm@uninova.pt

A. Cunha
TecMinho, TECM, Guimarães, Portugal

W. M. Mohammed
Tampere University of Technology, Tampere, Finland

K. Popplewell et al. (eds.), *Enterprise Interoperability VIII*, Proceedings of the I-ESA Conferences 9, https://doi.org/10.1007/978-3-030-13693-2_13

so-called intelligent factory whose main characteristics are adaptability and resource efficiency together with the possibility of integrating the complete value and supply chains, including customers, in business and value processes. To achieve this objective, industrial companies need to be capable of handling more complex and stricter requirements in a variety of fields (e.g., data collection, flexibility) while ensuring to maintain, or increase, the requested production capacity. Nowadays, companies are encouraged to think globally while acting and staying locally economically compatible. Similarly, inside the factory, enterprise level strategy must concerted with local actions at lower level (e.g., device level). In addition to this, the industrial manufacturing systems of this 4th industrial revolution intend to reposition several components, (e.g., Manufacturing Execution Systems (MES) and Enterprise Resource Planning (ERP) layers), so that these can be deployed in a web-based or cloud-based environment, in order to enable its adaptation to the concept of collaborative networks [1]. In this context, manufacturing systems need to be as generic as possible in order to answer the different needs of a variety of industries. Industry 4.0 paradigm stands as the baseline to answer these requirements, and data collection capabilities represent a major pillar in this strategy to enable real-time information on production system operation. The data collection needs to include not only the collection of data from shop floor devices or resources but also the manipulation of the enterprises information systems that handle orders and schedules. Thus, interoperability demands are extremely high. The C2NET project (Cloud Collaborative Manufacturing Networks project) intends to implement the Industry 4.0 vision aiming to provide a cloud-based platform for managing the company interactions and promoting enterprise interoperability. According to the architecture of C2NET project [2], the Data Collection Framework (DCF) consists of a dedicated layer designed for collecting and managing the data of the industrial networks resources. This data includes Internet of Things (IoT) data from the shop floor (e.g., sensors) and legacy systems data resources (e.g., ERP). This way, data definition appears linked to a set of resources that are managed via middleware systems, i.e., hubs. Regarding this research work, this paper presents the steps performed to solve interoperability challenges regarding collection and use of scattered and heterogeneous data in a specific context of a use case from a Portuguese metalworking SME. The collected data is used to detect the occurrence of abnormal situations that can lead to delays in production orders and affect the overall performance of the company. Next sections are organized as follows: Sect. 2 presents a brief state of the art on related areas, Sect. 3 provides an overview of developed Data Collection Framework (DCF), Sect. 4 describes the interoperability needs identified and how they can be tackled through the implemented use case, and Sect. 5 provides specific insight into results obtained for validating the approach. In Sect. 6, some conclusions are drawn and future work lines are presented.

2 State of the Art

2.1 Enterprise Interoperability

Enterprises are constantly seeking for solutions capable of optimizing their assets aiming to increase their business benefits and reinforce their competitiveness. The available ICT and connectivity solutions are attracting companies that intend to globally spread their operations, in investing in interoperability solutions. The so-called *inter-enterprise interoperability*, besides creating an easy communication channel with clients, enables the networking between different enterprises supported by ICT-based technologies. Moreover, similar processes are being applied within the same company (*intra-enterprise interoperability*) in order to facilitate the use of information being generated on different departments, systems and devices, and eliminating issues related with duplicate information, information formats, etc. This degree of interoperability is the focus of this paper in which the collection and use of scattered and heterogeneous data is explored. One of the major problems of this interaction is the heterogeneity of data coming from different sources [3]. When focusing on the technical aspects of interoperability, two levels can be defined: A low-level interoperability in which connection and communication are established and a high-level interoperability in which the objective is to enable system interaction and understanding.

Regarding the low level, systems use well-defined formats for exchanging messages and, in IoT environments, technologies for Wireless Sensor Networks (WSN) and Machine-to-Machine (M2M) are the most prominent. WSN can be implemented using different protocols (e.g., Bluetooth (IEEE 802.15.1), Near Field Communication (NFC), Radio-Frequency Identification (RFID), IEEE 802.15.4 wireless personal area network (WPAN), WIFI (IEEE 802.11), 3G/4G and ETHERNET (IEEE 802.3)) and the selection of the most appropriate one must take into consideration aspects such as desired speed and working environment [4]. Regarding M2M, several initiatives have been recently developed, from which the ETSI M2M Service Architecture [5] is an example. In addition, MQTT (Message Queuing Telemetry Transport—http://mqtt.org/) is a lightweight messaging protocol especially used when remote connections are required and bandwidth is not an issue.

Regarding the high-level interoperability, the main challenge is to find a common ground for the involved systems. To this end, ontologies and other model-based technologies have been extensively used as they provide a shared understanding of a specific domain. In Semantic Web's, "an ontology" is defined as a set of classes in a domain area that shows the properties and the relations between those classes [6]. To establish an ontology on a specific domain, we have to select an ontology language (OWL or RDS/RDFS) and an ontology engineering approach (bottom-up: starting from the specific concepts and, by generalization, builds a structure; top-down: starting from the generic concept and, by specialization, builds a structure; middle-out: identifying central concepts in each domain). Ontology development can be supported by ontology development life-cycle tools (from creation to maintenance and evolution). Some examples include: Protégé [7, 8] and the Topic Maps 4 E-Learning—TM4L [9]. Current developments in this area, integrating both low-level and high-level interoperability strategies, include cloud-based platforms that support enterprise interoperability (such as C2NET platform [10]). These solutions tend to be plug and play platforms that, although requiring some configuration work, enable the access to a set of new and improved services.

2.2 Enterprise Resources Virtualization

Smart enterprise is a term coined by entrepreneur and investor Joe Lonsdale, describing a new breed of computing companies focusing on "enabling knowledge workers to process and analyze massive amounts of heterogeneous data and to collaborate and monitor things." On a high level, this encompasses killing waste/inefficiencies, redesigning collaboration and surfacing untapped data in new potent ways for big, monolithic industries that exist today [11]. In addition to this vision, the term ubiquitous computing (proposed by Mark Weiser) [12] envisages a smart environment in which sensors, actuators, interfaces and other elements are seamlessly embedded into common objects which are connected with each other via a network. The transference of this vision to manufacturing (smart) environment [13] leads to the next development step, which addresses the fusion of both physical and digital/virtual world [14] to reach the so-called Smart Factory (SF). In a SF, activities like real-time data collection enable that the access to manufacturing relevant information, anytime/anywhere, become a reality. This is done without affecting other systems, maintaining their capabilities of accomplish their tasks based on information coming from physical and virtual worlds. In addition to this, SF is capable of reacting to disturbances in production using available information to build the context under which a specific situation is occurring and taking advantage of decentralized approaches (either on information and communication) for appropriate reaction [13]. This behavior requires a high degree of synchronization between digital and real world which is being implemented. New ICT developments are providing the needed real-time access to sensors and devices as well as the advanced networking and processing capabilities contributing for an active cooperation of all the components building a sort of factory "nervous system" [15, 16]. Some implementations have been developed to support interoperability of Internet of Things systems, and, more to it, to allow search and detect both IoT and real-world resources as well as their associations [17]. The virtualization of resources contributes for reducing the complexity of internal operations, as well as generates an agile environment through the possibility of implementing decentralized decision-making approaches [2, 18].

2.3 *Data Collection and Data Collection Frameworks*

In the context of real-world resources, data collection, also called data acquisition or data sensing, deals with the collection of data (actively or passively) from the device, system, or as a result of its interactions [19]. For data collection, critical information needs to be available at the right point in a timely manner, and in the right form. The main aspects of a data collection system are: (i) Communication with distributed devices: This can be done over wired or wireless links to acquire the needed data, and need to respect security, protocol, and application requirements; (ii) Nature of acquisition: It could be continuous monitoring, interval-poll, event-based, etc.; (iii) Frequency: This depends on, or is customized by, the application requirements (or their common denominator). In simple scenarios, due to customized filters deployed at the device, a fraction of the generated data may be communicated. In more sophisticated scenarios, data aggregation and even on-device computation of the data may result in communication of events (such as detection of faults), which can be detected based on a device's own intelligence and capabilities [20]. IoT devices interactions and cooperation capabilities are likely to create large amounts of data [21]. This type of platform can be classified considering three different areas, namely: communication, transformation and data storage: (i) Communication: depending on the types of protocols used for sending/receiving data; (ii) Transformation: related with the techniques used to enrich data (e.g., taking into consideration the context under which this data is being collected) and how to make it more meaningful; (iii) Data storage: defining the type mechanisms used to store structured and unstructured data.

The analysis of the state of the art allows concluding that having a consistent strategy for interoperability is fundamental to guarantee the proper operation of any system involving different components as it ensures that component interfaces and shared or exchanged data are completely understood, to be interpreted by other components or systems.

3 C2NET Interoperable Data Collection Framework

C2NET project aims at facilitating the collection and usage of scattered, heterogeneous, and sometimes ambiguous, data. For this reason, interoperability is a key feature of C2NET platform which is composed of different components that are dependent on other legacy software. The developed interoperability framework acts as a middleware between C2NET components and legacy systems, not only for data collection, but also to guarantee data exchange inside the platform and provide configuration features to ensure generic usage of components. The Data Collection Framework (DCF) is the domain module that acts as the C2NET entry point of data that is arising from companies' side (see Fig. 1).

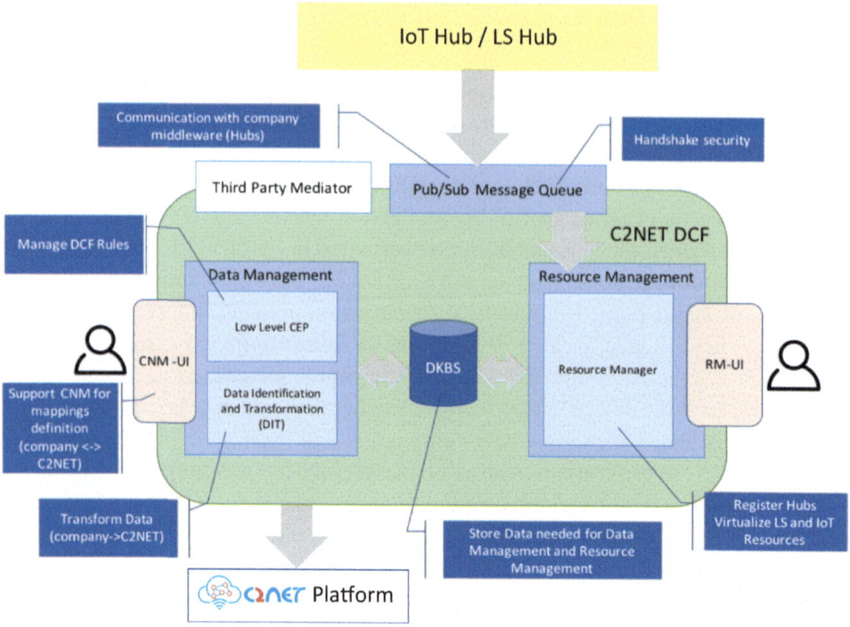

Fig. 1 C2NET data collection framework architecture

The company middleware composed by IoT and Legacy System (LS) Hubs provides the facilities required to scale the solution to any number of external resources and ensures interoperability with different data sources from ERP Systems, company databases, IoT devices, etc. To cover these aspects, three levels of interoperability have been defined for the C2NET DCF: (1) At Hubs level: where data are primary gathered from the company to the platform. Here, interoperability issues such as communication protocols, software interfaces, or data selection functions needs to be handled; (2) At Platform level: where the components are exchanging and using data from the Hubs. Here, interoperability issues such as data structures integration/mapping and transformation needs to be handled; (3) At User level: where the data can be presented to the user. This paper focuses on the first two aspects, namely:

- **Interoperability Support at the Hubs Level**: At the hubs level, there are two questions that interoperability tries to respond. The first is how to provide an interface to the hubs allowing plug and play and rapid deployment. The second is how data can be gathered intelligently to the platform, in a way that huge data flows and unnecessary information cannot affect the functioning of the system or abuse of system resources. The C2NET solution to these issues relies on the development of a secure layer able to interact in an easy and generic manner with the hubs in charge of data collection. In this way, all interfacing complexity with IoT and legacy data systems is hidden to C2NET platform. In what regards the C2NET Data Collection Framework architecture, the two main modules that are on charge of this level of interoperability are the PubSub module (responsible for the communications with companies' hubs) and the Resource Management module (provides the necessary configuration facilities for C2NET resources).

- **Interoperability Support at the C2NET Platform Level**: At the C2NET platform level, interoperability is handled from a classical enterprise information systems data interoperation problem. In technical interoperability, as the one implemented at the Hubs level, consensus between different vendors often relies on standards (e.g., communication protocols). In the case of data interoperability, standardization or the use of reference models can also be applied. C2NET applies the unified approach to interoperability. This approach consists in using a common format at a meta-level which provides a means to establish semantic equivalence, allowing consistent mapping between input and output models. Therefore, in C2NET, the unification requires a common structure at the meta-level for data to be exchanged between the different resources and C2NET platform (and its components). This is a big challenge for the modules implementing this feature because of the complexity of the task, which includes data processing and transformation. Data transformation aims to keep a common representation for collected data by providing a common structure called C2NET format.

4 C2NET to Detect Non-conformities and Reduce Waste

Production and product quality is an area of growing relevance for both manufacturers and consumers. Companies want to reduce the number of non-conformities to reduce the production costs, both by reducing the raw material that may be misused (and turned into waste), as well as to reduce the time spent in production. Additionally, increased public awareness on environmental sustainability and the rise of prices of non-renewable raw materials are contributing for market changes [22]. From a company point of view, in an ideal scenario, non-conformities and waste should be eliminated reaching a zero-defect production process. C2NET solution uses IoT in production lines to enable real-time quality control allowing the detection of non-conformities in an early process stage and contributing for waste reduction. The use case involves a Portuguese metalworking SME company that produces a huge variety of small pieces for different customers. Most of production data are scattered across different systems and in heterogeneous formats (manual registries are still common). To overcome these difficulties, the collection of data and its transformation in C2NET format were crucial to enable its usage. To support the implementation of this use case, a set of steps were defined (see Table 1).

Table 1 Use case steps

Nº	Name	Description	
0	Install company resources	Preparatory phase includes identification of data sources, additional instrumentation added (if needed), installation of hubs and setting up of IoT communication network.	
1	Company resource virtualization	Available resources and data collection rules are registered and configured in C2NET	
2	Define C2NET mappings	Data to be collected is mapped to C2NET data format - STables	
3	Define rules for non-conformity detection	Definition of rules that translate the quality patterns and allow the detection of non-conformities	
4	Upload Production Order	Production order is uploaded via Legacy System Hub	L
5	Start Production	Production start is detected via IoT devices which reaches C2NET platform through the IoT Hub	O O
6	Monitor the various stages of production	Production data is collected via IoT Hub	P
7	Detect Non-Conformities	IoT Hub collects data that triggers a non-conformity rule	
8	Check production order status	The status of the production order is stored (to be resumed later if production stops)	
10	Notify about Non-Conformities	C2NET sends message to production manager informing about the detected non-conformity	
11	Stop production (if needed)	In critical cases production may have to be stopped	
12	Setup Machine	These steps are performed out of C2NET	
13	Solve Non-conformity cause		
14	Separate non-conform products		
15	Catalogue non-conform products	The non-conformity is registered together with the actions developed to solve it	

Note that, steps 3 to 6 occur continuously up to the detection of a non-conformity. (LOOP). Also, steps 12 to 14 are not processed with the support of C2NET

5 Validation and Discussion

To support the evaluation of the business impact, a set of Key Performance Indicators (KPIs) were defined. These KPIs allow the confirmation of the benefits achieved by the involved company through the use of C2NET. KPIs are associated with a number of measurable indicators to facilitate the extrapolation of information. Table 2 presents the set of KPIs defined for this use case together with the measurable indicators that will be used to evaluate each one of them. KPIs were measured in the beginning of the project, and final measurements were made at the end of the project for comparison.

Table 2 Business KPIs for metalworking use case

KPI	Associated measurable indicator(s)	Initial value	Current value
Increase in machinery availability	Average No. of stops (per 10,000 units)	16	14
	Duration of stops (per 10,000 units)	1 h 23 m	1 h 15 m
Increase in machinery working hours	Production time (pieces/hour)	8604	9158
Decrease in non-conform product	Average % of non-conform products	0.90	0.87

Table 3 Functional/technical measurable indicators for metalworking use case

Functional/technical indicator	Initial value	Current value
No. of devices installed to collect production data[a]	2	13
No. of hubs installed at company to import data	0	2
Amount of data samples being collected through IoT devices[b]	0	~2.5 KB/min
No. of existing legacy system data files (ERP)[c]	3	5
No. of existing datasets	81	92[d]
No. of imported/mapped datasets (this use case)	0	13
No. of rules for automatic non-conformity detection	0	3

[a]Including: no. of products produced; temperature of painting chamber; quality of painting
[b]Measurement depends on the sending frequency being used. In this measurement, the IoT devices were configured to collect sensors data each 10 s and send the aggregated data package each minute (i.e., 6 samples/min/sensor)
[c]Containing data regarding production order; production recipes; available resources
[d]Including IoT datasets

The analysis of the associated measurable indicators allows us to conclude that there were significant improvements regarding the defined business KPIs.

In addition to business KPIs, a set of functional/technical indicators were also defined. These indicators enable the possibility to check the development status as well as to conclude about the usability of the proposed solution. Table 3 presents a compilation of the defined Functional/technical measurable indicators together with the available measurements.

The differences between initial values and current values allow detecting a consistent increase in the data being generated and used. Note that the table presents only the values related to detection of non-conformities in two specific production stations (the ones in which the testing phase is focusing). As the non-conformity detection is to be applied in other production stations, the number of rules will also increase. Moreover, and although this is parallel to the use of C2NET, the company has developed additional effort in structuring and organizing its legacy system data in order to take the most out of C2NET functionalities. An example is the growing number of legacy system data files available. These new files represent information regarding machine availability and material consumption, and the data being collected there will contribute for a decrease in the time spent in production planning.

6 Conclusions and Future Work

C2NET project is providing an interoperability solution through a cloud-based platform for managing the company interactions and promoting enterprise interoperability. The Data Collection Framework here presented for collecting and managing data is an example of the work developed. The collection and integration of data from multiple sources in a Portuguese metalworking SME are providing an interoperable solution for additional insight into the processes. The results being obtained, even in early stage, demonstrate clearly the potential of the approach.

Future work includes the addition of virtual resources to combine collected data being generated and generate additional information and knowledge about the production operation. Also, more features (e.g., negotiation and optimization) of the C2NET platform could be exploited to enrich the interactions between the company networks. Moreover, the expansion of the approach to cover all production stations of the company, as well as to implement it in other sectors, is also planned. Some ideas will be explored in BOOST 4.0 project.

Acknowledgements The research leading to these results has received funding from the European Union H2020 projects C2NET (FoF-01-2014—nr 636909) and BOOST 4.0 (ICT-15-2017).

References

1. Bohuslava, J., Martin, J., & Igor, H. (2017). TCP/IP protocol utilisation in process of dynamic control of robotic cell according industry 4.0 concept. In *2017 IEEE 15th International Symposium on Applied Machine Intelligence and Informatics (SAMI)* (pp. 217–222).
2. Ghimire, S., Melo, R., Ferreira, J., Agostinho, C., & Goncalves, R. (2015). Continuous data collection framework for manufacturing industries. In Ciuciu I. et al. (eds) *On the Move to Meaningful Internet Systems: OTM 2015 Workshops* (vol. 9416). Lecture Notes in Computer Science. Cham: Springer.
3. Govindarajan, N., Ferrer B. R., Xu, X., Nieto, A., & Lastra, J. L. M. (2016) An approach for integrating legacy systems in the manufacturing industry. In *2016 IEEE 14th International Conference on Industrial Informatics (INDIN)* (pp. 683–688).
4. IEC (International Electrotechnical Commission). (2014). *Internet of things: wireless sensor networks*, White Paper.
5. ETSI Technical Committee Machine-to-Machine communications. (2013). *Machine-to-Machine communications (M2M); Functional architecture*. Available at: http://www.etsi.org/deliver/etsi_ts/102600_102699/102690/02.01.01_60/ts_102690v020101p.pdf.
6. Musen, M. (1998). Domain ontologies in software engineering: use of protégé with the EON architecture. *Methods of Information in Medicine, 37*(4–5), 540–550.
7. Stanford Center for Biomedical Informatics Research. (2011). Stanford's protégé home page. Retrieved from http://protege.stanford.edu/.

8. Ontopia. (2015). About Ontopia. Retrieved from Ontopia: http://www.ontopia.net/page.jsp?id=about.
9. Dicheva, D., & Dichev, C. (2012). TM4L: creating and browsing educational topic maps. WinstonSalem State University.
10. Andres, B., Sanchis, R., & Poler, R. (2016). A cloud platform to support collaboration in supply networks. *International Journal of Production Management Engineering, 4*(1), 5–13.
11. Lonsdale, J. (2012). *What are some companies similar to Palantir Technologies in terms of challenge and work culture?* October, 2012. Available at: https://www.quora.com/topic/Smart-Enterprises.
12. Weiser, M. (1991). The computer for the 21st century. Special issue on *communications, computers, and networks*. Scientific American.
13. Lucke, D., Constantinescu, C., & Westkämper, E. (2008). Smart factory-a step towards the next generation of manufacturing. In *Manufacturing systems and technologies for the new frontier* (pp. 115–118). London: Springer.
14. Westkämper, E., Jendoubi, L., Eissele, M., & Ertl, T. (2006). Smart factory—bridging the gap between digital planning and reality. *Manufacturing Systems, 35*(4), 307–314.
15. Agostinho, C., Marques-Lucena, C., Sesana, M., Felic, A., Fischer, K., Rubattino, C., & Sarraipa, J. (2015). OSmosis process development for innovative product design and validation. In *2015 ASME IMECE*, Houston, USA.
16. Ko, J., Lee, B., Lee, K., Hong, S. G., Kim, N., & Paek, J. (2015). Sensor virtualization module: virtualizing IoT devices on mobile smartphones for effective sensor data management. *International Journal of Distributed Sensor Networks*. https://doi.org/10.1155/2015/730762.
17. IoT Open Platforms. Retrieved from http://open-platforms.eu/.
18. Marques, M., Agostinho, C., Zacharewicz, G., & Jardim-Gonçalves, R. (2017). Decentralized decision support for intelligent manufacturing in Industry 4.0. *Journal of Ambient Intelligence and Smart Environments, 9*(3), 299–313. https://doi.org/10.3233/AIS-170436.
19. Karnouskos, S., Villaseñor-Herrera, V., Haroon, M., Handte, M., & Marrón P. J. (2011). Requirement considerations for ubiquitous integration of cooperating objects. In *Proceedings of 2011 4th IFIP International Conference New Technology Mobile Security NTMS 2011*.
20. Karnouskos, S. (2009). Efficient sensor data inclusion in enterprise services. *Datenbank-Spektrum, 9*(28), 5–10.
21. Tan, L., & Wang, N. (2010). Future internet: The internet of things. In *2010 3rd International Conference on Advanced Computer Theory and Engineering (ICACTE)* (vol. 5, pp. V5-376-V5-380).
22. Peng, T., & Xun, X. (2017). An interoperable energy consumption analysis system for CNC machining. *Journal of Cleaner Production, 140.*

Manufacturing Software Units: ISO 16300-3 Main Guidelines for Interoperability Verification and Validation

H. Basson, M. Bouneffa, M. Matsuda, A. Ahmad, D. Chung and E. Arai

Abstract The ISO 16300-3 standard considers the required artefacts for interoperability verification and validation throughout the whole cycle of manufacturing application development. In this chapter, an important work of ISO series standards has been elaborated in the context of manufacturing application development. The current work presents the ISO 16300-3 main guidelines devoted to the verification and validation of software unit's interoperability for the manufacturing software units which can be considered as the most critical part of any automated manufacturing process. The various proposals of ISO 16300-3 are individually detailed to some extent showing its pertinence for a systematic verification and validation of manufacturing software units interoperability.

Keywords Manufacturing software units · Interoperability verification · Interoperability validation · ISO 16300 guidelines

1 Introduction

Manufacturing activities are increasingly automated through a growing set of manufacturing software units (MSU) in evolving smart industry. These activities control the execution of manufacturing processes. The MSU interoperability constitutes a central concern for obtaining targeted competitive products from implemented manufacturing chain, with the help of supervising the behavior of manufacturing devices,

H. Basson (✉) · M. Bouneffa · A. Ahmad
EILCO, LISIC, University of Littoral, Calais, France
e-mail: basson@lisic.univ-littoral.fr

M. Matsuda
Kanagawa Institute of Technology, Atsugi, Japan

D. Chung
Rockwell Automation, Cleveland, USA

E. Arai
Osaka University, Osaka, Japan

K. Popplewell et al. (eds.), *Enterprise Interoperability VIII*, Proceedings of the I-ESA Conferences 9, https://doi.org/10.1007/978-3-030-13693-2_14

hardware equipment and hosted operating processes. The MSU quality level and its different criteria, including the interoperability, are largely recognized as major requirement for satisfied processes executions and reaching targeted improvements of manufacturing processes capabilities. Most of evolving manufacturing environments share persistent goal of optimization of the manufacturing processes for their functional capabilities, resources consumption, as well as their ability to integrate rapidly the new market requirements. For the goal of innovating manufacturing applications, MSUs are regularly improved, retested and redeployed to increase the performance of manufacturing processes, by using the available digital technologies and tools [1]. In contrast to traditional manufacturing processes, today's manufacturing applications aim at a higher adaptability dealing with the rapid fluctuation of market expectations in terms of services, quality and cost of products. Subsequently, the development and evolution of intelligent reactive processes have become necessary to animate any "Smart Factory" environment [2, 3]. Nevertheless, the intelligent MSUs must also be capable of interoperating with an important range of embedded components related to key technologies such as robotics, 3D printing, intelligent materials, photonics, nanotechnologies and Internet of Things. Inside such heterogeneous contexts, the development of manufacturing applications requires a systematic verification and validation of both internal interoperability and external interoperability between MSUs along with the associated components related to mentioned key technologies.

The present work considers mainly the contributions of ISO 16300-3 [4] in terms of guidelines as part of road map for the verification and validation of the internal interoperability between MSUs. The wide typology of interoperability and the multiplicity of its various levels encourages the development of manufacturing applications in which the interoperability verification and validation are performed involving standardized references such as the ISO dedicated series. Hence, in this paper, we present the contribution of ISO 16300-3 in terms of provided guidelines to apply for MSU interoperability verification and validation.

In the rest of the paper, Sect. 2 introduces major elaborated guidelines of ISO 16300-3 for the verification and validation of MSU interoperability. It gives an architectural view of manufacturing application and its different levels of granularity. Section 3 details these contributions and corresponding targeted advantages. A global assessment of ISO 16300-3 contribution toward a systematic approach for interoperability verification and validation is given in Sect. 4. Section 5 presents the general profile of the required data schema for the interoperability verification and validation, whereas Sect. 6 presents the concerned interoperability mechanisms and the associated quality characteristics. The paper conclusion is given in Sect. 7.

2 Manufacturing Applications ISO 16100 Description

ISO 16300-3 [5] starts from the elaborated work of ISO 16100 series [6] for a generic description of a manufacturing system as being a set of applications, animated indi-

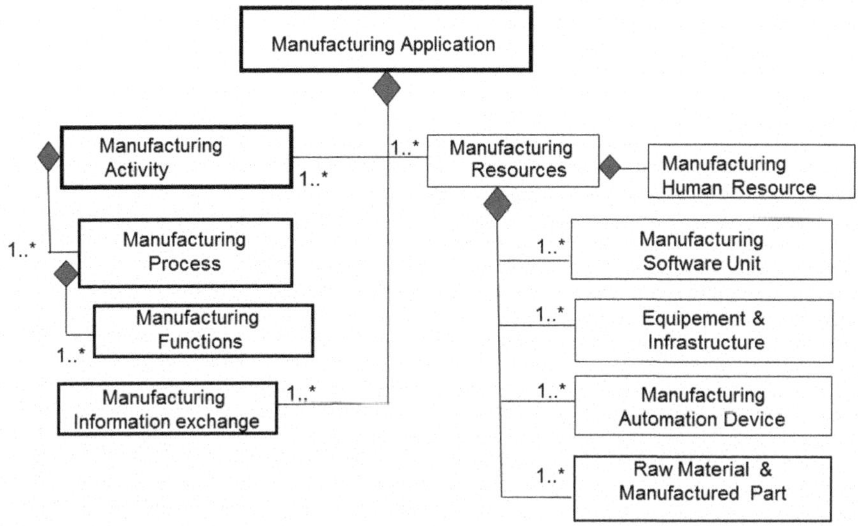

Fig. 1 A structured view of manufacturing applications

vidually by a set of activities (Fig. 1). A manufacturing activity according to its specified functional role within a manufacturing system is performed through a set of processes. These processes execute and interoperate according to their appropriate control flow and the corresponding required interoperability, respectively. Each process is associated with a set of functions activated according to an adopted time schedule. It can be triggered by a range of specific events. The functions associated with manufacturing processes are implemented through a set of manufacturing resources, mainly the MSUs. The MSUs are developed or selected, which are then configured to support their various information exchanges as well as various data flows with concerned manufacturing components, devices, sensors, actuators, etc. The whole system operates according to a scheduled sequence of manufacturing processes to accomplish manufacturing activities [1].

A manufacturing process often cooperates or coordinates with other processes according to its functional role. The respective functions of interacting processes are then considered to enable them to interoperate with each other. The invoked function requires a shared common set of criteria and a set of conditions under which the adopted interoperability mechanisms can be activated, in order to manage the cooperation or coordination. The software units that implement these functions should meet a related set of criteria and conditions for required interoperability.

The development of any manufacturing application starts by establishing manufacturing requirements specifications constituting a critical document: Manufacturing Application Requirements Document (MARD) [4]. It includes the architectural, functional, behavioral and qualitative specifications to be met by the concerned application. In respect of MARD, the application architecture is realized using adequate formalisms for various design artefacts. These mainly address the followings:

1. the application global structure,
2. its capability units including MSUs,
3. their interdependencies and
4. its configuration and deployment.

The manufacturing application processes shall be composed of designed and planned activities and operations of various types (human, mechanical, electrical, hardware, networking and/or computing, etc.). For each process, the manufacturing application design indicates its functional role inside the manufacturing application, its individual control flow as well as its underlying specific activities and functions. For the manufacturing processes implementation, the design shall specify the required manufacturing resources and their specific capabilities considered as necessary for the manufacturing execution. These manufacturing resources are of different types (mechanical, electrical, hardware, networking, software, etc.) [7], where corresponding capability units shall be described using the dedicated profile template.

3 Requirements for Interoperability Verification and Validation

ISO 16300 addresses requirements of users and suppliers of manufacturing software regarding the interoperability of software units devoted to the area of industrial automation [4]. User interoperability requirements include:

- integrating an automation application system by combining capabilities of a set of software components provided by various sources,
- substituting another software component in a software unit to provide an equivalent functional and/or qualitative capability required by the automation application system,
- integrating the capability of a software unit from one resource system platform to another platform,
- validating and verifying the capability of a software unit to meet the automation application system requirements.

ISO 16300 also addresses software interoperability services, which include:

- access to the description of a software capability to enable interoperability assessment,
- enabling the search and location of candidate software units and components, preferably automatically, using search engines,
- representing the interdependencies between software components for an automation application hosted on a particular system platform.

For the goal of interoperability verification and validation, ISO16300-3 specifies four sets of artifacts described below (Table 1) as necessary to reach the goal.

Table 1 Four major sets of required artefacts

Set of artefacts	Description
A	Composed of design schemas of expected activities of MSUs and associated interoperability mechanisms that shall be designed to meet requirements of data sharing, messages exchange, services invocation and exchange, or procedure call which may occur among MSUs
B	Composed of code parts implementing the effective capabilities of MSUs and working interoperability mechanisms permitting to concerned MSUs to accomplish associated activities
C	Composed of the quality model elements specifying the expected interoperability quality criteria as they shall be fulfilled by the implemented interoperability mechanisms and services. These criteria and corresponding characteristics, sub-characteristics and properties shall be specified according to ISO/IEC 25000 quality model with effective quality characteristics, sub-characteristics and properties of implemented MSUs interoperability
D	Composed of the quality reports providing the numerical values or ranking values of effective quality characteristics specified in the instantiated quality model

4 Interoperability Verification and Validation

As proposed in ISO 16300-3, the verification process concerns a set mapping checking between the two elements of the artefacts couple set (A, B), while the validation concerns the mapping checking between the two elements of the artefacts couple set (C, D). According to ISO 16300-3, the whole verification and validation processes are accomplished within the four major development phases of manufacturing application (Requirements Specification, Design, Implementation, and Testing).

The required artefacts for interoperability verification (see Fig. 2) include firstly the specification of functions to be provided by the MSU. They also include the design of required interoperability mechanisms to be implemented in order to meet the application MSUs requirements in terms of services exchange, functions call, messages communication, information sharing, database common access and updating. The invoked interoperability design is a specific part, which is composed of distinguished elements inside the whole design of manufacturing application MSUs. Subsequently, each required specific interoperability mechanism shall have its own specific design representation. In order to be realized, each required designed mechanism shall have one (or more) implementation solution(s) provided by the development environment. The concerned application development team can do the selection of an implementation solution of the interoperability. This selection is decided according to the required quality criteria of the current application in terms of reliability, performance, efficiency, security or other quality criterion.

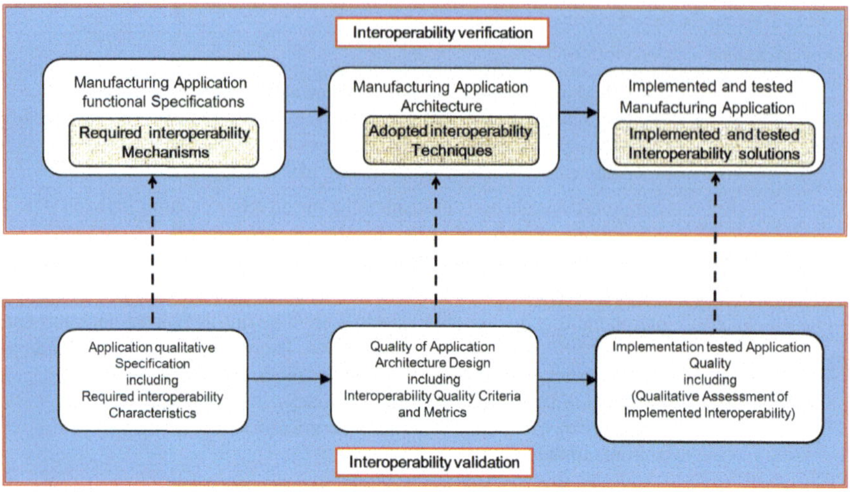

Fig. 2 Interoperability verification and validation processes

5 Data Schema for Interoperability Verification and Validation

In order to perform the interoperability validation and verification, a very large set of data is required. ISO 16300-3 proposes a generic schema (Fig. 3) to help to start a data schema design intended to contain all data elements necessary to perform the interoperability verification. The proposed schema can be adapted and then detailed to be a container of information which can be requested by users and dedicated tools to give adequate responses helping in verification and validation tasks. The given data schema concerns three development phases which should be extended to include realized test cases. These are the Requirement, the Design and the Implementation Phases. The data entities defined in a specified phase may be related to those related to another phase. This leads to a kind of traceability between the three different phases [8, 9].

A qualitative specification should be done and taken into account in the design and implementation phases, to each referenced chosen interoperability mechanism [10]. A precise qualitative specification through the two sets C and D (Table 1) is necessary for interoperability validation.

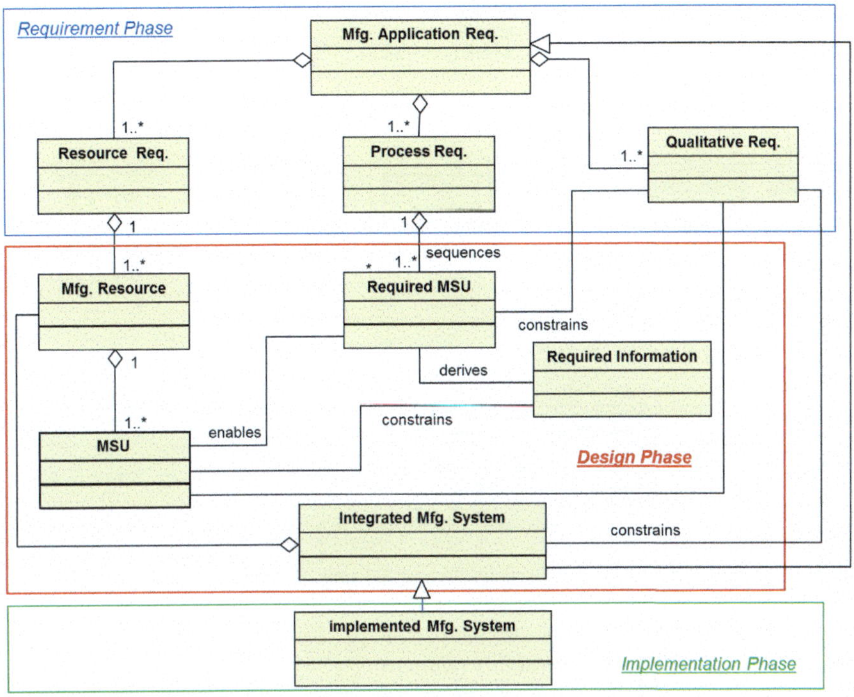

Fig. 3 Development data schema of manufacturing processes interoperability

6 Interoperability-Required Mechanisms and Associated Qualitative Characteristics

In order to identify the type of required interoperability, an interoperability typology is highly recommended. Figure 4 shows a useful example, not exhaustive, but it may illustrate the importance underlined by ISO 16300-3 of using some interoperability typology by level toward a detailed interoperability mechanism specification at the development phase of manufacturing processes design.

The specification of interoperability-required mechanisms starts by the list of couples of interoperating units inside the whole application architecture (see Fig. 4). The type of interoperability relationship is to be specified in reference to an interoperability typology. A small part of non-detailed typology is given below using a gradual tree-like enumeration of interoperability cases, which may occur between linked couple of software units (or components).

These can be illustrated as follows:

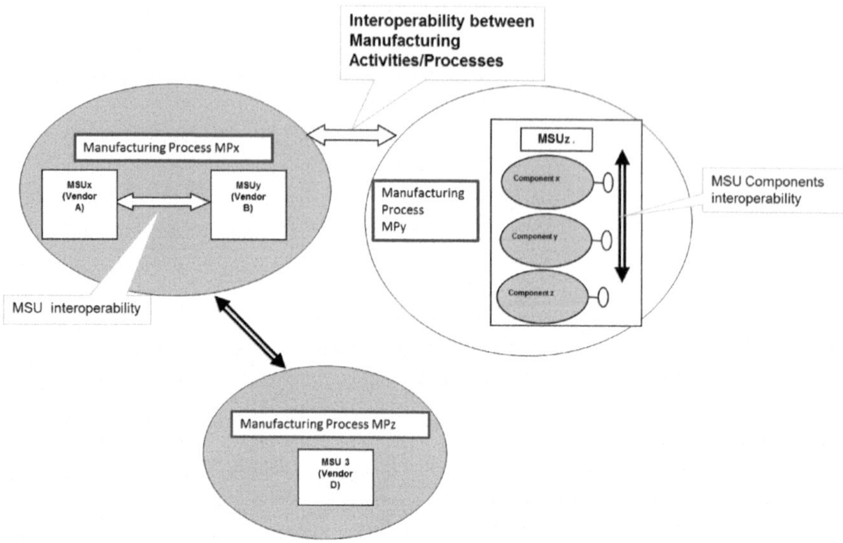

Fig. 4 Example of interoperability levels inside a partial view of application architecture

<MSU interoperability relationship> =>
<Inter-MSUs control relationships> |
<Inter-MSUs services exchange> |
<Inter-MSUs synchronization> | (1)
<Inter-MSUs Message based communication> |
<Inter-MSUs Events based communication> |
< MSU Data interoperability mechanisms>

<MSU Data interoperability mechanisms> =>
<Interoperability via shared variables> |
< Interoperability via shared simple files> |
< Interoperability via shared XML files> | (2)
< Interoperability via shared Tabular files> |
<Interoperability via share Databases>

<MSU Message exchange based interoperability> =>
<One direction connection initiation> |
< Request-Response interoperability> | (3)
<Bi-directional synchronous message flow> |
<One direction notification> |

Etc.

To each required interoperability mechanism, a set of quality characteristics and associated metrics are to be specified according to the prioritized quality characteristics of the currently developed manufacturing application.

7 Conclusion and Perspectives

The interoperability of software units, as an important resource of manufacturing application, is increasingly critical for the quality of manufacturing systems. The article presents the contribution of ISO 16300-3 in terms of provided guidelines to apply the interoperability verification and validation between manufacturing software units. These guidelines can serve to elaborate a systematic verification and validation of interoperability among software units. The manufacturing processes are associated with the activities with established roles, which triggers according to a time schedule. These processes also collaborate with other similar processes. It is significant for the implementing software units to verify their interoperability that may also require some shared common set of criteria. ISO 16300-3 proposes the verification of a set mapping between elements of design schema, software code parts and their associated quality model. This specification also highlights a prominent interoperability typology at different levels of specification, design and development of manufacturing processes. The ISO 16300-3 proposals have been individually detailed to some extent indicating their importance, since we target a systematic approach aimed at an integrated process of interoperability verification and validation for the developed or modified manufacturing software units.

References

1. Schweer, D., & Sahl, J. C. (2017). The digital transformation of industry–the benefit for Germany. In *The drivers of digital transformation* (pp. 23–31).
2. Khalfallah, M., Figay, N., Ferreira Da Silva, C. et al. (2016). A cloud-based platform to ensure interoperability in aerospace industry. *Journal of Intelligent Manufacturing, 27*(1), 119–129.
3. Wang, L., Chen, X., & Liu, Q. (2017). A lightweight intelligent manufacturing system based on cloud computing for plate production. *Mobile Networks and Applications*, 1–12.
4. ISO 16300-3: 2017. (2017). *Automation systems and integration—Interoperability of capability units for manufacturing application solutions—part 3: Verification and validation of interoperability among capability units*. Available at https://www.iso.org/obp/ui/#iso:std:iso:16300:-3:ed-1:v1:en.
5. ISO/IEC 1600-N (1…5). (2009). *Industrial automation systems and integration—Manufacturing software capability profiling for interoperability*.
6. Matsuda, M., & Wang, Q. (2010). Software interoperability tools: standardized capability-profiling methodology ISO16100. EAI2N 2010, 140–151.
7. ISO/TR 18161:2013. (2013). *Automation systems and integration—Applications integration approach using information exchange requirements modelling and software capability profiling*. Publication date: 2013–07. Available at https://www.iso.org/standard/60899.html.

8. Ahmad, A., Basson, H., & Bouneffa, M. (2017). Analyzing and modeling the structural and qualitative interdependencies of software evolution. In *7th International Workshop on Computer Science and Engineering, workshop of, The 5th International Conference on Information Technology and Science*, Beijing, China (25–27 June 2017).
9. Basson, H., Matsuda, M., & Chung, D. (2014). A contribution to Software interoperability of software application. In *I-ESA, INTEROP-VLAB, European Virtual Laboratory for Enterprise Interoperability*. Albi, France (24–28, March 2014).
10. Basson, H., Matsuda, M., Bouneffa, M., Ahmad, A., et al. (2016). Qualitative evaluation of manufacturing software units interoperability using ISO 25000 quality model. In *Enterprise interoperability VII* (pp. 199–209). Springer International Publishing.

Interoperability of Test Procedures Between Enterprises

Intermediate Representation for Test Procedure Exchange

Marco Franke, Volker H.-W. Meyer, Rainer Rasche, Andreas Himmler and Klaus-Dieter Thoben

Abstract The main objective of any test process is to check the functionality, reliability and operational safety of products. Recent developments resulted in a wide range of highly specialized test procedure languages. In addition to the heterogeneity between test procedure languages, the different versions of the same test language are also heterogeneous. To make test procedures exchangeable between enterprises and reusable, testers face the challenging task of effectively decoupling the test procedure descriptions from specific test tools as well as test benches. This paper presents an approach to achieve the interoperability of test procedure languages which envisages an intermediate representation. An appropriate intermediate representation is the enabler for the application of data integration solutions. The goal is to make test procedures interoperable and to enable transformation definitions for cross-compiling. The developed intermediate representation uses the State Machine Notation for Control Abstraction (SCXML) to represent all test procedure language features. The paper demonstrates how to transform a test procedure from Check Case Definition Language (CCDL) into SCXML. Moreover, the capabilities and restrictions of this approach are shown and summarized.

M. Franke (✉)
BIBA—Bremer Institut Für Produktion Und Logistik GmbH,
Hochschulring 20, 28359 Bremen, Germany
e-mail: fma@biba.uni-bremen.de

V. H.-W. Meyer
AIRBUS Group, Airbus-Allee 1, 28199 Bremen, Germany
e-mail: Volker.Meyer@airbus.com

R. Rasche · A. Himmler
dSPACE GmbH Paderborn, Rathenaustraße 26, 33102 Paderborn, Germany
e-mail: RRasche@dspace.de

A. Himmler
e-mail: AHimmler@dspace.de

K.-D. Thoben
Faculty of Production Engineering, University of Bremen, 28359 Bremen, Germany
e-mail: tho@biba.uni-bremen.de

K. Popplewell et al. (eds.), *Enterprise Interoperability VIII*, Proceedings of the I-ESA Conferences 9, https://doi.org/10.1007/978-3-030-13693-2_15

Keywords Standards for interoperability · Specification and modeling

1 Introduction

The main objective of any test process is to check the functionality, reliability and operational safety of products [1]. To meet the obligations, product manufacturers and system integrators execute a test process to certify the correct behavior of their product. The certification of safety-critical mechatronic systems like aircraft, cars and trains is based on test procedures which check the correct behavior in normal operation as well as in faulty operation mode. For this purpose, test procedures in hardware-in-the loop (HIL) test environments comprise the steps from setting up the system under test (SUT) to an initial state, to provoke the fault behavior and to monitor the system's reaction. Here, the SUT's reaction has to meet the relevant requirements, which also addresses safety-critical aspects. To enable a V-model-based development approach, the manufacturers and the system integrators each run through their own test process to certify the product at various test levels. While the manufacturers certify their product at the unit level, the system integrators focus on certification at least on the integration and system levels. Due to the different test purposes, different test system architectures, test benches, test procedure languages, company-specific guidelines, terminologies and standards are used. As a result, domain-specific and test-purpose-specific test procedure languages have been developed. Today, test procedure languages like Testing and Test Control Notation (TTCN-3), Automatic Test Markup Language (ATML) and the signal description definitions within XIL of the Association for Standardization of Automation and Measuring Systems (ASAM) are heterogeneous with respect to their syntax and semantics. Currently, there is no interchange format available covering all test procedure specific language features which is the enabler for applying data integration solutions.

The available data integration approaches to aggregate and exchange data are capable of transforming data formats ranging from CSV files, XML files, log files or models [2, 3] to an intermediate representation and back to a target data format. Thus, the available approaches overcome the data integration conflicts, which are mentioned by Wache [4]. In doing so, the considered approaches apply procedural transformation rules, which search for specific patterns in the source file and then add content in the target file by applying appropriate transformation rules. While the transformation works fine for property values or concepts, complete language features (like a loop) cannot be created on the basis of just a predefined loop statement in the source language.

Apart from the interoperability of language features, the insertion mechanism of current solutions is also an open issue. The current assumption [2, 3] is that the location and the amount of the insertions of the target content is source oriented. This assumption does not hold between different kinds of test procedure languages. The amount of insertion locations varies between languages. For example, the translation of a statement of an imperative language into an object oriented would need more than one location.

In addition to the test procedure language heterogeneity, a concrete test procedure includes more barriers to be solved. The consideration of the test bench and test process is mandatory to know: Which resources are available, which test bench-specific implementation guidelines must be met to be executable, and how the test procedure is included in the overall test process with respect to test automation, test analysis and test documentation. In conclusion, the available data integration solutions cannot handle the range of heterogeneity of syntax and semantics neither their transformation rules can manage resources and implementation restrictions which are test bench and test process specific.

A test process, which is established over the boundaries of a company, has to overcome the challenges of interoperability. Thus, the necessary interoperability between and within enterprises is considered by ATHENA Interoperability Framework on four levels [5]: data/information (for information interoperability), services (for flexible execution and composition of services), processes (for cross-organizational processes), enterprise/business, for collaborative enterprise operations. To achieve the interoperability of test processes, the fourth level must be completed. Therefore, it must be possible to exchange the content of test procedures between test procedure languages which consider the test procedure content as well as the overall test process properties. There are two methods for achieving the interoperability on enterprise level:

1. Directly transforming the syntax and semantics of one test procedure language into those of another language.
2. Using an intermediate representation to hold test procedure content, and enable two kinds of direct transformation:
   a. Transformations from a language to the intermediate representation.
   b. Transformations from the intermediate representation to a language.

The first method requires a set of transformation definitions for each quadruple ($language_i$, $hardware_i$, $language_j$, $hardware_j$). The complexity of transformation definitions varies according to the chosen languages and applied test system architectures. Moreover, the effort to create transformation definitions increases exponentially.

For these reasons, this paper presents an approach for creating an intermediate representation to overcome the syntactic, semantic and test bench-specific barriers of test procedure languages which is not available on the market. The objective is to represent the content of test procedure platform independent and to enable a transformation from one test procedure language into another one (cross-compiling). The general transformation process of a test procedure into an abstract representation is motivated in [6]. Following this process, creating an intermediate representation is one of the necessary steps to achieve the interoperability between different test processes of enterprises. The manufacturer can exchange his test procedures with the system integrator and vice versa. This would speed up failure detection in procedures where the test processes of manufacturers and the system integrators deliver different results.

Following, the authors present a state-based test procedures representation which all languages have in common. Then, the covered language features will be used to define the minimum functional requirements for an intermediate representation. SCXML is used to represent the minimum test procedure features needed to meet the requirements. The paper then presents the test-specific functions for HIL testing and explains how such functions can be implemented in the intermediate representation approach. Subsequently, the authors describe the transformation of a test procedure describing an aircraft-specific, realistic test from Check Case Definition Language (CCDL) [7] to State Machine Notation for Control Abstraction (SCXML) [8]. Finally, a summary and outlook is provided.

1.1 State-Driven Perspective of Test Procedures

Test procedures check the SUT behavior with respect to a specific aspect. A test procedure stimulates the SUT, provokes the aspect-related conditions and monitors the SUT's reaction. A formal process to model this behavior and to derive test procedures is provided by model-based testing (MBT) [6, 9]. MBT uses test models, e.g., variants of **activity diagrams** or **state machines**, to describe all the required test procedures for a specific test campaign. Therefore, a test model represents the possible test procedures and includes conditions to describe the respective test procedure variants. Each variant defines a set of states and transitions, thus describing a deterministic SUT behavior. In conclusion, a state-based perspective of a test procedure is possible for all test procedures. The state-based perspective is described in more detail in the following:

State. A state is defined by the information which is relevant at a given point in time. In test procedures, states are defined by the set of parameter values and signals of the SUT as well as the called test system functions. The concrete assignment of parameter values is language specific and differs between test procedure languages. While there are syntactical differences, the value assignment has similar semantics. For example, the assignment in common test procedure languages is performed by means of a keyword (e.g., 'set' or 'put') or through operators (e.g., '=' or ':'). In addition to this information, a state can also contain assertions which define its validity. Common constraints include timeouts and parameter value ranges that must be satisfied.

Test procedures can also use compound states to summarize the constraints that are valid for more than one state. Compound states are used to encapsulate states in logical blocks.

Transition. Transitions are used to model state changes. In a test procedure, each state change must have a reason. The most frequent reasons include events that are triggered, conditions that are no longer valid and elapsed time. Common variants of conditions are changes in parameter values. For example, the passing of time is

Table 1 Functional requirements of representing test procedure

Id	Requirement
State-specific requirements	
ST1	Assigning of parameters, including values and their units, to states
ST2	Assigning of active test system functions to a state
ST3	Assigning of specific conditions like timeouts, events and conditions
ST4	Creating of compound states to summarize states and enable cross-state conditions
ST5	Enabling parallel states to support concurrency
Transition-specific requirements	
TR1	Transition without restriction must be available
TR2	Transition with a timing restriction must be available
TR3	Transition with an event restriction must be available
TR4	Transition with conditions defined by a parameter values must be available
Test-specific functions	
SF1a	Stimulation or failure injection of a parameter/signal via predefined shapes (ramp, sine, pulse, etc.)
SF1b	Stimulation or failure injection by adding an offset to a parameter variable/signal
SF2	Failure injection via manipulation of hardware pins
SF3	Logging parameter values/signals during a specific interval
SF4	Enabling requirement traceability

implemented in test procedure languages through language-specific wait statements (e.g., ‘wait 5 ms,’ ‘Thread.sleep(5)’ and user-specific functions).

Test-specific functions. HIL test processes apply test-specific functions like bit-wise failure injection and tailored logging. These kinds of functions are test process specific and are grouped in one category. The most common test-specific functions are the support of logging according to parameter- and state-specific information, requirement traceability, direct access to hardware-specific functions, changing and observing simulation model variables, etc.

1.2 *Requirements for a State-Driven Representation of Test Procedure*

The objective of the intermediate representation is to represent the content of test procedures, which contains states, transitions and test-specific functions. Therefore, the most important requirement is for the intermediate representation to support these three groups. This section lists the detailed requirements for each group. The detailed requirements are extracted from the functionality provided by the test procedure languages ASAM (more precisely, XIL signal description definitions), ATML, TTCN-3, CCDL and FlatScript II. In addition, a questionnaire was published in which test engineers from the aviation sector could prioritize the available language features. Both outcomes have been applied to define the mandatory requirements for an intermediate representation which are listed in Table 1.

2 Approach

SCXML [8] has been chosen as intermediate representation to enable data integration. It is a standard by the World Wide Web Consortium (W3C) for representing Harel state charts in an XML-based data format. The integration of Harel state charts enables the representation of states, compound states and parallel states in a clear manner. Thus, it meets the requirements ST3-5. For each kind of sequential or parallel state, a corresponding SCXML model can be created. The integration of actions (e.g., <assign> or <send>) which change the internal state of the state diagram as well as the calling of test system functions (<invoke>) satisfies the state-specific requirements of ST1 (except for units) and ST2. Thus, all listed state-driven requirements are met.

A transition in SCXML connects two states. This means a transition postulates either no restriction or the occurrence of an event, of a condition, or of both [8]. The described variety of SCXML-specific transitions meets all transition-driven requirements (TR1–4). The languages for specifying the conditions and the corresponding data model are not set directly by SCXML, but SCXML offers appropriate placeholders. With the used SCXML version [8], the data model can be specified by any XML data structure and conditions are represented as strings. The data model and the condition are interpreted by concrete implementations that are part of an SCXML executor. The proposed approach envisages a specialization of the tuple-based data model (ID and expression) to hold the triple (ID, value and unit) as JSON. Therefore, the requirement ST1 is satisfied. The parameter IDs are used in the conditions of the transitions. The syntax and semantics of a condition use the parameter IDs as the subject and object of a condition. The assignable functions and operators can be defined by other standards, such as ASAM General Expression Syntax [10].

Except for the states and transitions, the test-specific functions cannot be implemented directly through SCXML statements. The required stimulation and failure injection (SF1a, SF1b, and SF2) can only be represented by triggering test system functions. In such a case, the SCXML <invoke> command would be used in such a way that the <invoke> command contains all necessary information to call the function on a test system. <invoke>-specific information includes the name of the function, the parameters, etc. The required logging (SF3) is supported by SCXML <log> statements and can be used to satisfy the requirements SF3 and SF4. The mapping (SF5) requires a framework to transform mapped values, such as labels, physical units and data types, so they can be used for the remote test bench site or the definition in the test procedure.

In sum, 13 of the 15 functional requirements are fulfilled when representing a test procedure as a general state diagram and as a specialized SCXML model. The missing test-specific functionality and the missing capabilities regarding condition modeling and units can be compensated by standards like XIL API.

3 Use Case

3.1 Description of the Test Procedure

Modern passenger jet wing designs are optimized for speed and efficiency during the cruise portion of flight. High-lift devices (slats and flaps) compensate for this design trade-off by adding lift at takeoff and landing, reducing the distance and speed required to safely land the aircraft, and allowing the use of a more efficient wing in flight.

The flaps are driven by a single actuator with two shafts. In case of a failure, a wing tip brake stops flap movements, e.g., if one shaft is jammed. Thus, the brake avoids the problem of asynchronous flap extension, which can cause significantly different lifts of both wings.

3.2 Overall SCXML Representation

The representation of a test procedure in SCXML is based on the SCXML implementation by Apache [11] and is implemented using a sequence of compound states, which is shown in Fig. 1.

In terms of the logical blocks of the use-case-specific CCDL script, the compound states *Initial Condition*, Test Step 1–4 (summarized as TestStep_i) and *Finally* are inserted into the compound state test-Sequence.

The highest block (*Initial Condition*) is the block that is executed first. The name of the first compound state is not set. The first state always includes the SCXML XML tag <initial> to mark it as the initial state. All following states are connected to each other by transitions. The transitions between *Test Step 1* and the transient *Finally* include events. The purpose of this kind of events is to realize time synchronization via absolute time conditions. In such a scenario, the objective of a state is to remain active for a specified absolute time frame. At the end of the time frame, the active state

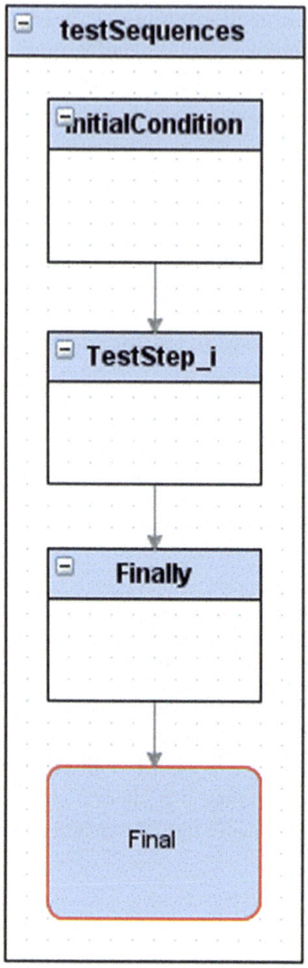

Fig. 1 Test procedure as SCXML

sends the appropriate event and the connected state machine can traverse to the next connected state. Within the compound state *testSequences* or in any other compound state, there is only one state active at a given point in time. To enable parallel active states, the compound state *testSequences* is a child of the parallel SCXML structure <parallel>. In the following, the representation of states, transitions and test system function-specific test procedure language features is shown on the basis of CCDL-specific code fragments. CCDL is used in aerospace industry to test the real-time behavior of mechatronic systems, because it is readability and provides real-time capabilities. The following transformation examples are transferable to other test procedure languages.

3.3 Representation of Simple Operations

Each test procedure changes and checks parameter values several times. An example of changing the parameter *HLSEnv.Output_Airspeed* is provided in Fig. 2. The change of a parameter value for a specific state is possible through the SCXML data model <data> which, in our implementation, uses JSON to store object-oriented data structures and the SCXML tag <assign>. Both SCXML elements can be used to change the value globally or just within a state. The set statements are used in the use case to enable simulator initialization and failure injection, for example.

Both the monitoring and the simple check of parameter values are realized in CCDL through an expect statement which is shown in Fig. 3.

The semantics of the expect statement specify to check the parameter value and to create a corresponding log message. The expect statement has no influence on the visited states/transitions. The representation of the semantics is possible through the SCXML elements <if> and <log>. The interpretation of the condition is not supported directly by SCXML, so external condition languages and interpreters are required. This functionality is available through the XIL API by using the GES 10.

CCDL

```
// ----- CAS - Computed Airspeed -----
set HLSEnv.Output_AirSpeed to 40 [kts]
```

SCXML

```
<data id="HLSEnv.Output_AirSpeed" expr="{"unit":"kts",
"value"=""}"/>
<assign location="HLSEnv.Output_AirSpeed{value}" expr="40"/>
```

Fig. 2 CCDL example of setting a value

CCDL

```
expect HLSS1.Output_SlatFault = FALSE (REQ:1822-1)
```

SCXML

```
<if cond="! HLSS1.Output_SlatFault == 0">
<log expr="REQ:1822-1: failed" label="Test Step1:
HLSS1.Output_SlatFault == 0"> </log></if>
<if cond="HLSS.Output_SlatFault == 0">
<log expr="REQ:1822-1: passed" label="Test Step1:
HLSS1.Output_SlatFault == 0"> </log></if>
```

Fig. 3 CCDL example of checking a value

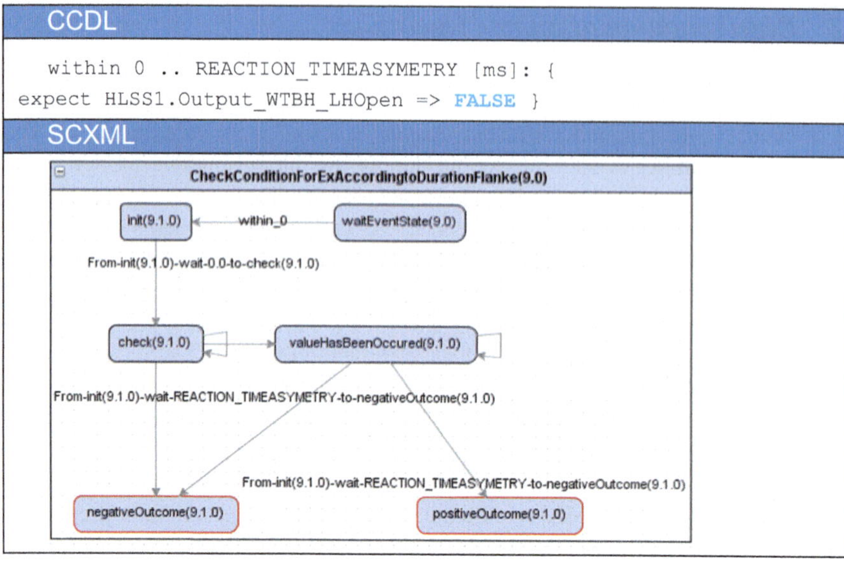

Fig. 4 SCXML compound state to represent a within statement

3.4 *Representation of Complex Operations, Including the Usage of Simple Ones*

Complex operations stimulate or observe specific parameters over a period of time. The example in Fig. 4 demonstrates the monitoring of the parameter *HLSS1.Output_WTBLHOpen* over certain a period of time. The period starts at 0 ms and ends after the constant time *REACTION_TIME_ASYMMETRY*.

The transformation of a within statement from a CCDL test procedure is realized via a compound state. The compound state handles the timing requirements and the monitoring of the parameter value. The outcome of a within or during statement is positive or negative. For each possible outcome, a final state <final> is created, which issues the corresponding log message. The result is shown in Fig. 4.

3.5 *Representation of Test-Specific Content*

Test-specific content belongs to functionality which is provided by the test system instead of the test procedure language. The semantics of such functions are defined inside the test system instead of the test procedure language. Such kind of structures is handled as black box. In SCXML, the <invoke> statement is used to represent the black box. The name of the <invoke> statements determines which specific function has been applied.

3.6 Evaluation

The evaluation based on a real test procedure in CCDL has been provided by AIRBUS. For that purpose, the developed prototype (Java RCP application) transformed the test procedure into a SCXML model. In doing so, the prototype transforms a test procedure from CCDL into a parse tree, subsequently into a timeline and finally into a SCXML model. This model is the base to translate it into the target test procedure language. For the evaluation, it was transformed into a test procedure in Python.

The original test procedure in CCDL was executed on a HIL test bench at AIRBUS Bremen, and the generated test procedure in Python was executed on Real-Time Testing Python-based test automation tool by dSPACE. Subsequently, the curves of the flaps angle, the flaps lever position and the brakes were compared between both executions. The expected system reactions with respect to the observed curves were slightly different, which was the result of the differences in the simulation models of AIRBUS and dSPACE. That means that the original test procedure, coded in CCDL for an AIRBUS test bench, was successfully translated into a Python test procedure, executable on a dSPACE environment. Therefore, the test procedure could be exchanged and the interoperability has been achieved.

In consequence, SCXML is an appropriate intermediate representation to enable both the transformation of a test procedure into SCXML model and the transformation from a SCXML model into a test procedure.

4 Summary and Outlook

Different kinds of test procedure languages are available and specialized for different application domains. The exchange of test procedures between different test procedure languages has to overcome the barriers of interoperability which was not possible yet. The paper has defined functional requirements for an interoperable data format. Then, the combination of SCXML and bijective transformation mechanism has been presented to take on the role of the interoperable data format. Hereby, SCXML is not test specific and the required test-specific language features have been defined in SCXML. Subsequently, the transformation of an avionics-specific test procedure into SCXML was presented in the use case which considered CCDL structures covering all defined functional requirements. A prototype for the transformation of a test procedure from CCDL into SCXML and into Python has already been implemented and evaluated. As a result, the test procedure in CCDL could be exchanged and the interoperability has been achieved between enterprises on ATHENA interoperability level 4.

Next steps are to finalize the specification of SCXML as an appropriate standard for interoperable test procedures. The translation of test procedures from other test procedure language than CCDL and Python must be continued. While CCDL and other test procedure languages provide a common test procedure structure, programming languages offer more freedom. This kind of freedom will challenge the transformation of a test procedure into the predefined structures of SCXML (Complex Operations). The ongoing research will analyze if the universal transformation will keep the current applicability of SCXML models or will result in a huge range of complex operations which are not transferable between test procedure languages.

Acknowledgements This research has been funded by the Federal Ministry of Economics and Technology of Germany (BMWi) in the project STEVE (project number 20Y1301G) and AGILE-VT (project number 20X1730D).

References

1. Franke, M., Gerke, D., Hans, C., et al. (2012). Method-driven test case generation for functional system verification. In *Proceedings ATOS*, Delft (pp. 36–44).
2. Shani, U., et al. (2017). Ontology mediation to rule them all: Managing the plurality in product service systems. In *Systems Conference (SysCon), 2017 Annual IEEE International*. IEEE.
3. Franke, M. et al. (2016). Semantic data integration approach for the vision of a digital factory. In *Enterprise interoperability VII* (pp. 77–86). Springer International Publishing.
4. Wacher, H. (2003). Semantische mediation für heterogene informationsquellen. *KI, 17*(4), 56.
5. Berre, A. J., Elvesæter, B., Figay, N., Guglielmina, C., Johnsen, S., and other. (2007). The ATHENA interoperability framework. In *Enterprise interoperability II* (pp. 569–580). London: Springer.
6. Franke, M., Klein, K., & Thoben, K.-D. (2017). Interoperable information exchange as enabler of NFF related TES. *Procedia CIRP, 59*, 240–245.
7. Razorcat, CCDL Whitepaper. https://www.razorcat.com/files/de/produkte/ccdl/Razorcat_Technical_Report_CCDL_Whitepaper_02.pdf. Last accessed 2018/01/04.
8. W3C. (2011). State chart XML (SCXML): state machine notation for control abstraction. *W3C Working Draft*, April 2011.
9. Utting, M., Pretschner, A., & Legeard, B. A taxonomy of model-based testing approaches. In *Software Testing Verification Reliability, 22*, 297–312. https://doi.org/10.1002/stvr.456.
10. ASAM expression—General expression syntax. (2011). Höhenkirchen, Germany: ASAM e.V.
11. Apache Common, Commons SCXML, http://commons.apache.org/proper/commons-scxml. Last accessed 2018/01/04.

Part VI
Ontology Modeling

Ontology for Continuous Learning and Support

Roua Elchamaa, Abir Mbaya, Nejib Moalla, Yacine Ouzrout and Adelaziz Bouras

Abstract Acquiring new knowledge and skills is the principal goal of any educational institution. Universities planned always to improve the learning methods to offer high academic competencies qualities for students. These competencies help students to respond to the needs of companies' missions during their internships. In some cases, some companies' missions required more competencies in a special domain activity. Therefore, we proposed our method based on a continuous learning ontology where a student can do his job with the support and assistance of a specific service or from his supervisor. This ontology will permit to take into consideration, in addition to competencies, the abilities of students to ensure interoperability between company and university by ensuring matching between the actual student skills and required mission skills. Thus, the proposed approach will establish an efficiency matching between the company needs and the students. Results show the benefits of such an approach to resolving the gap between industry needs and students' skills.

Keywords Education · Continuous learning · Skills

R. Elchamaa (✉) · A. Mbaya · N. Moalla · Y. Ouzrout
DISP, Univ. Lyon, University of Lyon 2, 69676 Bron, France
e-mail: roua.elchamaa@univ-lyon2.fr; roua.chamaa@net.usj.edu.lb

A. Mbaya
e-mail: abir.mbaya@univ-lyon2.fr

N. Moalla
e-mail: nejib.Moalla@univ-lyon2.fr

Y. Ouzrout
e-mail: yacine.ouzrout@univ-lyon2.fr

R. Elchamaa
Saint-Joseph University (ESIB), Beirut, Lebanon

A. Bouras
CSE Department, Qatar University, Doha, Qatar
e-mail: abdelaziz.bouras@qu.edu.qa

K. Popplewell et al. (eds.), *Enterprise Interoperability VIII*, Proceedings of the I-ESA Conferences 9, https://doi.org/10.1007/978-3-030-13693-2_16

1 Introduction

The skills learned and the skills acquired during life can be considered as the main strength of a student profile. For instance, future engineers are expected to have a wide range of skills in addition to math and science skills [1]. We can be more convinced by doing a simple search on the Internet; we can see that the profiles with high-quality skills are the profiles targeted by human resources and education organizations. Competency profiles are also used by ministers of education and teacher training institutes to prepare the educational programs and the criteria required for each teacher; as such, the Quebec Ministry of Education has organized a skills profile for principals [2]. The management and skill analysis of the profiles varies from one activity field to another. In the financial sector [3], organizations have found that the skill management can reduce their recruitment costs and training that may be needed to improve results. The goal is to be able to recruit candidates having a profile that includes knowledge and skills. These skills or competencies will allow them to work quickly after their recruiting. The objective is, therefore, to identify and analyze the required competencies of the candidate profiles. Several skills are necessary for this area, e.g., analytical skills, high level of knowledge in accounting and finance. For marketing, some consulting firms intend to share their expertise by citing the skills of their staff. Other companies guarantee their staff a regular updating of their knowledge. These companies have services that offer the staff knowledge management to enhance their expertise in a new business need. This control is carried out by tracking the current skills of employees and providing professional training to meet new demands on the company. The management of skills in human resources in some companies plays a valuable role in their marketing activities because their staff is considered as the most valuable element of the company [4]. In this period, we notice that universities are under pressure to produce graduates suitable for employment [5]. This is due to changes in education policies and the labor market. In fact, there are still debates and discussions regarding the exact elements required by the human resources and skills of the graduates necessary to be a favorable candidate [6]. Fitness for the job needs much more than the simple acquisition of generic skills present in the curriculum vitae of an employee. On the contrary, to achieve better economic and social outcomes, graduates must be dynamic in their work and able to self-manage the career development process.

In our model, we need to represent more desirable graduate abilities required by companies like self-management and lifelong career management. We aim to improve the visibility of the student profiles of IUT Lyon 2 and enhance their knowledge in the companies during their internships. The first step that we need to achieve is to support the learning process in IUT and present each semester of study at the IUT as an entity with theoretical or practical objective skills. Therefore, when the student attends courses, the acquisition of knowledge should be observed to well represent his expertise as competencies. The educational objectives of each semester are considered prerequisites for a future semester. So, in the student profile, it is interesting to describe these skills below three sub-concepts: current skills, prerequisite skills, and future skills. In addition, we need to present the abilities of student, such as his ability to create or design or manage.

This paper is organized as follows: Sect. 2 details different approaches using skills analysis and discusses various approaches that aim to match the profiles and company needs. In Sect. 3, we describe in detail the proposed model for learning in university Lyon 2. In Sect. 3, we describe the aim of the proposed ontology in competency management and matching profile. Finally, Sect. 4 summarizes the presented work and exposes some future work.

2 Related Work

Universities play a significant role in providing their students with basic skills. These necessary skills allow students to apply their knowledge in specific missions in an industrial field that is appropriate for these acquired skills. The management of knowledge and skills is a key factor in ensuring the performance of industrial processes [7]. To make an evident skill exchange between universities and companies, authors in [8] proposed at first steps to defining a concept for modeling skills, competencies, and abilities of a student who has a specific educational background based on courses taken at the university. This notion is named 'student profile' [9, 10]. A personalized 'student profile' is defined as the ability to provide content and services tailored to the individual based on the knowledge about his preferences and behavior [11]. On the other hand, there is a need for analyzing competencies in profiles to create a precise model representing skills that should be acquired by learners to ameliorate the reviewing of learners. Many research initiatives have approached competence representation. They describe competences in learning and working contexts. A classic analysis was performed in [12] to check the required computer competencies for university students. In [13], the proposed approach aims to illustrate skills in the field of logic programming in upper secondary education. The model proposed by experts based on different methods of detailed skills description is referring to an in-depth analysis. In [14], the author offered an improved competency model. A competency-based learning standard was developed and implemented by authors using RCN competencies. The treated problem in this article is the existing e-learning competency standards (IMS RDCEO, HRXML). It cannot model complicated competencies or draw the relations between them adequately. Furthermore, it cannot support comparisons of data between different communities and cannot follow the learner progress during learning process.

An ontology can be used to represent and clarify knowledge of a domain [15, 16] through a conceptualization to perform a useful ontological analysis of the field. So, if the analyses are weak, this leads to inconsistent knowledge. Moreover, the clarification of the terminology makes it possible to have coherent reasoning. Ontologies are proposed in [17] to represent a student model to extract the interesting points for describing the knowledge of student. This model is based on several ontologies, mainly the ontology of 'student profile' which represents the student's personal information, a 'state ontology' which tracks the progress of a student, and 'Objective Ontology' that defines the study objectives. We can find competencies ontologies in many fields, such as artificial intelligence where ontologies and multi-agent systems are used for a corporate semantic web [18], learning objects in mathematics in order to make competencies quasi-standard that enables interoperability and reuse [19], educational technology. The need for knowledge enhancement within an organization or university involves efforts to achieve improvement in the management of acquired and required skills. Automating the management of skills becomes an essential criterion for ensuring the smooth running of knowledge management activities. In [20], the author considers ontologies as a semantic infrastructure to perform skills management. Most of the research focuses on competence management especially in solving the management problems of human resource. From the industry point of view, to ensure the support of the recruitment process, competence is one of the first criteria to study. In [21], a competences semantic model is proposed to make a correspondence between the profiles postulating to an employment and the job offers introduced by the company. In [22], a competency ontology management has been developed for skills management; an application has been implemented based on this ontology. A description of competences and ontology was provided with functional analysis. The system was deployed for research purposes in a national subsidiary of Microsoft, the global IT services company, with a Microsoft.NET implementation which communicates with the competency ontology. The utility of modeling skills in 'student profile' is to be able to grant the internship to each student with skills closest to the needs defined by the company. In fact, this modeling makes it possible to visualize the experiences acquired in his diploma. Different solutions based on analysis of the skills profiles and the demands of a company aim to support the recruitment initiative and ensure the improvement of employability strategies for better matching and interoperability as well as a solid skills gap analysis [23, 24].

3 Continuous Learning and Assistance Model

3.1 Overview

To achieve the best alignment between the 'student profiles' of Institute of Technology of Lyon 2 and 'companies missions' for an internship, it will be necessary to analyze the 'competencies of the student profile' and the 'required competencies in the company.' The competency management can help universities to detect the competencies gap in student profile and check if the student is perfectly capable of responding to a mission defined by the company, or he may need some assistance or individual support to work properly especially if the accorded mission requires business need. So, it is necessary to be able to compare the competencies required in a mission with the knowledge acquired or being acquired by the learner (Fig. 1). This allows us to say that the student can continue to learn in the company where he is doing his internship. That is what one gets from knowledge management of student profiles.

3.2 Model Concepts

In the context of information systems, there are several ways to define and analyze the necessary data requirements. One of the processes is the data modeling that allows the definition and analysis of the required data.
The logical data model is a subset of a conceptual model that can be expressed using a particular technology. Another model is known by his description of the complete field, explanation of concepts, business processes, and description of the physical structures of the data stored in the database. This domain model is called ontology; all types of data can be considered as ontologies. The physical model can be viewed as an ontology of a particular database. In this case, we need to define several concepts in order to improve the knowledge in the domain of learning at universities and companies. Therefore, many concepts will be defined in our proposed 'learning ontology.' The principal concepts are as follows: competencies, student profile, company's missions and university semester objectives (Fig. 1). In the same figure we present one case of student profile where the student finished his first semester in Bac+1 and is continuing his second semester in Bac+1.

- Student Profile: The profile representation is a practical way to gather personal and academic information about the student. These data will be taken by observation of the learner competencies. The benefits of building ontology-based profile in our model are to match the needed competencies of companies and the skills in student profiles to ensure the interoperability. We need to follow the learning process of each student during of each semester until he does all courses of the six semesters at IUT in order to have his diploma. This profile contains all the information concerning the learner, name, address, his year of study at the IUT, the semester he is currently attending. Also, this profile must include all the abilities of the student.
- Competency: We defined an ontology based on a study of competencies of student's profile during the Competency Development Program at the university. The competences are linked to the 'student profile' in an indirect way. In our case, we will follow the path tracking of learners or in another word students during their studies at the IUT since they started their Bac+1, then their Bac+2, and finally

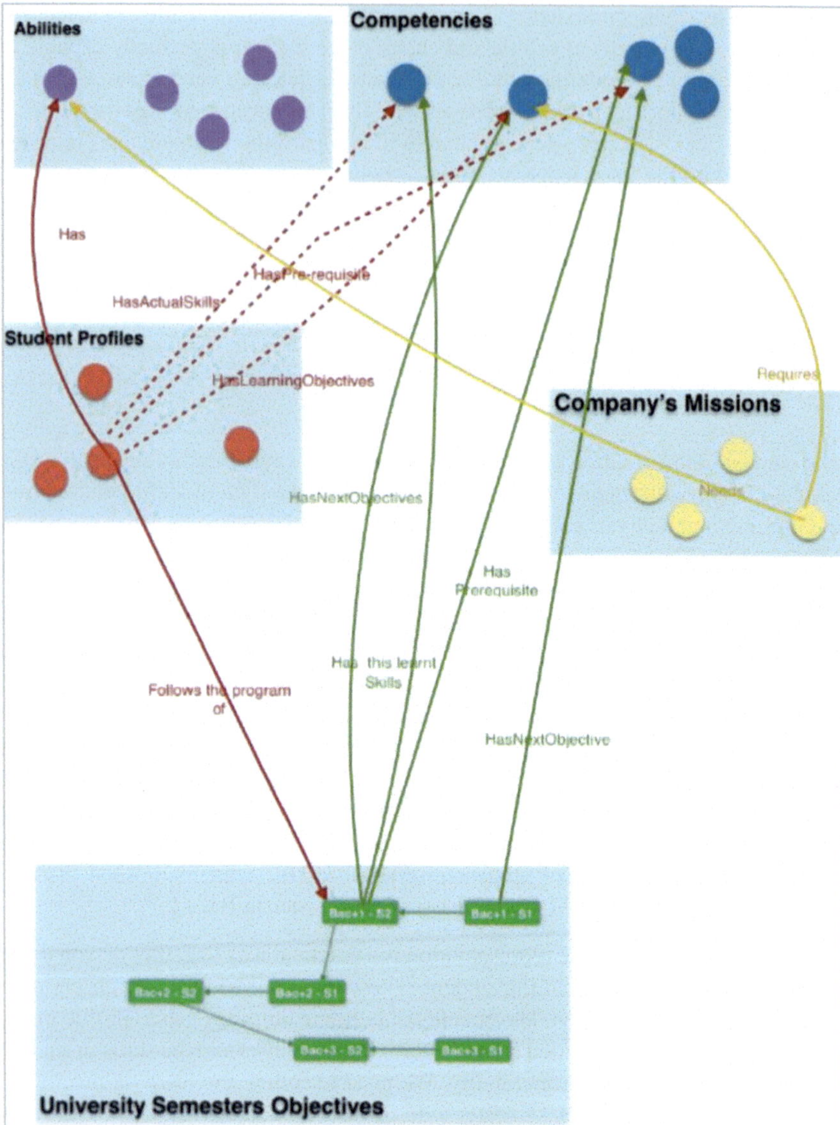

Fig. 1 Defined learning ontology concepts for IUT students

their professional degree LP. This tracking will be done semester by semester. We considered that the student begins to acquire knowledge since his first year of diploma (Bac+1), although in this year the student learns 'academic expertise' in the followed courses, more than the 'practical experience' acquired during the internship in this year. So, in this year university participates in building the 'student profile' with a very high percentage of academic expertise. Consider an example (Fig. 1): If the student has successfully completed his skills in Bac+1 in the second semester, then we can say that this student has some actual skills (academic or practical skills) acquired during this semester. And also, his profile should include the prerequisite knowledge of last semester (Bac+1 semester 1) and his objective learning skills that should be offered in Bac+2 semester 1. In this way, this representation of competencies for student profile clarifies to the companies what are the competencies already acquired, skills in progress and specially the competencies to receive in the next semester if the student follows a semester of alternation between the university and the company. In this case, the company will always be interested in the type of skills that the student will have during his work at the company.

- Difference between competencies and abilities: In concepts in student's profiles: 'competencies' and 'Abilities.' Competencies are the knowledge gained from courses at the university, whereas abilities are the capacities based on the analytical reasoning (technical and economic) or the acquired abilities outside the universities. The student can be able to manage, e.g., design some model, exchange ideas with his environment, organize and replicate. And he can be able to self-assess which is a good skill always required in companies. University semester objectives: First, the university needs to build a depth overview of its students' knowledge and skills in parallel with the company offers. This method allows the university to make their students' profiles more visible in companies. It also helps companies to find the adequate profile for the company's job offers.
- Company's Missions: Competence descriptions are indispensable for the definition of missions and learners. They specify what learners are expected to reach; competencies represent the basis for a company and university connection. They also help to offer support and assistance during learner internship. In Fig. 1, one case of student profile is represented. Student finished his first semester in Bac+1 with success, and he is continuing his second semester in Bac+1. We focus now

on the second semester (Bac+1_S2); this semester should have the competencies learnt in the first semester as prerequisite, learning competencies to succeed in this semester and objective competencies in the next semester. So, the students following this program will have in their profiles specific 'prerequisite skills, actual skills and learning objectives skills.'

3.3 Ontology Implementation

We will benefit from the reasoning in the built ontology after defining ontology relations, conditions, and restrictions. Furthermore, this ontology can provide a shared understanding of the learning domain which helps reuse the outcomes. Figure 2 includes all implemented classes in the 'learning ontology,' and the defined properties are presented in Fig. 3. In our approach, we are interested in achieving a representation of competencies within the IUT of Lyon 2. We consider that our university will give students different theoretical materials to be able to build the expected profile in his specific diploma. During studies at the universities, students do not only gain skills from the followed courses, but they also learn new knowledge through their projects implementation and internships well done in their organizations.

- Competencies Gap analysis: To make a proper matching between student profiles and company's missions, a competencies gap analysis will be necessary to detect any competencies lacks. In the gap analysis, in our ontology, we need to check the student profile competencies. In fact, this gap can be detected by minimizing the required mission competencies from the student profiles competencies. For example, if a student is following the program of the semester 1 (Bac+2) and he will do in the next semester his training (alternation semester), then it will be necessary to check all his competencies even the future competencies in the semester of alternation. In some cases, a student lacks some competencies considered as business skills. In this case, the company can offer him support during his internship to ameliorate some existing competencies in the student profile (Fig. 4).

  – Continuous Knowledge and Individual support: This part aims to improve the knowledge process for a student during his internship. This entity is the learning entity that continues even after the acceptance of this student profile in a mission of a company. Sometimes company accepts the student for a mission, although his profile lacks expertise in some business needs. In this case, the company knows the current capabilities of the student, but the goal remains to accomplish the mission's goal. So, the company offers him through one service the needed assistance or support during his internship to fill the gap with new knowledge acquired during this period. It can be like offering some document to help him understand some business needs.

Fig. 2 Implemented classes

Fig. 3 Defined properties

PROPERTIES
Need
Support
Propose
Has
Require
To ameliorate

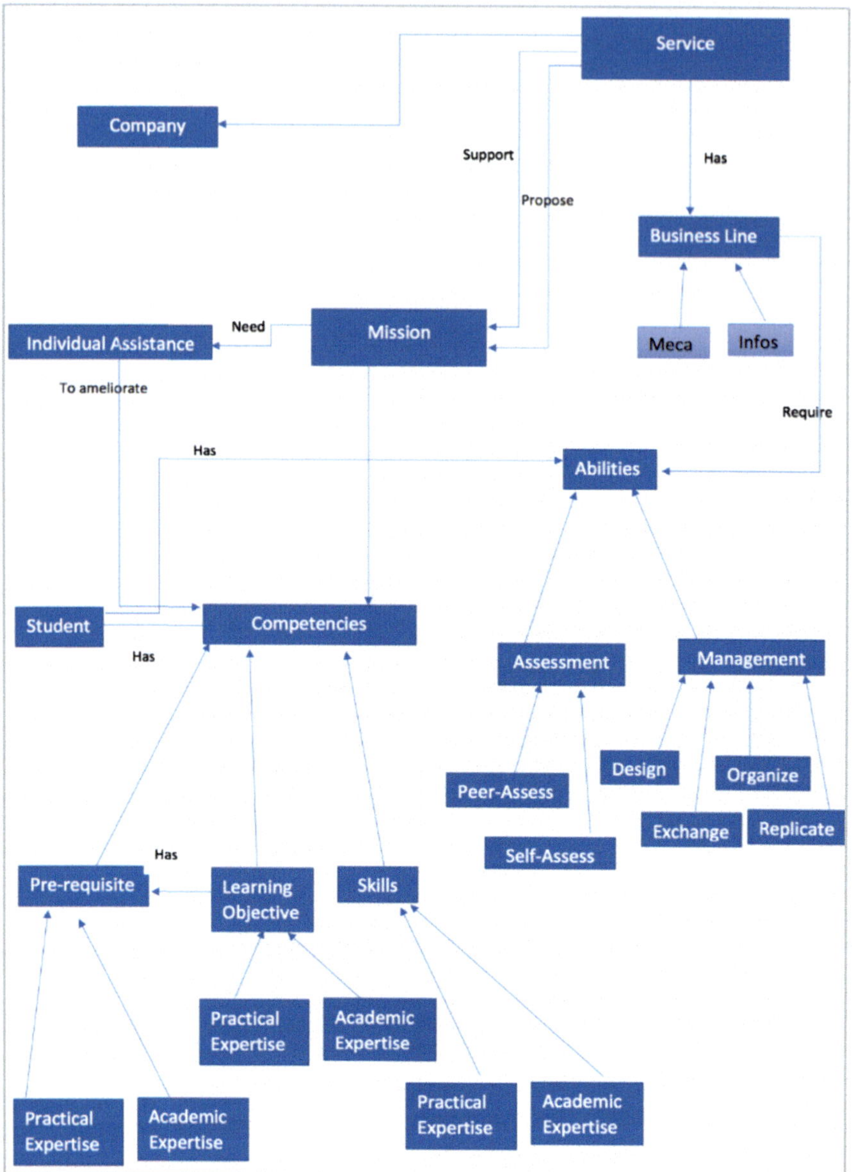

Fig. 4 Learning ontology

4 Conclusion

In this article, we have proposed a method able to offer the student the opportunity to learn continuously within his internship in a company. The proposed ontology aims to match the competencies required in companies' missions with the competencies of the students (prerequisite competencies, in-progress competencies, objective competencies). We take also into consideration the abilities of the students in these profiles. Based on many examples, we now know that competency management within educational institutions is a priority to ensure good results which will appear in learners' profiles. View as the main objective of these institutes is to offer skills through well-defined educational programs. Our method allows filling the competencies gap detected, and it allows to provide the environment of execution not just the knowledge acquired by this profile but also the situation to reach the objectives of acquisition of skills in the next semester. The proposed approach achieved an efficiency matching between the company needs and the students' skills. Results show the benefits of such an approach to resolving the gap between industry needs and students' skills. In future work, we intend to improve the ontologies representing the student profile and company missions with respect to our method which is primarily concerned with skills representation.

Acknowledgments This publication was made possible by NPRP grant\# NPRP 7-1883-5-289 from the Qatar National Research Fund (a member of Qatar Foundation). The statements made herein are solely the responsibility of the authors.

References

1. Haase, S., Chen, H. L., Sheppard, S., Kolmos, A., & Mejlgaard, N. (2013). What does it take to become a good engineer? identifying cross-national engineering student profiles according to perceived importance of skills. *International Journal of Engineering Education, 29*(3), 698–713.
2. Bouchamma, Y., Basque, M., & Marcotte, C. (2014). School management competencies: Perceptions and self-efficacy beliefs of school principals. *Creative Education, 5*(08), 580.
3. Rizvi, S., Teckchandany, S., & Ahuja, G. (2013) A study of competencies and skill sets needed by potential recruiters in the financial sector for management graduates. *Management Dynamics*.
4. Branch, R. (2016). The effect of human resource management practices of marketing sector on internal organizational social capital (case study: Manufacturing companies of Gilan province). In Saadat, M. & Ooshaksaraie, M. Department of Management, Department of Industrial Management. International Business Management.
5. Bridgstock, R. (2009). The graduate attributes we've overlooked: Enhancing graduate employability through career management skills. Higher Education Research & Development.
6. Bak, O., & Jordan, C. (2017). Linking industry and higher education: Assessing the skills requirements. *Logistics and Transport Focus, 55–56*.
7. Boucher, X., Bonjour, E., & Matta, N. (2007). Competence management in industrial processes. *Computers in Industry, 58*(2), 95–97.
8. Jean-Daubiasand, S., & Eyssautier, C. (2005) An environment helping teachers to track students' competencies. In *Workshop Le More (Learner Modelling for Reflection, to Support*

Learner Control, Metacognition and Improved Communication between Teachers and Learners) at AIED 2005 (12th International Conference on Artificial Intelligence in Education).
9. Melis, E., Faulhaber, A., Eichelmann, A., & Narciss, S. (2008). Interoperable competencies characterizing learning objects in mathematics. In *International Conference on Intelligent Tutoring Systems* (pp. 416–425). Springer.
10. Sieg, A., & Mobasher, B., & Burke, R. (2007) Ontological user profiles for personalized web search. In *Proceedings of the 5th Workshop on Intelligent Techniques for Web Personalization*, Vancouver, Canada, pages 84–91.
11. Liang, T.-P., Yang, Y.-F., Chen, D.-N., & Ku, Y.-C. (2008) A semantic expansion approach to personalized knowledge recommendation. *Decision Support Systems, 45*(3):401–412.
12. Furst-Bowe, J., Boger, C., Franklin, T., McIntyre, B., Polansky, J., & Schlough, S. (1995). An analysis of required computer competencies for university students. *Journal of Research on Computing in Education.*
13. Linck, B. (2008). Competence descriptions for informatics education using the example of logic programming. *Education and Information Technologies, 18*(2), 193–204. Melis, E., Faulhaber, A., Eichelmann, A., & Narciss, S. Interoperable competencies characterizing learning objects in mathematics. In *International Conference on Intelligent Tutoring Systems* (pp. 416–425). Springer.
14. Sitthisak, O., Gilbert, L., Davis, H.C., & Gobbi, M. (2007). Adapting health care competencies to a formal competency model. In Seventh *IEEE International Conference on Advanced Learning Technologies, 2007. ICALT 2007*. IEEE.
15. Chandrasekaran, B., Josephson, J. R., & Benjamins, V. R. (1999). What are ontologies, and why do we need them? *IEEE Intelligent Systems and their applications, 14*(1), 20–26.
16. Busse, J., Humm, B. G., Lubbert, C., Moelter, F., Reibold, A., Rewald, M., et al. (2015). Actually, what does ontology mean? *Journal of computing and information technology, 23*(1), 29–41.
17. Clemente, J., Ramírez, J., & De Antonio, A. (2014). Applying a student modeling with non-monotonic diagnosis to intelligent virtual environment for training/instruction. Expert Systems with Applications.
18. Gandon, F. (2002). Distributed artificial intelligence and knowledge management: Ontologies and multi-agent systems for a corporate semantic web. PhD thesis, Université Nice Sophia Antipolis.
19. Pan, J., Zhang, B., Wang, S., Wu, G., & Wei, D. (2007). Ontology based user profiling in personalized information service agent. In *7th IEEE International Conference on Computer and Information Technology, 2007. CIT 2007* (pp. 1089–1093). IEEE.
20. Sicilia, M.-A. (2005). Ontology-based competency management: infrastructures for the knowledge intensive learning organization. In *Intelligent learning infrastructure for knowledge intensive organizations: A semantic Web perspective* (pp. 302–324). IGI Global.
21. Bizer, C., Heese, R., Mochol, M., Oldakowski, R., Tolksdorf, R., & Eckstein, R. (2005). The impact of semantic web technologies on job recruitment processes. Wirtschafts informatik Springer.
22. Draganidis, F., Chamopoulou, P., & Mentzas, G. (2006). An ontology-based tool for competency management and learning paths. In *6th International Conference on Knowledge Management (I-KNOW 06).*
23. Schmidt, A. (2005). Bridging the gap between knowledge management and e-learning with context-aware corporate learning (Vol. 3782, p. 203). Lecture notes in computer science.
24. Szab´o, I., & Neusch, G. (2015). Dynamic skill gap analysis using ontology matching. In *International Conference on Electronic Government and the Information Systems Perspective*. Springer.

Enterprise Ontology for Service Interoperability in Socio-Cyber-Physical Systems

Alexander Smirnov, Tatiana Levashova and Alexey Kashevnik

Abstract Enterprises are complex systems comprising three fundamental components: a physical sub-system, a decision sub-system, and an information sub-system. Humans in the enterprises are not just resources but the first consumers and providers of data, information, and knowledge. The present research considers enterprise as a technical system embedding social component. Socio-cyber-physical systems naturally embody this idea. The research considers such systems as an enterprise component that combines the cyber aspects of computing and communications with the dynamics and physics of physical enterprise. The enterprise interoperability problem is treated as the problem of service interoperability in socio-cyber-physical systems at the service level and process level of an enterprise. The research proposes ontologies of enterprise and socio-cyber-physical system, a task-oriented service interoperability model, and a service communication mechanism.

Keywords Socio-cyber-physical system · Enterprise · Service interoperability · Service communication · Ontology

1 Introduction

Enterprises are complex systems comprising three fundamental components: a physical sub-system, a decision sub-system, and an information sub-system. Humans in

A. Smirnov · T. Levashova (✉) · A. Kashevnik
St. Petersburg Institute for Informatics and Automation of the Russian Academy of Sciences, 39, 14th Line, 199178 St. Petersburg, Russia
e-mail: tatiana.levashova@iias.spb.su

A. Smirnov
e-mail: smir@iias.spb.su

A. Kashevnik
e-mail: alexey@iias.spb.su

A. Smirnov · T. Levashova · A. Kashevnik
ITMO University, St. Petersburg, Russian Federation, 49, Kronverkskiy Pr, 197101 St.Petersburg, Russia

K. Popplewell et al. (eds.), *Enterprise Interoperability VIII*, Proceedings of the I-ESA Conferences 9, https://doi.org/10.1007/978-3-030-13693-2_17

the enterprises are not just resources but the first consumers and providers of data, information, and knowledge [1]. One of the enterprise definitions refers to it as a socio-technical system or socio-cyber-physical system (SCPS) where agents interact [2]. Agent interoperability is the key enabler of their interactions.

SCPSs are a kind of embedded systems. The SCPSs integrate computations with information and communication technology and with methods of intelligent planning and control. Embedding SCPSs into enterprises enables to unite the cyber dimensions of the SCPSs with the dynamics and physics of enterprise and to take the enterprises at a new level [3].

Research on enterprise interoperability distinguishes four enterprise levels at which interactions can take place. They are data, services, processes, and business [4]. The SCPSs consist of three spaces: cyber, physical, and mental [5]. These spaces are modeled by sets of resources. Resources interact to provide the required services. Interoperability in SCPSs is related to resource interoperability. Since resources expose their functionalities as services, the problem of resource interoperability becomes the problem of service interoperability. This problem obviously shows itself at the service level and the process level of enterprise. The both levels deal with enterprise functionality in terms of services. Service interoperability at these two levels is the center of attention of the present research.

The paper presents a research on ontology-based service interoperability in SCPS. SCPS is considered as a system embedded into enterprise. An ontology supporting this view integrates an enterprise ontology and a SCPS ontology. The enterprise ontology proposed deals with the concepts of resource, service, role, and task. The concept of resource is the key concept used to integrate SCPS and enterprise. Benefit for enterprises from such integration is twofold. At first, the enterprises gain cyber constituent, the advantages of which are well known. Secondly, in SCPSs, humans are not just objects served by cyber resources, but "collaborators" as well. The proposed ontology supports interoperability of the resources from different spaces. In particular, this concerns the interoperability of cyber resources and humans. Research in this direction is at an early stage.

The research contribution is the enterprise ontology integrated with socio-cyber component and a task-oriented service interoperability model which is feasible due to ontology-based task representation and task-oriented service communications.

The rest of the paper is structured as follows. Section 2 discusses the enterprise ontology in its part the research deals with. Section 3 focuses on the ontology of SCPS and the service interoperability problem. Some concluding remarks are summarized in Conclusion.

2 Enterprise Ontology

The ontology for resource self-organization in SCPSs [6] is the background for the enterprise interoperability ontology. The ontology shares the ontology view on an enterprise proposed in ORG-Master framework [7]. That framework introduces

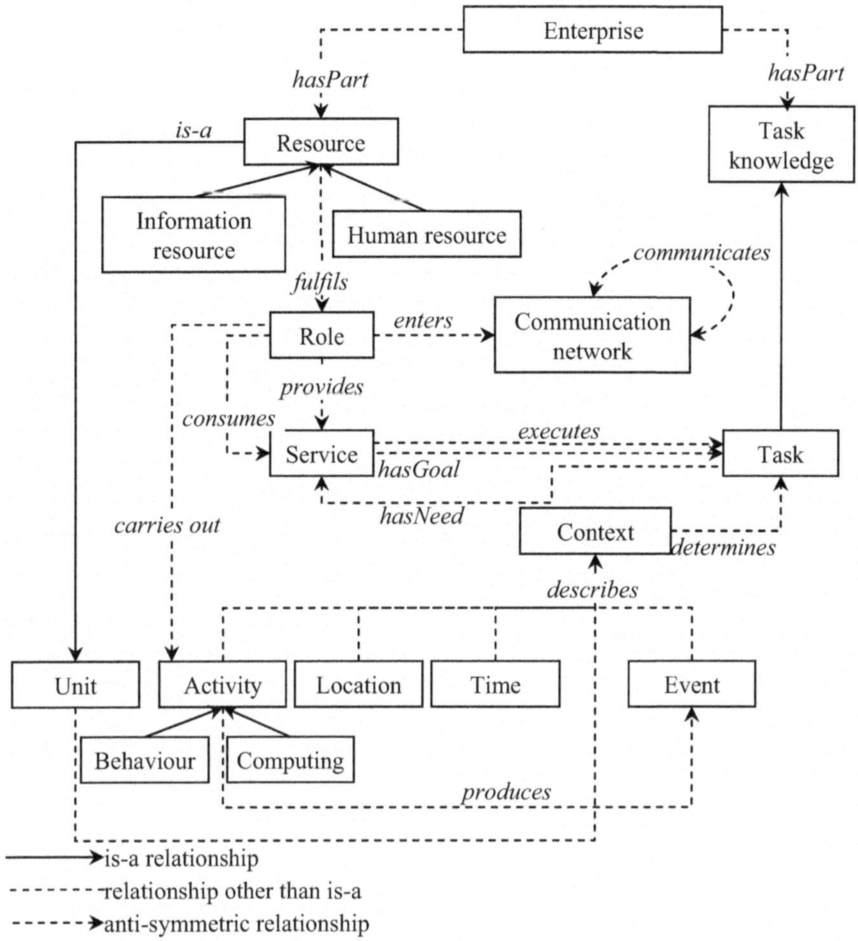

Fig. 1 Enterprise ontology

12 middle-level concepts to describe an enterprise: activities, production, performance management, norms, business, environment, information resources, material resources, financial resources, human resources, organizational structure, and infrastructure. Among the variety of resources distinguished in the ORG-Master framework, information and human resources organize the category of resources capable of providing services and participating in interactions. The enterprise ontology (Fig. 1) focuses on these two kinds of resources.

Resources provide *services* in accordance with *roles* that these resources *fulfill* in the current situation (context). Services provided by one resource are consumed by others. In the ontology, this idea is as follows. The resources fulfill *roles*; the roles *provide* or *consume services*. *Role* is a position that a resource can take in the context. The resources' roles may be changed in the process of executing *tasks*.

Service is some *activity* or effort that is done to satisfy a need or to fulfill a demand. From the SCPS perspective, the need or demand corresponds to a *task* to be *executed*. *Task* is a specific purpose to be achieved or a problem to be solved. Task knowledge is specified in the enterprise knowledge base. Services *execute* the tasks; specific services are needed to execute a particular task (*hasNeed* relationship). A service is designed to realize some goal (*hasGoal* relationship). Services can be involved in task execution if the fulfillment of their individual goals leads to the fulfillment of the entire task. That is, the services having the *hasGoal* relationship to the task can participate in the execution of this task. A service can be simple or complex. A simple service is provided by a single resource. A complex service requires collaboration of several resources. The proposed concept of service can be reconciled with the core reference ontology for services (UFO-S ontology) [8], which harmonizes different service perspectives. This is supposed to be done by introducing *service event* as a sub-concept for the concept of *event*.

Context determines the task to be accomplished in the current situation. In ubiquitous and pervasive environments, a widely adopted definition of *context* is any information that can be used to characterize the situation of an entity [9]. Categories of *individuality*, *activity*, *location*, *time*, and *relations* are used to model such context [10]. On the other hand, *context* is a situation, which could be seen as a course of events; this situation evolves organizing new relationships between the entities involved in it [11]. The proposed ontology unites the two perspectives above. In this ontology, context is characterized by categories of *unit*, *activity*, *location*, *time*, and *event*.

The *unit* category represents the entity itself. The name *unit* is introduced instead of *Individual* one so as not confuse with individuals used in Web Ontology Language OWL.[1] Resources are the units that carry out activities in the context. *Location* and *time* are used to model the spatiotemporal coordinates of the entity. *Event* is defined as occurrence happening at a determinable time and place. Entities or some factors can produce events. *Activity* is a sequence of actions or operations the entity carries out when executing a task. Events are instantaneous, activities last in time [12].

Resources carry out activities in accordance with their contextual *roles*. Two types of activities are distinguished: *behavioral* or *computing*. Behavioral activity refers to responses, reactions, or movements made by a resource in a situation. This activity does not suppose any computations. On the contrary, computing activity aims at performing computational operations like functions or procedures. As a rule, behavioral activity is inherent in humans, whereas information resources are responsible for computations. The consequences of activities are events. Computing activities produce service delivery event. Behavioral activities produce various types of behavioral events.

The category of *relations* is intended to represent any possible relations an entity may establish with other entities. In the present ontology, this category is omitted since relations are a typical ontology modeling component but not just a contextual category. All the relationships specified in the ontology can be classified as relations.

[1] https://www.w3.org/TR/owl-features/.

Resources use *communication network* to communicate. A *communication network* is a collection of resources linked so as to enable information exchange. The network provides predefined communication *roles* with which resources joint this network. This perspective complies with the Core Ontology for Semantically Interlinked Online Communities (SIOC) [13].

Below, the resource interoperability problem is treated as the problem of service interoperability. Service actions, operations, or functions refer to the actions, operations, or functions of the resources providing these services.

3 Service Interoperability

3.1 Service Interoperability Model

The service interoperability model (Fig. 2) relies on the earlier developed methodology for context-aware decision support in pervasive environments [14]. As provided by the methodology, context is an ontology-based model of the current situation. This model represents two types of knowledge. They are conceptual (domain) knowledge and task knowledge. The difference in types of knowledge corresponds to the classification proposed by Guarino [15].

The enterprise ontology is the source for context knowledge. The task knowledge and domain knowledge are interrelated in this ontology. The relationships between these two types of knowledge indicate which domain knowledge is needed for the task executions. As well, these relationships provide the tasks with the domain semantics. The context represents the tasks (and related to them domain knowledge) that are needed to be executed in the current situation. These tasks specify the expected service functionality.

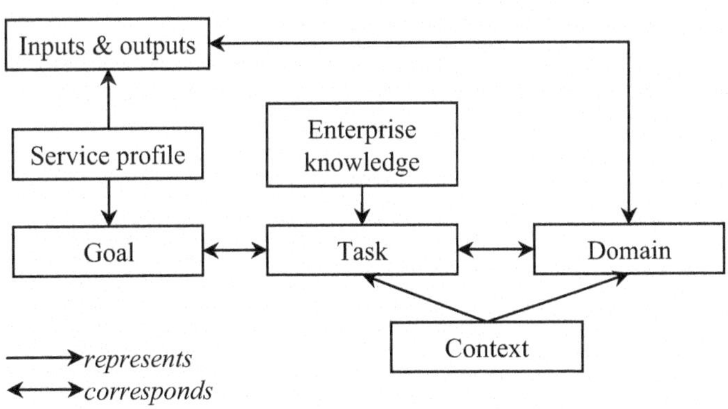

Fig. 2 Service interoperability model

Services are specified by their profiles. These profiles provide, among other things, a description of service functionalities in the form of the goals that the services are designed to and inputs/outputs of the services.

The interoperability among the services is supported through matching. The service goals are matched against the tasks represented in the context. The outcome of this procedure is a set of services available for the task execution. It means that the services from the set are suitable to execute the task jointly rather than each of these services can execute the entire task independently. For instance, the services are intended (or have goal) to assemble products, i.e., they can participate in the assembly task. At that, each service can perform subtasks corresponding to its individual goal. One service can deliver the product components, other service can orient these components, one more service can fasten the components, and so on.

The services configure a service network or a process (a process defines the sequence of the services (functions) according to some specific needs [4]) to execute the task. Configuring of the service network is supported by communication mechanisms.

4 Service Communication

Due to the fact that human resources are components of the enterprise, online communication of services is proposed. At present, online communications become common human practice. Popularity of such communications among people gave rise to the idea of applying such kind of communications to information resources.

The technology of *online communities* has been chosen to organize the service communication. *Online community* is a virtual community whose members interact with each other via the Internet. As opposed to social networks, an online community unites its members (resources here) based on some common interest or goal. The common goal or task underlies the task-oriented service interoperability model above. This fact determined the choice of online communities.

Services are proposed to communicate by messaging. The advantages of this are (1) an explicit form of information being communicated; (2) the common way of communication; and (3) no need to recognize different communication modalities. The proposed message format is:

$$< Type,\ Resource\_Send,\ Resource\_Recip,\ Task,\ Service,\ Content,\ Status >, \quad (1)$$

where *Type* is a message type (one of *Request*, *Replay*, or *Notify*); *Resource_Send* is a name of the resource (an instance of the concept Resource) sending the message; *Resource_Recip* (resource recipient) is a name of the resource for which the message is intended; *Task* is a task that the resource *Resource_Send* executes; *Service* is activity that the resource *Resource_Send* carries out to accomplish the task; *Status* is a status of the task execution (status can be one of *Ready*, *Failed*, or *Suspended*); and *Content* is some information specific to the particular task. The resource recipient

(*Resource_Recip*) may be represented by a name of the resource, a role name, or omitted. If this resource is represented by a name, then the message is addressed to the particular resource. If the field *Resource_Recip* contains an instance of the concept *Role*, this means that the message is addressed to a set of resources fulfilling the given role. If name of the resource is omitted, the message has no specific recipient and is sent into the community as a public message.

An example of a message for an assembly task is given to provide some idea of possible interpretation for the signature (1):

$$< Request,\ Robot\_Alpha,\ Consultant,\ X236,\ Carry, \{A17,\ 12,\ 24,\ ?,\ ?\},\ Suspended > .$$

The resource with the name *Robot_Alpha* sends a request message (the message type is *Request*) to anyone fulfilling the role of *Consultant*. *Robot_Alpha* is assembling the product *X236* (in the enterprise ontology, *X236* is an instance of the class product; the class product represents assembly objects). The ontology specifies the assembly task as a sequence of operations. One of these operations deals with transferring product components. *Robot_Alpha* is ready to perform this operation. In the task specification, the operation is defined as $Carry(Component, x_0, y_0, x, y)$, where *Carry* is the operation name and the bracketed expression specifies the list of input/output arguments of the operation. The arguments are as follows: *Component* is the product component that the resource should carry; x_0, y_0 are the coordinates of the point from which the component should be carried; and x, y are the coordinates of the point to which the component should be carried. Referring to the message, *Robot_Alpha* is going to carry the component *A17* from the point ($x_0 = 12$, $y_0 = 24$), but *Robot_Alpha* is unaware of coordinates for the destination point. In the message, the unknown coordinates are replaced with question marks. The operation is suspended. It is expected that someone who fulfills the role of consultant will provide the resource *Robot_Alpha* with the requested coordinates.

5 Ontology for Service Interoperability in Socio-Cyber-Physical System

The ontology of SCPS (Fig. 3) is based on the definition [5] declaring that SCPSs consist of three interrelated spaces (cyber, physical, and mental). The physical space comprises a variety of *physical devices* supplied with computing components. These components enable the devices to carry out computations, process data/information/knowledge, perform some activities, communicate, and provide services. *Information resources* represented by the enterprise ontology can be characterized in a similar way. But unlike the information resources of enterprises, some devices of SCPS are capable of action. These devices are classified as *actors*, where *actor* is anyone who acts and gets things done. The acting physical devices make up the *cyberspace*. The ontology represents this space through *cyber resources*. Cyber

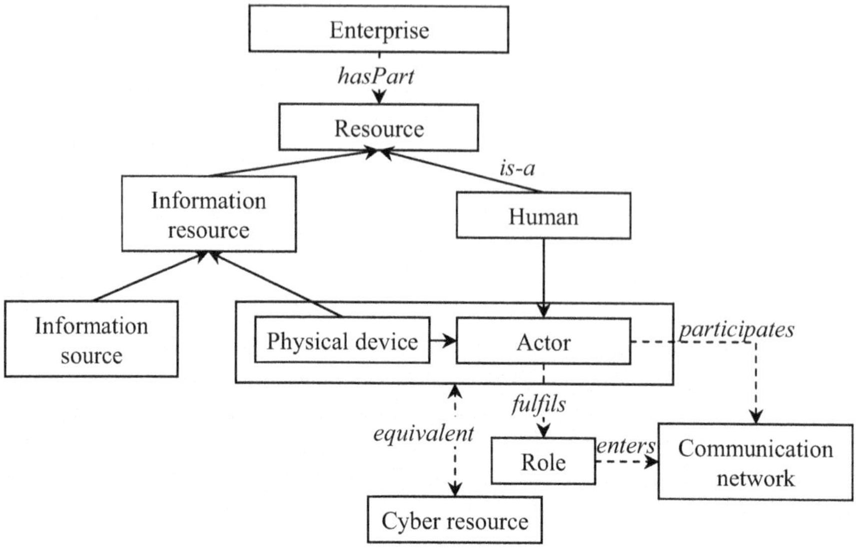

Fig. 3 Ontology of socio-cyber-physical system

resources are defined as physical devices capable to be actors. The equivalence axiom is used to define cyber resources as stated above:

$$(is - a\ Physical\,device)\ and\ (is - a\ Actor) \equiv Cyber\,resource.$$

Due to the introducing of cyber resources in the enterprise ontology as a subcategory of information resources, the category of *information sources* has been added. Information sources are any electronic resources that provide data/information/knowledge but do not act. They are databases, Web sites, sensors, etc.

Humans with their knowledge, mental capabilities, and sociocultural systems make up the mental space. Like the cyber resources, human resources can act and therefore are classified as *Actors*. The cyber resources and humans can perform both behavioral and computing activities (not shown in the figure).

A piece of an ontology model of a cyber resource (Fig. 4) is presented to illustrate applicability of the proposed ontologies. This model is developed to model coalitions of mobile robots that jointly solve a task. The model relies upon definitions and abbreviations proposed by IEEE-SA Standards Board [16]. High-level classes of the ontology include 27 classes with three types of relationships specified between them: "*is_a*," "*has*," and "*associate with*." The most abstract ontology classes (besides the class "*Thing*") are "*Physical*" and "*Abstract*." The class "*Physical*" describes physical entities and processes in the physical space while the class "*Abstract*" describes virtual entities. The class "*Physical*" is subdivided into classes "*Object*," "*Environment*," and "*Process*." The class "*Object*" describes *phys-*

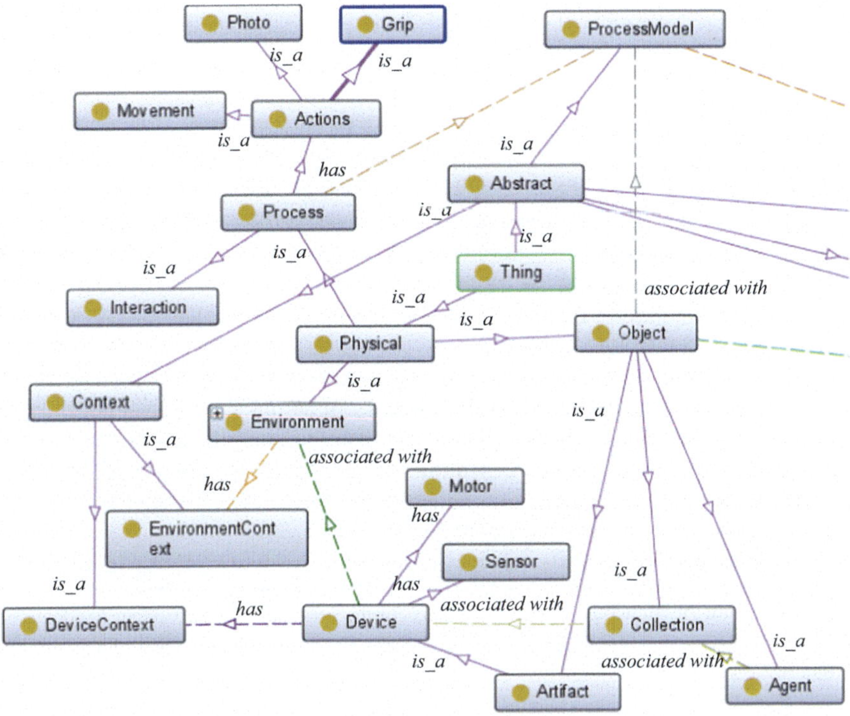

Fig. 4 An ontology model of a cyber resource (a piece)

ical devices, their *collections*, and *agents* that control them. The class "*Process*" is subdivided into the classes "*Actions*" and "*Interaction.*" Action is a movement, gripping, taking a photograph, and other actions that a physical item can implement. The class "*Abstract*" is subdivided into classes "*Process Model,*" "*Configuration,*" "*Policy,*" "*Context,*" and "*Competence Profile.*" According to the presented model, a cyber resource is defined as an artificial physical device associated with an environment (in the context of this paper, the enterprise) according to a process model.

6 Conclusion

The paper presents the research on service interoperability in socio-cyber-physical systems embedded in enterprises. The enterprise interoperability problem is treated as the problem of service interoperability in socio-cyber-physical systems. In this direction, an approach enabling to unite the conceptual models of enterprise and socio-cyber-physical system is proposed.

The enterprise ontology provides, among other concepts, the concept to specify tasks. Such a view gave a rise to a task-oriented interoperability model. This model supplies the services with the semantics of the problem domain and makes them aware of the task expected to be executed in the current situation. The services configure a service network to execute this task.

Configuring of the service network is supported by a communication mechanism. It is proposed to use online communities for message-based service communications. The main advantage of using message-based communication is that the messages represent explicit information/knowledge. The paper offers a message format for task-oriented messaging. An exemplified message from the product assembly domain illustrates the applicability of the format.

The ontology of socio-cyber-physical systems introduces into the enterprise ontology the concepts enabling to take into account specific of cyber resources, which are not just information resources but actors like humans. Consideration of cyber resources and humans belonging to the category of actors allows ones not only to organize joint actions of cyber resources, but also to combine the efforts of cyber resources and humans while task executing. An ontology model of a cyber resource is developed to model coalitions of mobile robots that jointly solve a task.

Acknowledgements The projects funded through grants 16–29–04349, 16–29–12866, 17–29–07073, 17–07–00247, and 17–07–00248 of the Russian Foundation for Basic Research, the research Program 29 “Actual problems of robotics” of the Presidium of the Russian Academy of Sciences, and Grant 074-U01 of the Government of Russian Federation provided support for the research.

References

1. Romero, D., & Vernadat, F. (2016). Enterprise information systems state of the art. *Computers in Industry, 79*(C), 3–13.
2. Weichhart, G., Guédria, W., & Naudet, Y. (2016). Supporting interoperability in complex adaptive enterprise systems: A domain specific language approach. *Data & Knowledge Engineering,* 105(Sept.), 90–106.
3. Frazzon, E. M., Hartmann, J., Makuschewitz, T., & Scholz-Reiter, B. (2013). Towards socio-cyber-physical systems in production networks. *Procedia CIRP, 7,* 49–54.
4. Chen, D. (2006). Enterprise interoperability framework. In *Proceedings of the Open Interop Workshop on Enterprise Modelling and Ontologies for Interoperability. CEUR Workshop Proceedings.* http://ceur-ws.org/Vol-200/19.pdf.
5. Liu, Z., Yang, D.-S., Wen, D., Zhang, W.-M., & Mao, W. (2011). Cyber-physical-social systems for command and control. *IEEE Intelligent Systems, 26*(4), 92–96.
6. Smirnov, A., Levashova, T., Shilov, N., & Sandkuhl, K. (2014) Ontology for cyber-physical-social systems self-organisation. In *Proceedings of the 16th Conference of Open Innovations Association FRUCT* (pp.101–107). IEEE.
7. Grigoriev, L., & Kudryavtsev, D. (2011) The ontology-based business architecture engineering framework. In H. Fujita & T. Gavrilova (Eds.), *New trends in software methodologies, tools and techniques: Vol. 231 Frontiers in artificial intelligence and applications* (pp. 233–252). IOS Press, Amsterdam.
8. Nardi, J. C., de Almeida Falbo, R., Guizzardi, G., Pires, L. P., Sinderen, M. J., Guarino, N., et al. (2015). A commitment-based reference ontology for services. *Information Systems,* 54(December) 263–288.

9. Dey, A. K. (2001). Understanding and using context. *Personal and Ubiquitous Computing, 5*(1), 4–7.
10. Zimmermann, A., Lorenz, A., & Oppermann, R. (2007). An operational definition of context. In B. Kokinov et al. (Eds.) *CONTEXT 2007: 4635. LNAI* (pp. 558–571). Springer-Verlag, Berlin, Heidelberg.
11. Baumgartner, N., Gottesheim, W., Mitsch, S., & Retschitzegger, W. (2010). BeAware!-Situation awareness, the ontology-driven way: Editorial. *Data & Knowledge Engineering, 69*(11), 1181–1193.
12. Sanfilippo, E. M., Borgo, S., & Masolo, C. (2014). Events and activities: Is there an ontology behind BPMN? In P. Garbacz & O. Kutz (Eds.), *Formal ontology in information systems* (pp. 147–156). Amsterdam: IOS Press.
13. Berrueta, D., Brickley, D., Decker, S., et al. (2010). SIOC core ontology specification. http://rdfs.org/sioc/spec/. Last accessed January 15 2018.
14. Smirnov, A., Shilov, N., Levashova, T., & Kashevnik, A. (2008). Context-aware operational decision support in decentralised environments: A methodology. In K. Engemann & G. Lasker (Eds.), *Advances in decision technology and intelligent information systems* (Vol. IX, pp. 26–30). Windsor, Canada: The International Institute for Advanced Studies in Systems Research and Cybernetics.
15. Guarino, N. (1997). Understanding, building and using ontologies. *International Journal Human-Computer Studies, 46*(2/3), 293–310.
16. IEEE 1872–2015 Standard Ontologies for Robotics and Automation. *IEEE Standard, 60* (2015). DOI: https://doi.org/10.1109/ieeestd.2015.7084073.

Application of Allen's Temporal Logic to Ontological Modeling for Enterprise Interoperability

Alena V. Fedotova, Karl A. Hribernik and Klaus-Dieter Thoben

Abstract The problems of creating lifecycle ontologies for enterprise interoperability are addressed in this paper. Enterprise interoperability and enterprise integration are essential components of enterprise engineering (EE). A few definitions and viewpoints on EE are examined in the first hand. An original discipline for EE is considered. A generalized ontological approach to enterprise engineering is developed on the basis of combination of the lifecycle modeling, knowledge management and ontological engineering. It calls for the modeling and coordination of at least three lifecycles: enterprise lifecycle, knowledge lifecycle and product lifecycle. A general representation of lifecycle knowledge graph by a mind map is given. Particular emphasis is put on granular lifecycle upper ontology and meta-ontology. The lifecycle representations being discussed include both visualized and abstract ones. Allen's logic is used to construct principle temporal relations between stages and phases of lifecycle.

Keywords Enterprise interoperability · Enterprise engineering · Ontological modeling · Product lifecycle management · Allen's logic · Information granulation · Fuzzy interval

1 Introduction

Enterprise integration, interoperability and network operation are a portion of the real trends that are empowering organizations to enhance communication and collabora-

A. V. Fedotova (✉)
Bauman Moscow State Technical University, 2nd Baumanskaya 5, 105005 Moscow, Russia
e-mail: afedotova.bmstu@gmail.com

K. A. Hribernik · K.-D. Thoben
BIBA—Bremer Institut für Produktion und Logistik GmbH, Hochschulring 20, 28359 Bremen, Germany
e-mail: hri@biba.uni-bremen.de

K.-D. Thoben
e-mail: tho@biba.uni-bremen.de

K. Popplewell et al. (eds.), *Enterprise Interoperability VIII*, Proceedings of the I-ESA Conferences 9, https://doi.org/10.1007/978-3-030-13693-2_18

tion in the best way [1]. The capacity for an undertaking to interoperate with others is not just a quality and favorable position for picking up competition in the market yet in addition an issue of survival for some organizations [2]. In fact, to decrease the cost, abbreviate deferrals and propose ceaselessly new product available, endeavors call for more interoperations during the whole product lifecycle and in a networked organizational condition [3].

There are numerous ideas identified with interoperability. The interoperability barrier is a crucial idea in characterizing the interoperability area. Numerous interoperability issues are particular to specific application areas. Conceptual barriers are the primary barriers to interoperability. Conceptual barriers are implicated about syntactic, content and semantic incompatibilities of information to be exchanged. These issues concern modeling at an upper level of abstraction (for instance, an organization's enterprise models). Semantic incompatibility is viewed as a vital boundary to interoperability as the information and knowledge in many models or programming have no unmistakably characterized semantics which would accommodate an unambiguous comprehension of the importance of the pertinent information. At the present stage, the most understood method for solving this issue is semantic comment and compromise utilizing ontology [2].

Recognition of the fact that enterprise integration is a fundamental part of enterprise engineering (EE), which concerns the arrangement of strategies, models and instruments that one can use to analyze outline while accomplishing integration [1]. Enterprise integration can be seen in various integration levels as indicated by the interest of the study. Since the end of the 1990s, enterprise integration approaches have needed to address the developing emphasis on enterprise interoperability. Enterprise interoperability is a key part of enterprise engineering. Enterprise interoperability is accepted to be more versatile (less cost and faster implementation) in adaptable, decentralized, organized framework situations for product and process plan conditions including production [1].

From the EE point of view, the development of enterprise is a kind of engineering activity. In addition, it has a tendency to look at every part of the enterprise, including information flows, business processes, organizational structures and resources [4].

The customary thought of enterprise as a group of business processes may break its foundational integrity; here different methodologies are required, for example, developing summed up enterprise models by utilizing agent-oriented technologies [5] and ontologies for industrial enterprise [6].

2 An Ontological Approach in Industrial Enterprise Engineering

Different perspectives on EE are introduced in Sect. 2. Some essential disciplines of EE are viewed, and the comparative visual pyramid portrayal is delineated. A generalized a transdisciplinary, synergistic ontological approach to enterprise engineering

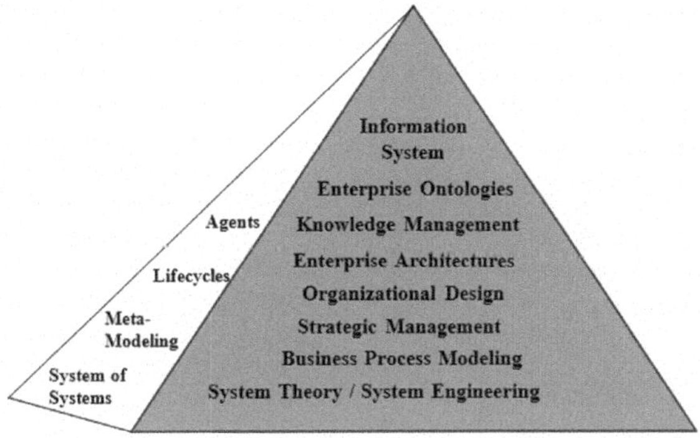

Fig. 1 Enterprise engineering activities pyramid

is developed on the basis of the combination of knowledge management, lifecycle modeling and ontological engineering.

Three fundamental objectives of EE are specified in [7]: intelligent manageability, organizational concinnity and social dedication.

In his turn, Martin [8] centers around seven EE disciplines assembled around a value system: strategic visioning saw as continuous cycle of significant positioning; venture upgrade—broken change in value definition; value stream reevaluation—discontinuous change in value offering; methodology update—irregular rehash of value creation; full quality management—proceeding with change in value creation; data innovation advance (persistent value enablement); hierarchical and social advancement—nonstop value development.

As per Vernadat [9], EE is the skill of comprehension, characterizing, indicating, breaking down and actualizing business processes for the whole enterprise's lifecycle, so that the enterprise can accomplish its goals and additionally be financially savvy and more focused in its market condition. Two essential disciplines for EE, namely enterprise integration and enterprise modeling, are allocated.

It [10] was noted that the two largest projects of enterprise engineering are the projects of TOVE [11] and enterprise project [12], started back in the 1990s. Today is popular Industrial Ontologies Foundry (IOF) [13]. Interoperability breakdowns are a noteworthy obstruction to coordinated product lifecycle and supply chain management. The Industry Ontology Foundry is an activity intended to address this issue by NIST and by various industrial and academic partners. The methodology is to form a group of digital manufacturing. Modules under thought include: core product model, functional basis, product lifecycle, materials and material attributes [13].

Our own pyramid of enterprise engineering activities (EEA-pyramid) is shown in Fig. 1.

Our way to deal with EE is established on the incorporation of lifecycle modeling, system of systems concept [14], ontological engineering and knowledge management [11]. It assumes the detail and coordination of no less than three lifecycles: enterprise, knowledge and product lifecycles [15] (see Fig. 2).

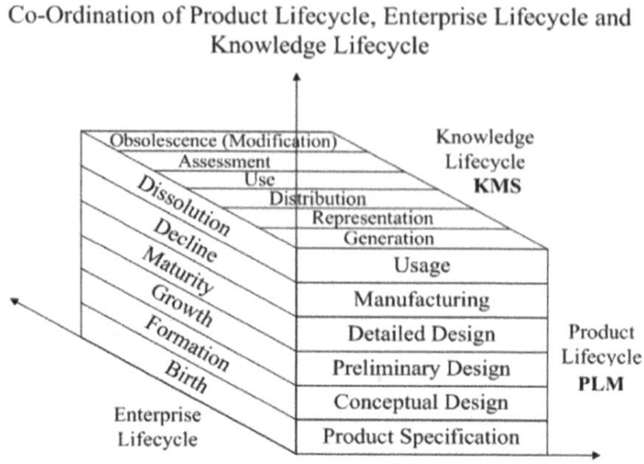

Fig. 2 The integration of PLM and KM as a generalized lifecycle management

From one viewpoint, a computer-based combination of knowledge lifecycle and product lifecycle prompts the combination of knowledge management (KM) and product lifecycle management (PLM) technologies. The idea of lifecycle exemplifies an essential execution of systemic approach to deal with complex objects that consists in visualizing their state changes for a temporal interval. The term "product lifecycle" denotes the possibility of the dissemination of artifacts between the areas of design, manufacturing, use (utilization) and transfer or reuse. Product lifecycle management is the process of managing a whole product lifecycle (see, for instance, [16–18]). It combines data, personnel, processes and organizations to give a product data and knowledge core for networked enterprises. The improvement of PLM systems demands lifecycle engineering and modeling. It implies incorporating an assortment of key lifecycle values of product into the most basic creation and usage time interims. Be that as it may, the PLM system just furnishes us with new chances to pick up competition. To expel insufficiency inside the product lifecycle and boost its focal points, Kiritsis presented the idea of closed-loop PLM [19]. Taking the idea of closed-loop PLM into account helps organizations raise the quantity of new product introductions in the market driving to a lessening in the time to market and subsequently to a shortening of the product lifecycle itself [20].

Then again, the interest of an enterprise in unions or consortiums, and additionally the arrangement of broadened, virtual or intelligent enterprises (see [21, 22]) prompts the prolongation of the best phases of enterprise lifecycle, for example, enterprise maturity and growth.

Knowledge management is frequently characterized as the way to apply a systematic way of dealing with the capture, structuring, dissemination and utilizing of knowledge all through an enterprise so it can work quicker, reuse best practices and decrease costs from project to project [11]. It is clear that KM turns out to be increasingly important to knowledge of lifecycle on account of virtual enterprises.

The ontological approach to enterprise knowledge management is of unique concern here, and PLM systems are more appropriate as the center for next IT hybrids and synergistic intelligent technologies for enterprise interoperability. These systems generate and bolster a unified knowledge and information space through the span of the product lifecycle (Fig. 2).

3 A Key to Enterprise Engineering—Lifecycle Ontologies

Time is a principal asset in lifecycle management. A particular feature of lifecycle is its heterochronous character, that is, irregularity identified with contrasts in temporal criteria and limitations on different phases. Actually, we attempt both to quicken design and production time and in addition to increase usage time. For example, at the design stage, the fundamental criterion is to diminish design time, for instance, by utilizing concurrent design strategies [23]. Conflictingly, at the utilization step we try to keep or increment the reglementary period, for instance, by enhancing the maintenance and support systems. Below the lifecycle knowledge graph is shown by a mind map (Fig. 3). Such knowledge graph specifications as its goal (lifecycle modeling, integration of stages, lifecycle knowledge management, learning), role (knowledge representation method, integration method of heterogeneous information, providing interoperability between enterprises), language (informal, formalized, formal), representation form (symbolic, graphic), type (by generality level (meta-ontology, top level ontology, domain ontology), by depth of study (task ontology, application ontology, dictionary, constraint system, lightweight ontology, heavyweight ontology) and basic relations (taxonomy (cycle stage phase), mereology (whole part), genealogy (parent–child) are of special concern. The attention here is on different ways of representing ontologies of lifecycle. A cycle all in all is portrayed by the nearness of limited and recurrent parts at temporal intervals; the key parameter is duration.

«Time arrow» and «time wheel» are two well-understood time representations bringing about lifecycle consequent and circular time models. From one viewpoint, linear consequent models declare time properties, for example, irreversibility, requesting facility and course, and then again, circular models of time stress variations, rhythms, emphases and self-supporting procedures.

The idea of lifecycle can be investigated from different perspectives; diverse variations for indicating its activities and stages are recommended. The point of cyclic product definition is to realize products, processes and economic arrangements that are more intelligent and better by incorporating the lifecycle theory into economy and technology.

The ontological way to deal with lifecycle knowledge engineering, we suggesting, assumes the construction visual and formal models of lifecycle ontologies. Formal models depend on Maltsev's [24] idea of the algebraic system which is a characteristic mathematical reason for determining meta-ontologies. Below we review the idea of the algebraic system and extend it to consider fuzziness and granularity. Visual representations encompass the concept of closed-loop PLM of Kiritsis [19]. To remove ineffectiveness within the product lifecycle and maximize the advantages, Kiritsis introduced the concept of closed-loop PLM. Closed-loop PLM depicts an approach which closes information loops between singular periods of the product

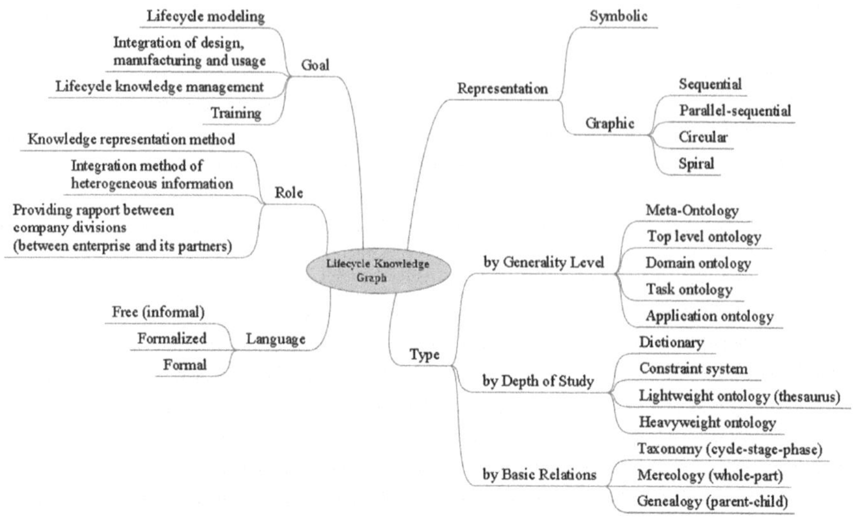

Fig. 3 General lifecycle representation by a mind map

lifecycle. It plans to accomplish an inescapable accessibility of applicable product information anytime in the product lifecycle. Moreover, the idea manages closing information loops between various IT layers, from data obtaining, through middleware and knowledge transformation layers to the business application layer [25]. Kiritsis [26] suggested an ontology-based approach for PLM. Basic formal ontology (BFO) was used on the top level of an ontology for PLM. BFO is a small, upper-level ontology that is designed for use in supporting information retrieval, analysis and integration in scientific and other domains. BFO is a genuine upper ontology. BFO is interesting for functional modeling by its ontological rigor (no overlaps in terminology, strict universal scope), its treatment of processes and the prospect of a first-order logic implementation [23].

In this work, we paid specific consideration to lifecycle ontology regarded as an upper ontology for enterprise engineering. We likewise present the idea of granular lifecycle meta-ontology. This depends on such ideas as granule, level, hierarchy and relations between levels (see [27, 28]). The requirement for lifecycle modeling in granular meta-ontology (contra to conventional singular one) is self-evident. It merits focusing on that no ideal granulation level exists. Sizes of granule are issue oriented and rely upon examination setting. Lifecycle stages can be considered in a more particular way and others, less completely, considering modeling objectives. Here, lifecycle stages are coarse-grained parts that are generally partitioned into lifecycle phases, fine-grained parts, where each phase relates to a particular system's state. Time is an essential asset in lifecycle management. A particular lifecycle feature is its heterochronous character, i.e., unevenness identified with the distinction in temporal criteria and requirements at different stages.

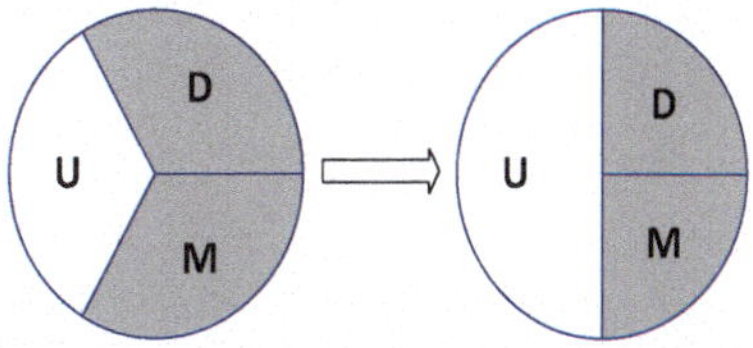

Fig. 4 Product lifecycle circular representation: an illustration of decreasing lead (design and manufacturing) time and incrementing usage period

Above all else, we show the stages of lifecycle in the system of the set-theoretic approach as granules received by segmentation. In general, the product lifecycle consists of three main stages: beginning of lifecycle (BOL), middle of life (MOL) and end of life (EOL) [26]. BOL features design and production phases; MOL contains logistics (distribution), use, service and maintenance phases; and EOL comprises of inverted logistics (collecting), remanufacturing (disassembly, refurbishment, reassembly, etc.), reuse, recycle and disposal phases. We introduce natural symbols for complex systems's lifecycle: D—design; M—manufacturing (production); U—MOL (including logistics (distribution), use, service and maintenance); R—EOL (including reverse logistics, remanufacturing, reuse, recycle, and disposal). We have

$$\mathrm{LC}_1 = \mathrm{D} \cup \mathrm{M} \cup \mathrm{U},\ \mathrm{D} \cap \mathrm{M} = \varnothing,\ \mathrm{M} \cap \mathrm{U} = \varnothing,\ \mathrm{U} \cap \mathrm{D} = \varnothing \tag{1}$$

or

$$\mathrm{LC}_2 = \mathrm{M} \cup \mathrm{U} \cup \mathrm{R},\ \mathrm{M} \cap \mathrm{U} = \varnothing,\ \mathrm{U} \cap \mathrm{R} = \varnothing,\ \mathrm{R} \cap \mathrm{M} = \varnothing. \tag{2}$$

The LC_2 structure (2) communicates the "ecological imperative" of present-day manufacturing being firmly identified with Kimura's lifecycle reversal idea mentioned previously. The primary lifecycle partition LC_1 (1) can be illustrated utilizing the circle segments (see Fig. 4).

It merits focusing on that the lifecycle representation by partition is fairly oversimplified and does not express numerous current interrelations and cooperation connections between partially overlaying phases. Also, this concurrent task empowers exceptionally important features. For instance, at the junction of use and design system specification is appeared. Production technologies should be talked about on the edge of design and manufacturing, while maintenance requires the joint effort of users and manufacturers. Accounting for such factors, we get a circular model of lifecycle with fuzzy limits. For such cases, granulation of lifecycle depends on covering (see Fig. 5). In this way,

$$\mathrm{LC}_1' = \mathrm{D} \cup \mathrm{M} \cup \mathrm{U},\ \text{but}\ \mathrm{D} \cap \mathrm{M} \neq \varnothing,\ \mathrm{M} \cap \mathrm{U} \neq \varnothing,\ \mathrm{U} \cap \mathrm{D} \neq \varnothing. \tag{3}$$

By and large, our approach depends on a relational model of time and interim time primitives. We take a fuzzy extension of Allen's [29] well-known temporal logic to show the connections between lifecycle phases (or lifecycle phases and stages) and

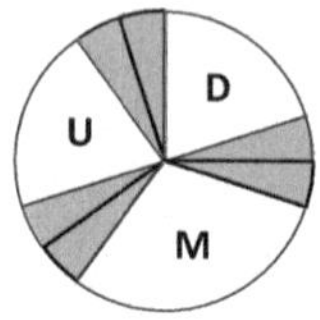

Fig. 5 Circular representation of lifecycle based on covering: the presence of collaborative tasks and fuzzy boundaries between stages

Table 1 A crisp model of main temporal relations between lifecycle phases and stages

Notation	Relations and their inversion	Illustrations	Examples
$r_{1/2}$	Phase a/b precedes (is performed before)/later phase b/a		Design phase is performed before logistics phase
$r_{3/4}$	Phase a/b is adjacent to (immediately precedes)/suits phase b/a		Use phase is performed immediately before service phase
$r_{5/6}$	Phase a/b partially overlaps phase b/a	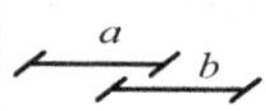	Design phase partly overlaps with production phase
$r_{7/8}$	Phase a resides inside stage s/stage s comprises phase a	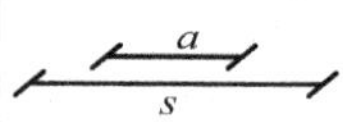	Production phase lies inside BOL stage
$r_{9/10}$	Phase a resides inside stage s, so that their beginning points coincide/Stage s comprises phase a, so that their beginning points coincide	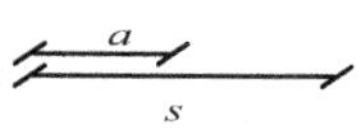	Logistics phase lies inside MOL stage, so that their starting points coincide
$r_{11/12}$	Phase a resides inside stage s, so that their finish points coincide/stage s comprises phase a, so that their finish points coincide	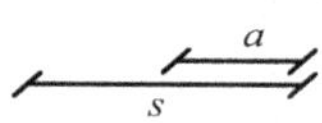	Maintenance phase lies inside MOL stage, so that their endpoints coincide
r_{13}	Phase a coincides with phase b	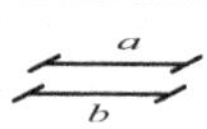	BOL stage coincides with design and production phase

model them. There are primarily two kinds of relations: overlaying and consequence relations (see Table 1).

Fuzzy quantity is defined as a fuzzy arrangement of the real line. Fuzzy quantities are more suitable to describe flexible requirements on lifecycle parts duration.

Let us present a formal model of lifecycle ontologies ONT_{LC} as a fourfold.

$$\mathrm{ONT}_{\mathrm{LC}} = \langle C_{\mathrm{LC}},\ R_{\mathrm{LC}},\ W_{\mathrm{LC}},\ T_{\mathrm{LC}} \rangle, \tag{4}$$

where C_{LC} is the arrangement of concepts refers to lifecycle, R_{LC} is the arrangement of relations between these concepts, Ω_{LC} is the arrangement of actions over concepts and/or relations and T_{LC} is the arrangement of temporal features for lifecycle.

Phases and stages are the fundamental concepts of any lifecycle; thus, the triad below can be taken as the systemic core of lifecycle.

$$\mathrm{ONT}_S = \langle S,\ R_s,\ \Omega_s \rangle, \tag{5}$$

where S is the arrangement of lifecycle phases (stages), R_s is the arrangement of relations between these phases (stages) and Ω_s is the arrangements of actions used on these phases (stages).

It is worth noting that every lifecycle phase can be seen as an interval entity $s = [a^-, a^+]$, where a^- is a beginning point and a^+ is a finish point of the interval. A fuzzy interval extends the understanding of an interval: It is a special type of fuzzy quantity which is represented by means of a convex fuzzy real line subarrangement. As a special instance, we have

$$\mathrm{ONT}_{S1} = \left\langle S, <_f, \approx_f \right\rangle, \tag{6}$$

where $<_f$ is a fuzzy strict linear order relation which is non-reflexive, transitive, linear and asymmetric and $\approx_f$ is a fuzzy concurrent relation, i.e., fuzzy symmetric, reflexive relation. Speaking more generally, it is possible to use the linguistic variable "Time" with a linguistically ordered term arrangement as {almost simultaneously, a bit later, later, much later, very much later}.

This portrayal of product lifecycle disentangles the examination of concurrent engineering issues and the improvement and adjustment of fitting artificial intelligence methods and instruments. Dubious and loose knowledge on a products' structure and its production technology, impaired design and simulation models requires a rehashed going of manufacturing stages took after by utilization benchmarks keeping in mind the end goal to confirm how starting specifications are fulfilled.

4 Conclusion

A very important way of developing EE paradigm consists in implementing lifecycle knowledge management on the basis of ontological approach for further enterprise interoperability. Using ontology allows breaking an important semantic incompatibility barrier to interoperability. Here, specific consideration is paid to lifecycle ontology regarded as an upper ontology for enterprise engineering. For the most part, our approach depends on a relational time model and interim time primitives.

We take an Allen's temporal logic fuzzy extension to model the connections between lifecycle phases and stages.

This work has been supported by DAAD and Russian Ministry of Education (Project 9.10005.2017/5.2).

References

1. Panetto, H., Zdravkovich, M., Jardim-Goncalves, R., Romero, D., Cecil, J., & Mezgar, I. (2016). New perspectives for the future interoperable enterprise system. *Computers in Industry*. Special Issue: "Future Perspectives on Next Generation Enterprise Information System: Emerging Domains and Application Environments" *79*, 47–63. Elsevier.
2. Chen, D., & Vernadat, F. (2004). Standards on enterprise integration and engineering—a state of the art. *International Journal of Computer Integrated Manufacturing (IJCIM), 17*(3), 235–253.
3. INTEROP. (2007). Enterprise interoperability-framework and knowledge corpus—Final report, Research report of INTEROP NoE, FP6—Network of Excellence—Contract n 508011, Deliverable DI.3.
4. Liles, D., Johnson, M. E., Meade, L. M., & Ryan, D. (1995). Enterprise engineering: A discipline? In *Society For Enterprise Engineering Conference Proceedings*. (vol. 6).
5. Tarassov, V. B. (2002). *From multi-agent systems to intelligent organizations*. Moscow: Editorial URSS. (in Russian).
6. Dietz, J. (2006). *Enterprise ontology—theory and methodology*. Berlin: Springer.
7. Dietz, J., Hoogervorst, J., et al. (2013). The Discipline of enterprise engineering. *International Journal of Organisational Design and Engineering, 3*(1), 86–114.
8. Martin, J. (1995). *The great transition: Using the seven principles of enterprise engineering to align people*. New York: Technology and Strategy. American Management Association.
9. Vernadat, F. (1996). *Enterprise modeling and integration: Principles and applications*. London: Chapman and Hal.
10. Guryanova, M. A., Efimenko, I. V., & Khoroshevsky, V. F. (2011). Ontological modeling economy of enterprises and branches of modern Russia: Part2. In *World research and development: An analytical review. Preprint WP7/ 2011/08 (part 2)*. State University Higher School of Economics, Moscow.
11. TOVE ontology project, http://www.eil.utoronto.ca/enterprise-modelling/tove/, last accessed 2017/11/30.
12. Uschold, M., King, M., Morales, S., & Zorgios, Y. (1998). The enterprise ontology. *The Knowledge Engineering Review, 1*(13), 31–89.
13. http://ncorwiki.buffalo.edu/index.php/Industry_Ontologies_Foundry:_ASME_Workshop_2017. Last accessed 30 November, 2017.
14. System of Systems engineering (2008). Innovations for the twenty-first century. In M. Jamshidi (Ed.). Wiley, New York.
15. Fedotova A. V., Tarassov V. B., Mouromtsev D. I., & Davydenko I. T. (2016). Lifecycle ontologies: background and state-of-the-art. In *Proceedings of the 5th International Conference on Intelligent Systems and Applications (INTELLI'2016, Barcelona, Spain, November 13–17, 2016)*, (pp. 76–82). IARIA XPS Press, Copenhagen.
16. Kimura, F., & Suzuki, H. (1996). Product life cycle modeling for inverse manufacturing. In F. L. Krause & H. Hansen (Eds.), *Proceedings of IFIP WG 5.3 International Conference on Life Cycle Modeling for Innovative Products and Processes (PROLAMAT'95, November 29-December 1, 1995)*. (pp. 81–89). Berlin: Springer.
17. Saaksvuory, A., & Immonen, A. (2008). *Product lifecycle management*. Berlin: Springer.
18. Stark, J. (2011). *Product lifecycle management: 21st century paradigm for product realization* (2nd ed.). London: Springer.

19. Jun, H.-B., Kiritsis, D., & Xirouchakis, P. (2007). Research issues on closed-loop PLM. *Computers in Industry, 58,* 855–868.
20. Kadiria, S., Grabotb, B., Thoben, K.-D., Hribernik, K., Emmanouilidise, C., Cieminski, G., et al. (2016). Current trends on ICT technologies for enterprise information systems. *Computers in Industry, 79,* 14–33.
21. Camarinha-Matos, L. M., & Afsarmanesh, H. (2007). A comprehensive modeling framework for collaborative networked organization. *Journal of Intelligent Manufacturing, 18,* 529–542.
22. Tarassov, V. B. (2001). Special session on intelligent agents and virtual organizations in enterprise. In Z. Binder (Ed.), *Proceedings of the 2nd IFAC/IFIP/IEEE Conference on Management and Control of Production and Logistics 2000 (MCPL'2000, Grenoble, France, July 5–8, 2000).* (vol. 2, pp. 475–478). Amsterdam: Elsevier Science Publishers.
23. Tarassov, V. B., Kashuba, L. A., & Cherepanov, N. V. (1994). Concurrent engineering and AI methodologies: Opening new frontiers. In *Proceedings of the IFIP International Conference on Feature Modeling and Recognition in Advanced CAD/CAM Systems (Valenciennes, France, May 1994)*, (Vol. 2, pp. 869–888).
24. Mal'tsev, A. I. (1973). *Algebraic Systems*. Berlin: Springer.
25. Hriberni, K., Cassina, J., Rostad, C.C., Thoben, K.-D., & Taisch, M. (2012). Potentials of Item-level PLM and Servitization in the Leisure Boat Sector. In *Proceedings of the 5th International Conference on Through-life Engineering Services (TESConf 2012).*
26. Matsokis, A., & Kiritsis, D. (2010). An ontology-based approach for product lifecycle management. *Computers in Industry, 61,* 787–797.
27. Tarassov, V. B., Fedotova, A. V., Stark, R., & Karabekov, B. S. (2015). Granular meta-ontology and extended allen's logic: Some theoretical background and application to intelligent product lifecycle management systems. In *Proceedings of the 4th International Conference on Intelligent Systems and Applications (INTELLI'2015, St.Julians, Malta, October 11–16, 2015),* (pp. 86–93). IARIA XPS Press, Copenhagen (2015). ISBN: 978-1-61208-437-4.
28. Zadeh, L. A. (1997). Toward a theory of fuzzy information granulation and its centrality in human reasoning and fuzzy logic. *Fuzzy Sets and Systems, 90,* 111–127.
29. Allen, J. F. (1983). Maintaining knowledge about temporal intervals. *Communications of the ACM*, *26*, 832–843.

Part VII
Block Chain and Decentralized Approaches

The Development of Smart Contracts for Heterogeneous Blockchains

Henry Syahputra and **Hans Weigand**

Abstract The advanced development of blockchain technologies at present is to have external applications that interact with the native blockchain. A smart contract allows us to insert business logic which can add transactions and share them with multiple parties. This paper discusses the development process of a smart contract platform that aims to generate smart contracts for heterogeneous blockchain technologies. We start the process of creating a blueprint design, a datalogical stereotype that employs the ontology design from a Resources-Event-Agent (REA) perspective (McCarthy in The Accounting Review, LVII(3) (1982) [1]). With the modeling approach we are using in this paper, Unified Modeling Language (UML) and Object Constraint Language (OCL), we implement the workflow and algorithm in a supply chain demo sample [2].

Keywords Smart contracts · Blockchain · UML · OCL · Resources-Event-Agent (REA)

1 Introduction

Contracts, transactions, and records are among the defining structures in our economic, legal, and political systems [3]. They protect assets and set organizational boundaries. However, these tools and the bureaucracies formed to manage them have not kept up with the economy's digital transformation. Blockchain promises to solve the problem by recording the transactions between parties efficiently in a verifiable and permanent way [4]. The ledger itself can also be programmed to trigger transactions automatically [5] with what we know as smart contracts.

H. Syahputra (✉)
Yishun Industrial Street 1, Singapore 768160, Singapore
e-mail: henry.syahputra@gmail.com

H. Weigand
Tilburg University, P.O. Box 90153, 5000 LE Tilburg, The Netherlands
e-mail: h.weigand@uvt.nl

K. Popplewell et al. (eds.), *Enterprise Interoperability VIII*, Proceedings of the I-ESA Conferences 9, https://doi.org/10.1007/978-3-030-13693-2_19

Although we share the enthusiasm for the blockchain potential, many business practitioners worry about this hype. It is not just security issues [6] that concern them. All the various technology stacks of a blockchain [7], various blockchain technologies, and the distributed governance [8] involve an ecosystem of agents, policies, services, and this makes it challenging to develop blockchain applications and smart contracts in particular. The direct impact of one code into valuable assets stored in blockchain makes it vulnerable to manipulation attempts, especially if the networks openly operate. A restricted operating network, however, has also unexpected failures [9, 10]. Incidents, malfunctioning, and unexpected results of smart contracts have been reported [11–13].

The design of smart contracts aims to provide developers with a model for creating flexible and reusable applications. Security, evaluation, and contribution are among the critical aspects of smart contract development [14]. Blockchain from the commitment-based ontology perspective [15, 16] differentiates into three levels, Essential, Infological, and Datalogical. In this approach, smart contracts at least contain (1) goals, (2) commitments, (3) conditions, (4) actions, and (5) timing. We formulate smart contracts into a standard modeling approach with MDA to perform the generation of smart contracts implementation with predefined parameters [15].

In this paper, we propose a workflow and framework to develop a smart contract in a platform-independent way. We take the use case of the heterogeneous environment and scenarios which resulted in how to come up with the ideal composition of smart contract from a practical point of view. This enables the developers to adopt a blockchain model-driven approach, focusing on the logical design of a blockchain application rather than the complexity of technical implementation of the various blockchain technologies and stacks.

To achieve a better understanding, in the Appendix section [2], we use a supply chain example to explain the process of creating smart contract [17]. We illustrate (Fig. 2) how a company is transformed into blockchain by employing our stereotype. We generate the code skeleton for both Ethereum and Hyperledger fabric as our blockchain target platforms by using the workflow and framework we have developed in this paper.

2 Background

In the software development cycle, models are being used and models are the primary key in Model Development Architecture [18]. The software development process starts from the definition of a business process which later transforms into a business process model while creating traceability links to the next process, analysis and design. The latter process will define how the code generation work can be started (Table 1).

We will map the developed MDA design [2] into a runnable smart contract code [19] in our development environment using (1) Unified Modeling Language that provides standard mechanisms [20] and simplified metamodel for developers [21].

Table 1 Comparison of Ethereum and Hyperledger

Characteristic	Ethereum	Hyperledger
Platform description	Generic blockchain platform	Modular platform
Governance	Ethereum developers	Linux foundation
Mode of operation	Permissionless, public, or private	Permissioned, private
Consensus	Mining based on proof of work Ledger level	Allow transaction-level consensus types

(2) OCL, a formal language used to describe expressions on UML [14, 22] with pre- and post-meta-level conditions. (3) Acceleo, a code generator implementation of the OMG's model-to-text (M2T) specification [11]. (4) Blockchain Network, Fabric Hyperledger and Ethereum that are highly flexible [3, 23].

3 Conceptual Collaboration Context

A stereotype is a class that defines how a metaclass is being extended as part of a profile [24]. A Stereotype has properties that are referred to as tag definitions. Model element's value properties that are applied by stereotype are referred to as tagged values. In class-based redundancy [25], stereotypes qualify specific classes and define a component of a redundant structure. Tagged values and parameters that are assigned to elements of the UML can be prescribed by stereotypes which are later to be associated with UML elements.

The complete list of tagged values required for all types of elements can be found in [26]. Our UML diagram uses MDA with commitment-based ontology [15] approach to defining a PIM. Once the PIM-to-PSM transformation is realized, the specific code of the target platform is generated by using Acceleo M2T (Model to Text).

3.1 Essential Interface

The Essential or business layer is concerned with what is created directly or indirectly by communication [15] and is similar to CIM. To get the general abstraction for the smart contracts, we adopt the previous research [16], Enterprise Ontology with the Business Ontology of which the essential interface is as depicted in Table 2. The developed REA model [1] is used as a domain ontology for accounting.

The Operations of the essential interface (Table 2) consist of (1) totalSupply(): an operation to get the total token supply from the related blockchain. (2) balanceOf(), the operation that aims to get the account balance related to a given blockchain address. (3) transfer(): the operation to send economic events to a given destination that relates to one's address. (4) transferFrom(): the operation to get an economic

Table 2 Essential interface

Operations	Tagged values	Details
totalSupply(...)	*_supply: uint256*	Get the total token supply
balanceOf(...)	_owner: address *balance: uint256*	Get the account balance of account with an address
transfer(...)	_to: address _value: uint256 *success: bool*	Send value amount to an address
transferFrom(...)	_from: address _value: uint256 *success: bool*	Send value amount from address to an address

event from a given source of origin's address in the blockchain. We differentiate the transfer into origin and destination as a way to capture the pre- and post-condition of an address in a real-time manner.

3.2 *Infological Interface*

The infological interface is concerned with value transfer between objects [15]. Logics of value transfer functions agreed by contract participants are the foundation of commitments at infological layer. The representation of business exchange [27] contains (1) goals, (2) commitments, (3) conditions, and (4) actions. While [15] added timing, the infological interface here uses a changed(...) operation as a tagged values pointer to record the time changes. The changed() operation refers to the blockchain platforms API, Ethereum [28] and Hyperledger [29] that handle the time for the block creation.

In this infological interface (Table 3), we define several operations: (1) reserve(): to use the infological, we have to reserve an account for it. (2) owner(): once an account is reserved, this operation will set who owns the account. (3) transfer(): to transfer assets, we change it by the account of the owner with this operation. (4) address(): this is the primary address associated with an account that is similar to an A record in traditional DNS. (5) content(): operations related to information associated with the owner. (6) registrar(): any associated parties/peers related with sub-registrar handled by this operation. (7) disown(): to release current control of an associated account. (8) changed(): a value pointer to remark the time changes. The owners of a contract have mutual commitments, and it has to be reserved by pointing to an account owner first. Ownerships are transferable by referring to account owner. To record changes in commitments, we use changed() operations. The commitment has to be balanced [30] and since it is part of the smart contract abstraction, ownership and disowns are very crucial for this particular reason.

Table 3 Infological interface

Operations	Tagged values	Details
reserve(...)	_account: string _success: bool	Reserves an account and sets its owner to you if it is not yet reserved
owner(...)	_account: string _ref: address	Get the owner of a particular account
transfer(...)	_account: string _newOwner: address	Transfer ownership of an account
addresss(...)	_account: string _address: address _ref: address	The primary address associated with an account (similar to an A record in traditional DNS)
content(...)	_account: string _content: bytes _ref: bytes	Associated content of an owner
registrar(...)	_account: string _subRegistrar: address _ref: address	Sub-registrar associated with the given account
disown(...)	_account: string	Released current control of an associated account
changed(...)	_account: string _hash_account: bytes	Pointer to remark changes

4 Datalogical Profile

The smart contracts context is constructed at the datalogical level [31] where it is created, operated, and mapped into a targeted blockchain platform. The process starts from model initiation to code generation that will be running on the targeted blockchain. In Table 4, we define ≪Datalogical≫ as a UML profile that provides generic extension for customizing UML models. This profile consists of stereotypes that have their tag definitions applied to UML specific model elements.

Each row in the ≪metaclass≫ column correlates with ≪stereotypes≫ column defined in the right column. The ≪metaclass≫ and the corresponding ≪stereotypes≫ are:

- **Operation**: Applicable stereotypes with this element are: (1) Event, (2) Function, (3) Modifier, and (4) Constructor. All these stereotypes relate to the targeted programming language and the smart contracts code.
- **Class** model element manages objects and features classifications as formal template parameter. Applicable stereotypes are: (1) Struct, (2) Contract, and (3) Library.
- **Constraint** element manages the semantics of elements by applying ParameterModifier stereotype.
- **Enumeration** element imports all non-private members belonging to a package from the target blockchain by applying FunctionType stereotype.

Table 4 Datalogical property

≪Metaclass≫	≪Stereotype≫	Tagged values
Operations	Event Function Modifier Constructor	Specify an operation can be exposed as a formal template parameter and provided as an actual parameter in template binding
Class	Struct Contract Library	Classify set of objects and features that characterize the structure and behavior of that objects
Constraint	ParameterModifier	Declare the semantics of Elements
Enumeration	FunctionType	Return modification of functions
Interface	Datalogical	Declare coherent services by BehavioredClassifiers via InterfaceRealizations
PackageImport	Import	Imports all non-private members of a package into namespace
Parameter	Indexed	Passing information from/to invocation of BehavioralFeature
Property	Delegate GetterSetter Const	Represent GetterSetter, Delegate and Const attribute including association end
ValueMapping	Mapping KeyMapping	Represent Key and Value mapping

- **Interface** element connects with APIs of blockchain platforms (i.e., Hyperledger Fabric API, Ethereum API) by applying the Datalogical stereotype.
- **PackageImport** element connects with different kinds of packages by applying import stereotype and importFilter tagged values.
- **Parameter** element passes information from/to invocation of behavioralFeature by applying the indexed stereotype.
- **Property** element changes the semantics of how the property behaves. By applying Getters and setters stereotype with property element, we can allow different access levels of public properties while still being protected.
- **ValueMapping** element represents key and value from the various target blockchain platform by applying Mapping and KeyMapping stereotypes.

5 Model Framework Creation (Model to Text)

Model transformation is a compilation process which transforms a source model into a target model [32]. We adopt Obeo's acceleo model transformations [33] and code generation script tools that make the code generation trivial by template editors [34].

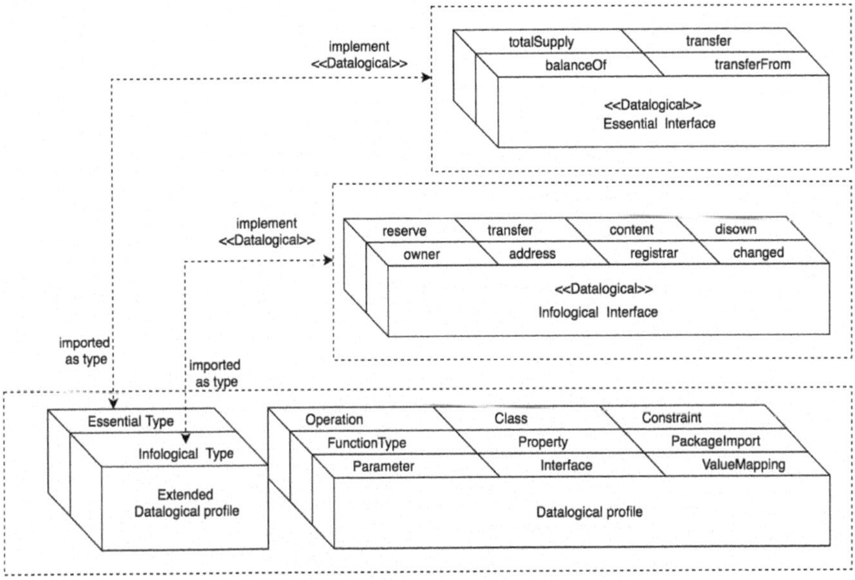

Fig. 1 A blockchain toolbox framework [19]

The top-level structure shows the model transformations of class and interface in the Datalogical profile (Fig. 1). In this diagram, we can see the connection between each interface class: Both the Essential and Infological interfaces are connected to the core Datalogical by applying the ≪Datalogical≫ profile. Simultaneously, to access these interfaces, we apply the Essential and Infological types from the extended Datalogical profile. In Appendix [2], we implement these relations in the supply chain demo example wherein the Quotation contract, the ≪Essential≫ type is applied by TransactionID and FreightTerms, while the other attributes IssueDate and PaymentTerms implement ≪Infological≫ type.

5.1 *Service Interface*

When we create a class diagram to model a solution, we have classes, interfaces, properties, and operations. Service interface aims to classify and associate these classes with Essential, Infological, and Datalogical. In algorithm 1, The SERVICE_INTERFACE pseudocode supplies an interface for transforming UML Model into targeted blockchain code by using Acceleo scripts tools.

We adopt the UML Model as an input responsible for populating dependencies according to relationships usage and association at each particular class. By applying this pseudocode, we take parameters, operations, and attributes to generate a structure composed by the attributes and pointers at the associated classes. For all related members of the datalogical stereotype (Table 4), the pseudocode will iteratively generate the respected stereotype interface through all classes inside the UML model.

In the case where the GENERATE_BLOCKCHAIN_PACKAGE and GENERATE_BLOCKCHAIN_INTERFACE templates are called (algorithm 2 and 3), it will create the PSM-related code implementations.

```
Algorithm 1: Interfacing UML Service
    template SERVICE_INTERFACE
        for all class ∈ ≪Datalogical≫ do
            If all in parameters not empty  then
               GET_ESSENTIAL()
               GET_INFOLOGICAL()
            end if
        end for
        for all class ∈ Datalogical  do
           GENERATE_DATALOGICAL_STEREOTYPES()
       end for
```

```
Algorithm 2: Generating Blockchain Package
    template GENERATE_SMART_CONTRACT_PACKAGE
        for class ∈ packageOfContracts do
           GENERATE_CONTRACTS_DEPLOYER()
           GET_ALL_PUBLIC_CLASS_FUNCTIONS()
           GENERATE_DIRECTORY_STRUCTURE()
        end for
```

```
Algorithm 3: Generating Blockchain Interface
     template GENERATE_SMART_CONTRACT
         for classObject ∈ ClassStereotype do
            GENERATE_STEREOTYPE_INTERFACE()
         end for
```

5.2 *Code Generation for Target Blockchain Smart Contracts*

The main purpose of code generation is to generate (1) the smart contract package that consists of source code with deployment script and (2) smart contract platform-specific implementation code. In algorithm 2, the GENERATE_BLOCKCHAIN_PACKAGE template shows the pseudocode transformation for the targeted blockchain platform into its targeted language. The template processes iteratively the inputted UML model diagram; if a class is part of package contracts of the targeted blockchain platform, it will create factory class by calling the PRINT_CLASS_FACTORY. Moreover, it will create the script for the smart contracts package configuration, GENERATE_BUILD_SCRIPT and RUN_BUILD_SCRIPT. This script will be executed by peers and is accessible via REST client (implemented by i.e., Ethereum and Hyperledger as client code) created by GENERATED_REST_CLIENT.

The GENERATE_SMART_CONTRACT template produces source code that interacts with the core targeted blockchain. In algorithm 3, the operation parameters will be examined with (1) GET_INTERFACE_PREFIX(), (2) GET_NETWORK_CODE() and (3) GET_UML_INTERFACE_SERVICE_CODE, that deals with the targeted blockchain operations (Fig. 2).

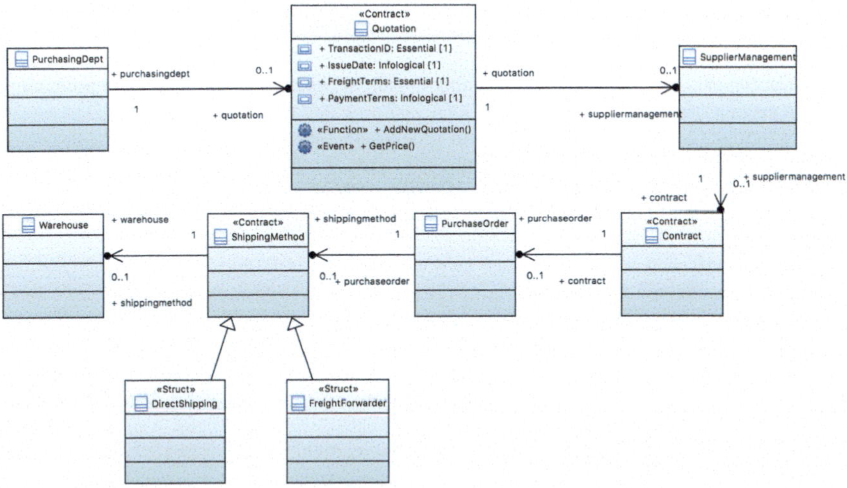

Fig. 2 A supply chain smart contract sample for heterogeneous blockchain [2]

6 Running the Generated Model

We want to use the generated source code at the target blockchain platform. Here, we employ two different blockchain platforms as our target code generation, (1) Ethereum: Generic permissionless blockchain platform and (2) Hyperledger: Modular permissioned blockchain platform. In the previous section, we have applied the model-to-text transformation that generates source code for Ethereum and Hyperledger fabric. We run the Ethereum-generated.sol code using solidity remix browser. As for the Hyperledger fabric, the generated codes are packaged into maven structured directories [2].

7 Discussion

A methodology workflow for UML modeling to create smart contracts for heterogenous blockchains, using REA ontology, commitment-based smart contracts, and MDA guidelines, is illustrated in this paper. Compilable targeted blockchain platform code resulted from the detail process of model transformation. We are using open-source tools, Eclipse modeling tools that include Papyrus and Acceleo in the compilable targeted blockchain platform code processes. Thus, we propose algorithms and framework (Algorithm 1, 2, 3) to simplify the development process for heterogeneous blockchain platforms. In an upcoming paper, we expect to involve the study of orchestrating the modular blockchain network and incorporate it into the methodology workflow presented in this paper.

References

1. McCarthy, W. (1982). The REA accounting model: A generalized framework for accounting systems in a shared data environment. *The Accounting Review*, *LVII*(3).
2. https://gitlab.com/Syahputra/SupplyChain-Blockchain-Demo-Tilburg.
3. Lansiti, M., Lakhani, K., The Truth about blockchain. *Harvard Business Review*, 118–127.
4. Weigand, H. (2009). *Value encounters—Modelling and analyzing co-creation value.*
5. Swan, M. (2015). Blockchain thinking: The brain as a dac (decentralized autonomous organization). In *Texas Bitcoin Conference* (pp. 27–29).
6. Bitcoin's $460 million disaster. https://www.wired.com/2014/03/bitcoin-exchange/.
7. Herrera-Joancomartí, J., & Pérez-Solà, C. (2016). *Privacy in bitcoin transactions lecture notes in computer science* (Vol. 9880). Cham: Springer.
8. Norta A. *Establishing distributed governance infrastructures for enacting cross-organization collaborations.*
9. Corbett, J. C., et al. (2013). Spanner: Google's globally distributed database. *ACM Transactions on Computer Change*, *31*(3), 8.
10. Baker, J. et al. (2011) Megastore: Providing scalable, highly available storage for interactive services. In *CIDR* (pp. 223–234).
11. EtherDicesmartcontract is down for maintenance. http://bit.ly/2ikzAsM.
12. KingOfTheEtherThrone smart contract. http://bit.ly/2jwzAJN.
13. GovernMental's 1100 ETH payout is stuck. http://bit.ly/2zEC3rP.
14. Luu, L., Chu, D., Saxena, P., Olickel, H., & Hobor, A. *Making smart contracts smarter.*
15. De Kruijff, J., & Weigand, H. (2017). An ontology for commitment based smart contracts.
16. De Kruijff, J., & Weigand, H. (2017). *Towards a blockchain ontology.*
17. Omran, Y., Henke, M., Heines, R., & Hofmann, E. (2017). *Blockchain-driven supply chain finance: Towards a conceptual framework from a buyer perspective.*
18. OMG-MDA. (2003) MDA Guide revision 1.0, OMG.
19. https://gitlab.com/Syahputra/BlockchainToolbox-Eclipse-Update-Site.
20. Bondavalli, A., Majzik, I., Mura, I., *Automatic dependability analysis for supporting design decisions in UML.*
21. Cariou, E., Marvie, R., Seinturier, L., & Duchien, L. *OCL for the specification of model transformations contracts.*
22. Pollet, D., Vojtisek, D., & Jezequel, J. M. *OCL as a core UML transformation language.*
23. Valenta, M., & Sandner, P. *Comparison of Ethereum, Hyperledger Fabric and Corda.*
24. UML Stereotype. https://www.uml-diagrams.org/stereotype.html.
25. Xu, J., Randell, B., Rubira-Calsavara, C. M. F., & Stroud, R. J. (1994). *Towards an object-oriented approach to software fault-tolerance*. PDCS-2 Technical Report.
26. Bondavalli, A., Majzik, I., & Mura, I. (1998). From structural UML diagrams to Timed Petri Nets. European ESPRIT Project 27439 HIDE, Deliverable 2, Section 4.
27. Szabo, N. (1997). Formalizing and securing relationships on public networks. *2*(9).
28. Ethereum Web3 J: http://web3j.readthedocs.io/en/latest/transactions.html?highlight=time.
29. Hyperledger: https://godoc.org/github.com/hyperledger/fabric/core/chaincode/shim.
30. Chopra, A. K., Singh, M. P., Oren, N., Miles, S., Luck, M., Modgil, S., et al. (2011). Analyzing contract robustness through a model of commitments. *LNCS, 6788,* 17–36.
31. De Kruijff, J., & Weigand, H. (2017). Understanding the blockchain using enterprise ontology. In *29th International Conference on Advanced Information Systems Engineering*, (pp. 29–43). Cham: Springer.
32. De Oliveira Rodrigues, A. W., Guyomarc'H. F., & Dekeyser, J.-L. (2011). *An MDE approach for automatic code generation from marte to openCL RR-7525*, INRIA. 2011, p. 27.
33. Acceleo: https://wiki.eclipse.org/Acceleo.
34. Obeo (2010). Acceleo—Model to text transformation.

Blockchain in Decentralized Local Energy Markets

Andy Hermann, Tobias Teich, Stephan Kassel, Daniel Kretz, Tim Neumann, Sven Leonhardt and Sebastian Junghans

Abstract Within the context of implementing decentralized energy markets, it is necessary to integrate existing structures into this process of change and let established participants play an active role in it. The problem of the conversion is the way how the current, and sometimes still rigid, accounting systems can be adjusted to a flexible and adaptable energy market. It should be noted that today's ordinary consumer must have the ability to feed energy into the grid, thereby taking a prosumer role. This article describes a concept developed within the context of the WindNode project to address this problem. The concept includes an approach focused on billing processes to realize a decentralized energy market. Attention was paid to scalability, flexibility, business process modeling and security. The result describes an approach to how any participant can be integrated into such a network, whether as an ordinary consumer or as a prosumer.

Keywords Blockchain · Decentralized energy market · Energy exchange · Ubiquitous networks · Ubiquitous infrastructures

1 The WindNode Project

1.1 Context of Blockchain Usage

Since 1 January 2017, the Westsächsische Hochschule Zwickau (WHZ) participates in the WindNode project consisting more than 50 partners and more than 30 associated partners. The biggest challenge in switching to a predominant share of renewable energy, generated by wind and sun, is to use and store it in a useful way. The feasibility of such a realization has to be demonstrated as part of the WindNode project within north-eastern Germany.

A. Hermann (✉) · T. Teich · S. Kassel · D. Kretz · T. Neumann · S. Leonhardt · S. Junghans
Westsächsische Hochschule Zwickau, Dr.-Friedrichs-Ring 2A, 08056 Zwickau, Germany
e-mail: Andy.Hermann@fh-zwickau.de

K. Popplewell et al. (eds.), *Enterprise Interoperability VIII*, Proceedings of the I-ESA Conferences 9, https://doi.org/10.1007/978-3-030-13693-2_20

The requirement for such an enterprise is an efficient and capable information and communication infrastructure. This provides the participants of such a network with the tools to exchange information and thus perform network stabilization. Therefore, active participants, such as heat and energy storage, electric vehicles, cold stores or smart home applications, can contribute to shape and enable the energy revolution.

Within nine sub-projects, innovative applications are implemented as demonstrators at all levels of energy systems. After testing, the various subsystems should be merged into an overall model. Thus, the energy system is presented in a tangible way and those who are interested in this topic can be convinced by the future-oriented system. With the support of the involved federal states' politicians, this project shall take the pioneering role of the region in terms of energy supply [1].

Within the professorship of "Electrical Energy Technology/Regenerative Energies," the Faculty of Electrical Engineering deals with the implementation of a practically usable development of energy storage systems, smart grid solutions as well as the simulation and evaluation of transient processes in electrical energy supply networks [2].

Among other things, the WHZ is working on the implementation of a decentralized energy market with energy storage and linked e-mobility within the context of the project. Another aspect of the subproject at the WHZ is the implementation of energy exchange. The question that had to be asked was "how can the billing pro-cesses be presented and how can these be tailored to a decentralized energy market?"

For this purpose, a concept based on blockchain technology and smart contracts was developed. The challenge was the way the technical components of the blockchain technology could be applied to the application-oriented infrastructure of energy consumers and producers.

1.2 Introduction to Blockchain

The most known blockchain applications are in the field of digital currencies. The properties of blockchain make the technology reliable and safe, especially in such a critical area. However, blockchain does not come alone into play on this mission. It requires even more technical principles, which in their combination makes success. When designing the decentralized energy market, therefore, only the required technologies are explained and their purpose is justified.

The fact that the use of blockchain can be transferred to the energy market and the associated energy and value shifts are also shown by the fact that RWE is also active in this area, especially with regard to the billing of charging stations [3].

One of the most important features is the verification of the accuracy of the data based on their integrity. In a blockchain network, there are several subscribers, and it is not known who or what exactly such a subscriber is, and each of these subscribers has an account called node in usual blockchain applications. These nodes contain all the data. The data are transactions or transaction history. The content of these transactions is, for example, a sum of money. In a certain period of time (e.g., every minute), all transactions made are summarized, checked for consistency, and a so-called block is formed. This block is provided with a hash value, so-called a checksum, which verifies the correctness of the data. So everyone can read the data, and the hash value is generated automatically. Now, if a block is presented whose hash value can't be confirmed by a node owner, it is wrong data. Once such a block has been acknowledged as correct, it is added to the transaction history in each node. The special feature of the hash method is that the hash value of an earlier block is included in the hash value calculation of the current block. This means that the current hash value is indirectly linked to all preceding ones, hence the term blockchain. Due to the redundancy of the data, each node has a representation of it, and it is almost impossible to fake it. At least 50% of the nodes would need to be compromised simultaneously, which in this project is hampered by the fact that an existing VPN network exists. In addition, all data would have to be replaced from the date from which the data are forged [4].

There are different types of nodes. There are those who have only one account and can initiate transactions and those who generate the blocks. In a decentralized energy market, the latter allowed them to authorize transactions to be made, to check them, to generate blocks and then present the status of the accounts to the network.

The generation of transactions, for example, the transfer of money to another account, can only be performed in the network by verified subscribers. For this purpose, signatures are used. Thus, the sender is always identifiable. Based on the transaction history, in which each participant can view and check, it is understandable whether the sender of the transaction even has the necessary funds, i.e., the money, to transfer someone a corresponding amount. If all of these conditions are met, the transaction can be added to a new block.

2 Usage of Blockchain

2.1 Technical Components

Each participant in the network, which can be assigned to four essential types, receives a node. The different kinds of participants are consumers, owners of power-generating devices, owners of energy storage devices and network operators. Because of practical reasons, the real implementation within the project shall be based on logical nodes. The administrative responsibility lies with the network operator, who is the only one entitled to generate blocks. Thus, the real redundancies only exist in instances of the network operators nodes.

Referred to the decentralized energy market, the transaction values are energy quantities, which are offset against real monetary values at a specific point in time. Each real physical participant must logically possess a node to generate transactions.

For proper transaction processing, the following data should be contained in a transaction:

- Transaction ID
- Name of the sender node
- Role of the sender node
- Name of the recipient node
- Role of the recipient node
- Transaction Type
- Tariff at the time of the transaction
- Start time or end time of the transaction
- Meter reading
- Type of energy source (CHP, PV, …)
- Additional optional data depending on transaction type and medium
- Additional data for smart contracts

For the logical generation of energy in the network, there must be an administrative system. By generating blocks, it credits the amount of energy to the respective instances of the energy-generating nodes. The administrative system must be associated with technical devices, especially electricity meters.

By defining the system just once and having a central administrative instance for the design of smart contracts, such a network basically forms a virtual power plant (VPP). This VPP is arbitrarily scalable and therefore remains open to additional participants in the network at a later time. Furthermore, it is adaptable and can be transferred to other urban areas (Fig. 1).

The administrative system acts as an interface between the outside world and the blockchain network. In this way, a reaction to signals from the outside and a legally binding settlement of purchase and sales processes can be implemented without any problems. Thus, the business logic is in the smart contracts and can be adjusted depending on the conditions (Fig. 2).

2.2 *Practical Applicability*

If energy from an energy-generating device is fed into the grid, this must be reported to the administrative system, which credits this energy to the corresponding node. This means that the new amount of energy is in the network and has a concrete owner. Physically, this energy is conducted in an energy store. Therefore, depending on the billing model, a transaction is created, which gives the owner of the node, which is representative of the energy storage, remuneration for the storage of the electricity.

Mobile energy storage is treated like an ordinary node, which has an owner. If energy is supplied to the network, corresponding transactions are generated, which place the amount of energy to the node of the recipient and credit the sender appropriate compensation. If energy is withdrawn, opposite transactions are carried out correspondingly. From a logical point of view, it is irrelevant from where the mobile memory executes these energy transfers.

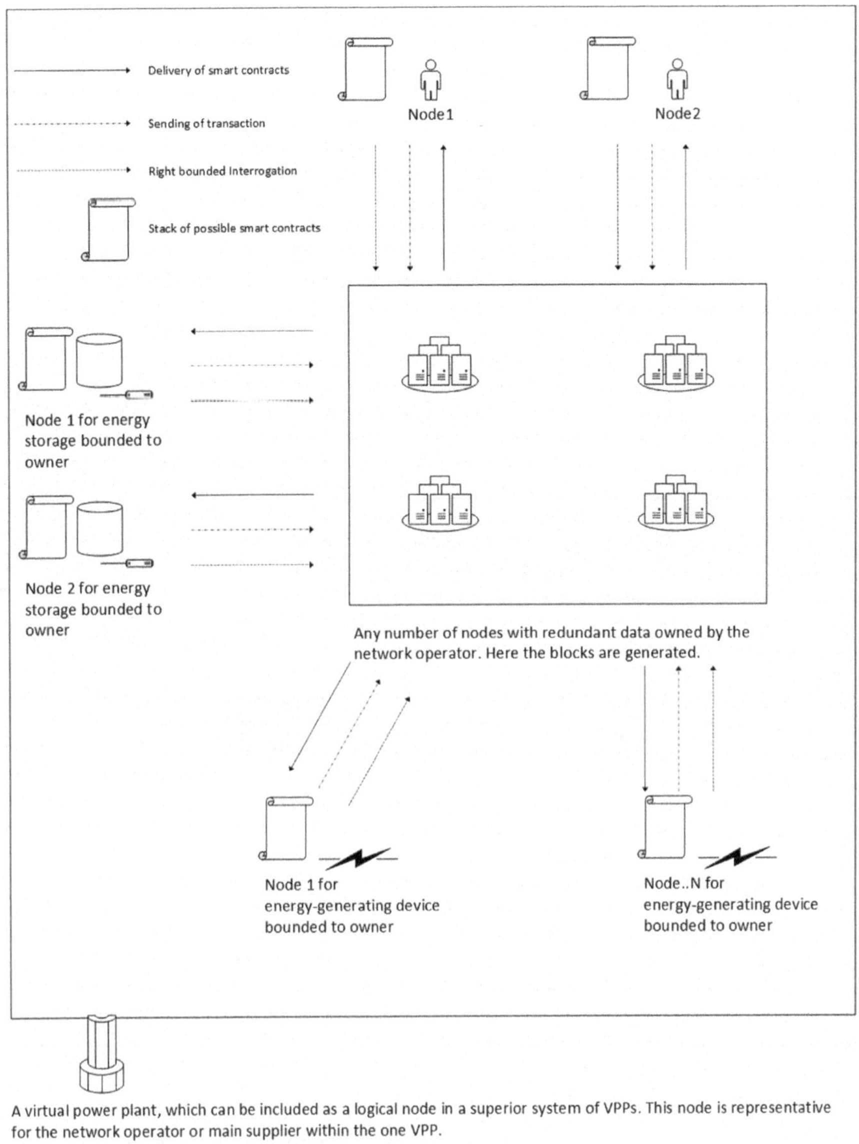

Fig. 1 Communication structure on the basis of blockchain

The logical abstraction of the transaction medium allows a transfer to other media. Thus, this concept could be transferred to thermal energy. Therefore, the entire supply structure could be mapped and billed in one and the same network, which also greatly facilitates the coupling of the two media, and thus combined contracts.

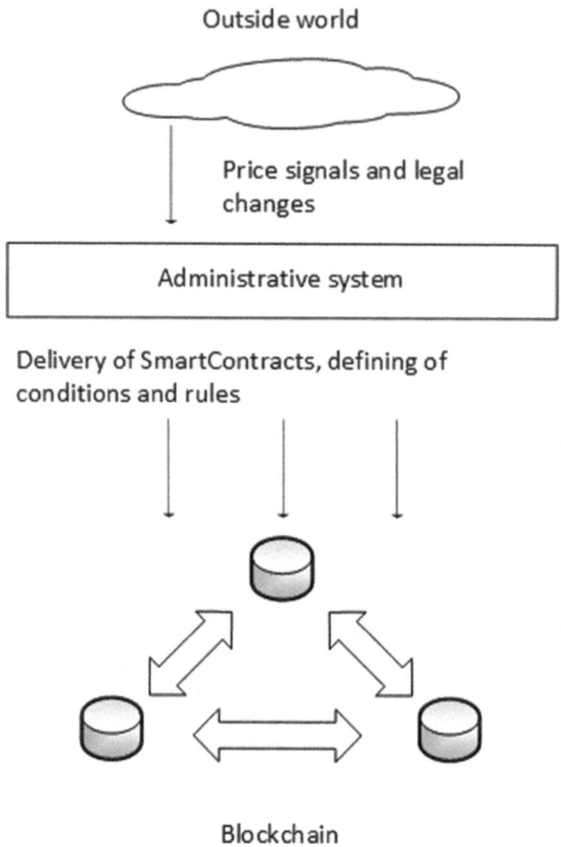

Fig. 2 Treatment of external influences and business logic

The administrative system is responsible for logically bringing new energy and money into the network and in addition to the removal of energy in terms of consumers. Furthermore, it is responsible for the withdrawal of money with respect to the owners of energy storage and energy-generating devices.

Depending on the payment model, each participant can be credited with a lot of money at a certain time, for example, during the monthly payment of the electricity flat rate. In this way, a wide variety of payment models are possible within a network. In this way, a participant can use a monthly flat rate and still consume more power. The compensation is then made once a year. At the same time, another participant might ask to receive as much power as he has paid for.

In particular for energy-producing node owners, access to their own account is of interest. Thus, the energy in possession can be sold at any pre-defined time and any pre-defined conditions. Smart contracts can be used to set the time of sale as a condition, thereby automating the sale.

Basically, all business processes are completely flexible and so are the billing processes. Conversely, changes by law can easily be implemented in the form of smart contracts, which can be distributed within the network. This would make obligations feasible in a very short time.

The administrative system and the conditions and rules for the smart contracts enable a quick reaction to the current energy market. The business logic is formulated by previously defined rules via smart contracts. These can include scope for variables representing the prices within the energy market. Thus, buying and selling decisions can be defined with certain price limits, which then automatically occur according to the current conditions [5].

The data of the transactions are designed so that different tariffs can be developed, which are within the context of the circumstances thereby allowing flexible pricing. For example, energy transfers may be more expensive at times of high network load than at times of low load.

Economically speaking, such a virtual power plant can also be considered as a separate system. This fact allows the creation of microscopic energy exchanges. On this basis, participants of an energy exchange can trade with each other. For purchases and sales beyond their own economic system, additional fees may be required.

3 Realization in Ubiquitous Networks and Infrastructures

For the implementation of the defined usage of blockchain in local districts, we need a ubiquitous grid. Originating from the Latin term "ubique," the term "ubiquitous" can be translated to the English word "everywhere." Often and in different contexts, especially in scholarly discussions, the term omnipresent is used. The substantive meaning of the term is explained in biology or biogeography. The so-called ubiquitist describes an animal or a plant species which is found in large numbers of severely differentiated habitats. Adapted to the discipline of computer science, especially to the information technology, the term "Ubiquitous Computing" is widespread. According to Weiser, it describes the computer of the twenty-first century as: "… the omnipresent computer that invisibly and unconstructively supports the people in their activities and releases them most far-reaching from annoying routine tasks" [6].

In reference to intelligent information and supply networks, it means that such networks of different players of varied domains are supporting invisibly and unconstructively. Therefore, the ubiquitous network or rather the ubiquitously available infrastructures are found in all areas such as energy, health, mobility or living. In the analysis of the term spreading of "ubiquitäres Netz "ubiquitäres Grid" or "ubiquitous Grid," it is conspicuous that they are not considerably spread for the purpose of the meaning ascribed here. However, delineation at this point should have occurred to few scientific contributions to the topic ubiquitous grid. The distribution of calculations that require a large amount of computing power is described in those contributions within the bounds of grid computing [7]. Furthermore, the term of the ubiquitous power grid is used in scientific literature. This can be linked to the definition of the ubiquitous network described here. In the context of the contributions to the ubiquitous power grid, however, a clear restriction to the electrical engineering view can be seen. In particular, the integration of electric vehicles as energy storage in a grid is defined as ubiquitous [8].

The ubiquitous nature of intelligent infrastructures is initially distinguished at the atomic level by the communication capability of available sensors as well as the flexible control of existing actuators via different communication paths and bus systems. Especially standardized bus systems and communication protocols, such as KNX [9], as well as uniformly established data paths in the fieldbus area or ethernet or RS485 communication, allow direct and therefore a decentralized control of the hardware among each other. In addition to the drastic reduction in complexity, this decentralized communication variant has the distinct advantage that if one subcomponent fails, as a rule other components can continue to operate unchanged [10].

A decisive disadvantage is the basic requirement of central communication within a closed subsystem of defined granularity for the realization of an overall view and optimization. A complete subsystem depending on the considered domain and objective may be, for example, a single room, a whole apartment, the entire property, a neighbourhood or continuing a complete municipality. For specific application scenarios such as intelligent heating control with individual room control based on user profiles or an inactivity detection based on heterogeneous transspatial sensor data is a local data collection including a pragmatic decision logic to determine the requirements and optimized decision-making not enough anymore. The disadvantage of a central communication component is that its failure may be responsible for the collapse of a complete system. Therefore, a hybrid approach to ensure the basic supply and functionality is to be preferred. This implies that a central communication and decision node within a subsystem is responsible for decisions for overall optimization. Consequently, the individual actuators and sensors should be linked primarily to the neighbouring central decision node as a fallback strategy. Because, if they fail, the decentralized communication continues. To realize an information and communication solution (ICT solution) based on these aspects, different challenges should be faced. The overall problem is the question of a heterogeneous system world based on physically connectable hardware that can serve as a data source and origin for the decision and control logic. Therefore, a holistic solution is crucial. In this context, there are challenges induced by the coupling of different bus systems and communication protocols. Another problem is the heterogeneity within equal product types. Devices with identical intended purpose from two different manufacturers can be very different, regarding terms of functionality and the way in which they communicate. For the implementation of a preferably flexible ICT solution, abstracted, device-specific and manufacturer-specific artefacts for integration in an overall solution have to be abstracted. From the concept of a decentralized communication environment, the communication and decision-making node can be understood as a concept that is capable of communicating in case of a partial failure of the entire system, for example, due to a technical defect. For the various levels of analysis, this demands a self-sufficient communication module that can generally operate decentralized and is influenced by cross-system communication and decisions. The consideration of data protection and security is another issue. For example, information at the level of a municipality can only be used anonymously if there is sensitive information within a housing or real estate area that may not be disclosed and can only be used for temporary processing. The requirements for its individual components with regard to the information processing and data load due to the aggregation are very specific. For an apartment, a single-board computer (SBC) or embedded computer is sufficient for coping with information, while the flood of information in a city area requires a powerful server system to handle it [11].

These structural conditions are a particular challenge for the ICT solution and especially the software solution is required for this. Consequently, it has to be flexible, scalable and dynamically configurable, depending on the respective area of application and the required functionality. The distribution aspects also require middleware to address these needs. Depending on the respective function and configuration of the system environment, the option of connecting heterogeneous systems is additionally required. Besides the hardware and fieldbus connection, it applies to systems such as ERP systems or provider-specific billing and evaluation systems. The integration of such systems offers an approach to the establishment of innovative, digitized services in the different domains. The selection of the operating systems depends on the problem of the distributed computer network and different granularity of thus hardware requirement. Especially the SBC solutions need an embedded system with minimal resource requirements, while in the server sector, for example, value-added services with complex graphical requirements and sophisticated analyses has to be provided to the end user. In general, it is not possible to make a blanket statement regarding the operating system. Microsoft Windows, for example, is seen as an established end-user sector due to its popularity. By contrast, Linux systems are preferred due to license costs, functionality in the embedded and server areas. Some SBC solutions are limited to special platform ports that are customized for this hardware. Thus, forcing flexibility in choosing the operating system or not allowing selection [12].

Based on the scalability and flexibility requirements for the computer technology used, the operating system selection and the programming environment these requirements are identical for the software solution. Like the efficient reuse of existing source, the code represents a basic requirement of software development. Is the provision of maximum reusability of software components and flexibility, for example, through configuration capabilities decisively in this application context? The dynamic assembly of complex applications at runtime from individual components has already been coined under the term service-oriented architectures (SOA) [13].

A commonly used technology in this context is the Web service. The communication can take place, inter alia, via the hypertext transfer protocol (HTTP) or extensible mark-up language (XML) and is thus programming language and platform independent. OSGi (open service gateway initiative) is an SOA-like approach designed specifically for use with a JAVA virtual machine. OSGi offers a solution to develop highly modular JAVA-based applications that meet the different scaling requirements. It provides at the same time a broad technological spectrum for the development of specific software building blocks based on the JAVA Framework, which can thus be dynamically assembled into a complex system. This means that changes or updates to the configuration can be made at runtime without having to shut down or restart the entire system. This aspect is critical to the development of services especially for devices with low memory availability, which are thus widely applicable within a ubiquitous grid.

References

1. WindNode Project Homepage. http://www.windnode.de/konzept/. Last accessed 15 November, 2017.
2. WindNode Project Homepage. http://www.windnode.de/partner/. Last access 15 November, 2017.
3. https://www.heise.de/newsticker/meldung/E-Autos-Innogy-entwickelt-Ladestations-Plattform-mit-Blockchainzahlungen-3702960.html. Last accessed 15 November, 2017.
4. Hofmann, E., Strewe, U. M., & Bosia, N. (2017). Supply chain finance and blockchain technology the case of reverse securitisation. SpringerBriefs in Finance, St. Gallen, Switzerland, pp. 35–40.
5. Morabito, V. (2017). Business Innovation Through Blockchain The B^3 Perspective. *Springer International Publishing AG 2017*, Cham: Switzerland. S, 102–105.
6. Mattern F. (2003). *Ubiquitous Computing—Die Vision von der Informatisierung der Welt, Zitat entnommen mit Referenz auf Weiser* (1991), Online: http://www.vs.inf.ethz.ch/publ/papers/UbicompLogin.pdf. Last accessed 15 Febuary, 2017.
7. Sample N., Keyani P., Wiederhold G. (2002). Scheduling Under Uncertainty: Planning for the Ubiquituous Grid. In *Proceedings of the 5th International Conference on Coordination Models and Languages*. Springer:Berlin.
8. Ota, Y., Taniguchi, H., Nakajima, T., Liyanage, K. M., Baba, J., & Yokoyama, A. (2010). Proposal of smart storage for ubiquitous power grid—Autonomous distributed vehicle-to-grid of electric vehicle, Institute of Electrical Engineers of Japan. *IEEJ Transactions on Power and Energy, 130*(11).
9. Merz, H., Lansemann, T., & Hübner, C. (2009). *Building automation: Communication systems with EIB/KNX*. LON and BACnet: Springer Verlag.
10. Mahmoud, M. S. (2011). *Decentralized systems with design constraints: stability analysis and design*, Springer.
11. Teich, T., & Igel, W. (2010). *Energieeffizienz in Wohngebäuden* (Vol. 1). GUC.
12. Bender, K. (2009). *Embedded systems—qualitätsorientierte Entwicklung*, Springer.
13. Masak, D. (2007). SOA? *Serviceorientierung in Business und Software*, Springer.

Toward Large-Scale Logistics Interoperability Based on an Analysis of Available Open Standards

Wout Hofman

Abstract Data sharing is important to optimize and manage logistics chains. Open standards have been developed and implemented by various logistics stakeholders. These implementations have led to closed solutions with barriers to change and implement innovative concepts like synchromodality. This paper analyzes the implementation of open standards by providing an overview of available open standards based on different implementation strategies for logistics. It reflects the state of the art of business-to-business (B2B) and business-to-government (B2G) interoperability in international trade and logistics. An analysis of the underlying problem and a proposal for a solution to create a system-of-systems for data sharing in supply and logistics with key elements of the Industrial Data Space are presented in this paper.

Keywords Open standards · Implementation guides · Data sharing infrastructure

1 Introduction

Universal connectivity as in 'super-fast connectivity, always on, on the move, roaming seamless from network to network, where we go—anywhere, anytime, with any device' [1] is mentioned as one of the most important aspects of the Physical Internet [2]. Such a radical innovation like the Physical Internet or Industry 4.0 supported by the Industrial Data Space [3] is mostly described at business level referring to technology like the Internet or Application Programming Interfaces (APIs). These visions lack the understanding of data standards and their implementation. Data sharing is required to achieve state awareness [4], also known as situational awareness [5]. Conceptual interoperability [6], which is currently not implemented by supply and logistics stakeholders [7], is required for large-scale implementation of supply and logistics innovations [2, 4]. This paper will argue that an open infrastructure supporting all logistics stakeholders and law enforcement agencies to access and share data to support their processes is required. The underlying problem is that of

W. Hofman (✉)
TNO, P.O. Box 23, 3769 ZG Soesterberg, The Netherlands
e-mail: wout.hofman@tno.nl

K. Popplewell et al. (eds.), *Enterprise Interoperability VIII*, Proceedings of the I-ESA Conferences 9, https://doi.org/10.1007/978-3-030-13693-2_21

the current implementation of open standards. Firstly, an overview of open standards for B2B and B2G interoperability in international trade and logistics is provided, secondly the current situation of interoperability is analyzed, and finally, a solution based on the key elements of the Industrial Data Space [3] is proposed. Conclusions will complete this paper.

2 Interoperability in Supply and Logistics

There are different models to represent the various aspects of interoperability, each with its levels of interoperability. For instance, the European Interoperability Framework (EIF, [8]) distinguishes four levels of interoperability, namely technical, semantical, organizational, and legal. Others like the Levels of Conceptual Interoperability Model (LCIM, [6]) identify six levels of interoperability. These models can be mapped to each other, for instance, the technical level of the EIF can be mapped to level 1 and 2 of the Levels of Conceptual Interoperability, the semantic levels are identical, and the organizational level of EIF addresses the levels 4 up to 6 of the other model. The Levels of Conceptual Interoperability does not distinguish the legal level, although it is of great importance for international supply and logistics, as it involves authorities like customs.

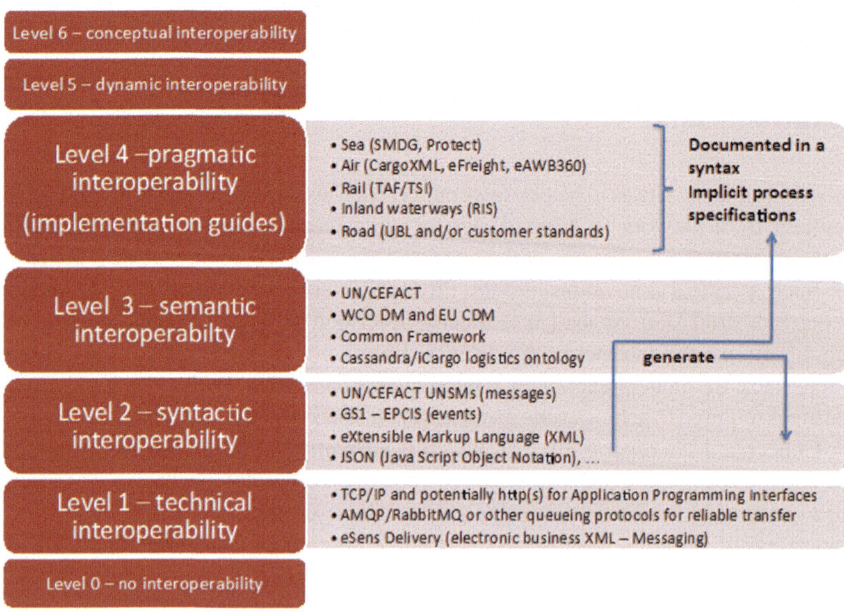

Fig. 1 Overview of relevant logistics standards

Figure 1 illustrates that semantic models are applied to (1) generate syntactical standards and (2) are used to create Implementation Guides (IGs). These IGs are either represented syntactical or documented in free format text. Level 1, technical interoperability, will not be discussed further. It consists of standard protocols like TCP/IP, AMQP (Advanced Message Queuing Protocol), and eSens Delivery (www.esense.eu) for secure and reliable message exchange over the Internet. These protocols can be used to support any level 2 standard.

The levels of interoperability are the following for supply and logistics:

- **Level 2 syntax**—structure of the data during exchange. A distinction needs to be made into syntax and technical protocols utilized to share data. Where a syntax only structures data, the technical protocols specify the mechanism to share data

Table 1 Technical protocols for data sharing

Technical protocol	Features	Syntax
Messaging	A sender duplicates data to a recipient	EDI, XML, or any other
Event Driven Architecture (EDA) [9]	IT systems of organizations can subscribe to events published by sensors (IoT—Internet of Things), IT applications, or platforms. An event either has sufficient data, e.g., the location and speed of a truck are given by the event, or refers to data available in an IT application. Mostly combined with a subscription mechanism. In case of generation of new events, EDA is positioned at level 4	XML, GS1 EPCIS
Data Crawling [10]	Periodic (controlled) access to a data source. The period between crawling can differ per source, whereas a source can specify any access restrictions to data	Any (not EDI based)
Link evaluation [10]	Data links to other data with a URI (Uniform Resource Identifier). Evaluating links directly accesses data	XML, RDF
Service-Oriented Architecture (SOA) [9, 11]	Capabilities of IT applications are published as services consisting of a service call and a response with Web Service Definition Language (WSDL) or REST APIs	XML

Table 2 Pragmatic standards for modalities

Modality	Standard	Semantics	Syntax	Roles
Road	ISO standards for paperless transport (www.tln.nl) or GS1 solutions (XSDs)	IGs	XML	Customer–carrier
	Booking and ordering		EDI	Forwarder–carrier
	Pickup and drop off of containers		EDI	Terminal–carrier
	Datex II—traffic management	UML Class diagram	WSDL/XML	Carrier–infrastructure manager
Rail	Telematic Applications for Freight/Technical Specifications for Interoperability (TAF/TSI)—track optimization	IGs (rail specific)	XML	Carrier (railway undertaking)–infrastructure manager
Air	Cargo Interchange Message Procedures (CIMP)	IGs (air specific)	XML	
	Cargo XML supporting CIMP		XML	Airline–airport
	e-AWB360—electronic Air Way Bill (AWB)		XML	Freight forwarder–airline
Sea	Booking and ordering	IGs	EDI UNSMs	Shipper/consignee–forwarder; forwarder–carrier
	Vessel operations (SMDG—smdg.org)		EDI UNSMs	Shipping line–terminal
	Formalities for handling vessels in ports (Protect—protect-group.org)		EDI UNSMs	Shipping line–port authority
Inland waterways	IGs for Formalities for barges on inland waterways (River Information Services—RIS)	IGs	EDI UNSMs	Skipper–infrastructure manager
	Corridor management		XML	Skipper–infrastructure manager

(Table 1). Particular syntaxes are based on a particular technical protocol, e.g., EDIFACT is supported by a messaging-based technical protocol.
Traditionally, supply and logistics utilize EDIFACT, EDI for administration, commerce, and transport [12] for booking and ordering by messaging (UNSMs), based on basic generic building blocks, called segments, a common dictionary, the United Nations Trade Data Elements Directory (UNTDED; [13]), and code sets specified by the United Nations Economic Commission for Europe (UNECE). XML (eXtensible Markup Language; [14]) and XML Schema Definition (XSD; [15]) are the most common used for sharing data between different applications over the Internet. The rail-, air-, and retail (GS1—Global Standards One) industry apply for instance XML Schema Definition for structuring their data. JSON (JavaScript Object Notation), a key-value pair syntax for sharing data, is a defacto standard used in software development. RDF, Resource Description Framework, has been developed specifically for representing linked (open) data. Both are not applied in supply and logistics.

- **Level 3 semantic**—interpretation of data by semantics represented by meta-models like Unified Modeling Language (UML; [16]) class diagrams and Ontology Web Language (OWL; [17]). Others are Object Role Modeling (ORM; [18]). UML is applied to develop a semantic model for data sharing in logistics in EU-funded projects, the Common Framework [19] for booking and ordering in logistics. Ontologies for logistics have for instance been developed by other EU-funded projects [20].
The World Customs Organization (WCO) and UN/CEFACT apply a proprietary meta-model for semantic models [21]. The WCO data model supports EDIFACT and XSDs, is applied by various customs authorities, and European Commission, Directorate for customs and taxation (EC DG Taxud) has developed its own subset of the WCO data model, the EU Customs Data Model. UN/CEFACT has developed a generic model for data sharing in logistics and published as free text by UN/CEFACT.
- **Level 4 pragmatic interoperability**—so-called Implementation Guides (IGs) of open standards for process alignment of organizations. These IGs are documented as free format text documents for UNSMs or as XSDs, potentially based on semantic models (Table 2).

3 Analysis of Interoperability in Supply and Logistics

This section further analyzes the state of the art of interoperability in supply and logistics from different perspectives, namely development of B2B and B2G standards for data sharing, implementation typologies underlying specific situations, and its application in supply and logistics.

3.1 B2B and B2G Interoperability

The generic perspective refers to documentation and development of open standards for B2B and B2G interoperability:

- IGs (level 4) are documented at syntax level and free format text (especially, the process aspects), which makes implementation of these IGs cumbersome.
- Semantic models (level 3) represent data aspects of EDI UNSMs and are used to generate XSDs or IGs (see also previous remark), either from a proprietary meta-model or UML Class diagram.
- Open standards, semantic models, and IGs replace business documents (paper forms). To be able to cater with exceptions and support resilience and agility, process specifications (level 4) are documented as part as textual descriptions to the IGs.
- There are no level 5 or 6 open standards to support data sharing in a more agile and resilient supply and logistics network.

Basically, open standards and their IGs reflect data aspects; process aspects are only specified as free format text. Open standards for specifying process aspects, e.g., Web Service Modeling Ontology (WSMO; [22]), semantic annotations for Web Services (SAWSDL) or business process modeling (BPM), are not applied. Level 5 and 6 standards are not available. IGs are not available in a machine-readable format, which leads to different interpretations and thus different implementations.

3.2 Three Typologies for Implementing Open Standards

Since syntactical open standards like EDI are generic, organizations have developed IGs along one of the three following typologies [23]:

1. **Electronic Monopoly**: Large, dominant players like customs authorities, infrastructure managers, large retailers, LSPs, or large carriers develop and impose IGs to their suppliers for optimization of their processes. IGs of different dominant players are incompatible and contain specific process optimization rules.
2. **Multilateral Inter-Organizational Information Systems**: Communities develop IGs to optimize for instance port operations. (Port) Community Systems often support these IGs. Community standards differ per community, which implies that members participating in more than one community have to implement different standards.
3. **Electronic Dyads**: Two organizations optimize their processes, mostly in the context of a framework contract, and develop specific IGs. To implement another framework contract requires redrafting the IGs, which results in a lock-in (Fig. 2).

Each of these typologies results in a closed system and does not provide universal connectivity. Small and medium-sized enterprises (SMEs) as suppliers or carriers

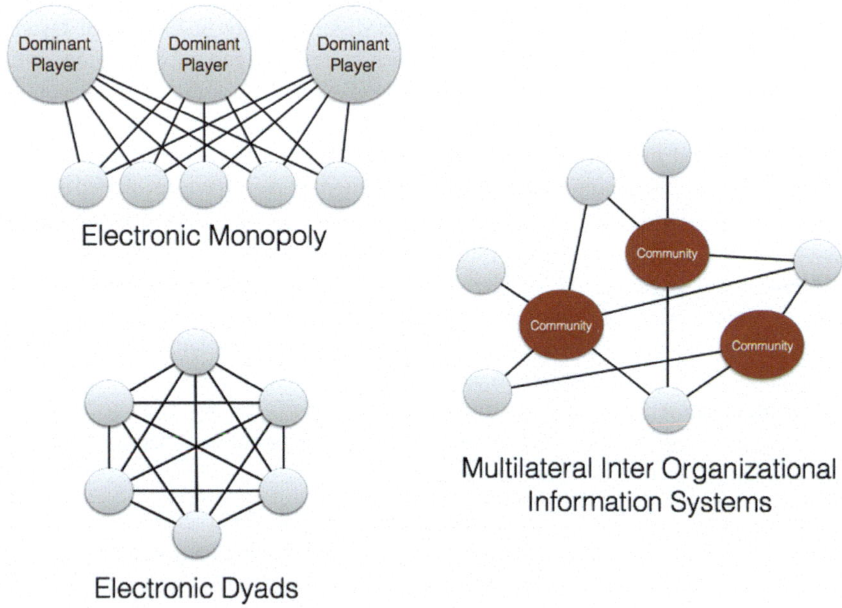

Fig. 2 Three typologies for implementing interoperability

to several large retailers have to deal with a variety of IGs, which results in the Electronic Dyads, where interoperability is differently implemented for each dyad. Commercial Integration Service Providers (ISPs) can act as an intermediate to hide this complexity.

3.3 B2B and B2G in Supply and Logistics

With respect to interoperability for supply and logistics, the following analysis is made:

- **Air**. The International Airline Transport Association (IATA) has initiated programs to fully replace paper with data. Although IATA has involved airlines in development, they have no power to stimulate implementation and community standards have to align airlines as dominant players, leading to limited electronic data sharing.
- **Sea**. Ports have developed community standards and PCSs to increase competition and being able to handle increased goods flows, integrated with carriers (SMDG and Protect, see before). Data sharing between ports or single points of customer interaction is only based on commercial booking systems like INNTRA.

- **Road**. This modality consists of a lot of SMEs, resulting in Electronic Dyads and lack of electronic data sharing. Infrastructure managers have created open standards for optimization of road utilization (DATEX II).
- **Rail**. Based on an EU Directive for an open market, the infrastructure managers have developed messages for infrastructure utilization (TAF/TSI) supported by systems of Rail Net Europe. Additionally, dominant players have developed closed community solutions, e.g., RailData (train composition data of limited number of carriers) and CESAR (booking system of a number of Railway Operators).
- **Inland waterways**. The EU RIS Directive led to harmonization of optimization of inland waterway optimization, e.g., by reporting formalities of skippers to infrastructure managers. Since Member States have different implementations of the Directive, there are differences in applying these standards.
- **Customs authorities** (the major authority in trade facilitation). They act as dominant player with national IGs based on the WCO data model reflecting national implementations of EU Directives.

Thus, IGs are either enforced by dominant players (infrastructure managers or customs) based on Directives with their national implementation guide or community standards for particular hubs. Where dominant players are missing, there is no Directive, or no clear business incentive to increase volumes of goods flows, interoperability implementation is lacking. The general conclusion is thus that implementation of open standards leads to closed solutions.

4 Toward an Innovative Solution

The previous illustrates availability of open standards not necessarily leads to a solution meeting for instance requirements of the EIF [8] or the Industrial Data Space [3]. Firstly, this section identifies the requirements for interoperability, and secondly proposes a new implementation typology for conceptual—and dynamic interoperability (level 5 and 6).

4.1 Requirements for B2B and B2G Interoperability

These EIF and the Industrial Data Space formulate key principles that are applicable for supply and logistics:

- **Organizational network**. The solution should be available to all organizations against reasonable costs, with a choice to connect to any system that adheres to the agreed set of services. Any solution should also be technology neutral [8].
- **Economies of scale—uniformity of services**. Stimulate innovation (open source, ISPs, etc.) and connecting once allows trading with all that participate in the

system of systems. Inclusion of all SMEs (and citizens) should be addressed with this point.

- **Data sharing**. Trust, controlled data access, agreed rules for data (re-)use, and data provenance have to be addressed. 'Open' data for all authority data that has no restrictions, implying a particular governance structure [8].

In addition to these key principles, the EIF provides recommendations regarding governance and (semantic and technical) interoperability, without prescribing any solution or making any choices. The EIF and Industrial Data Space do not specify the unified services, nor do they prescribe an implementation typology. By lack of such a typology, stakeholders will revert to any typology they are familiar with, leading to closed solutions.

4.2 Proposed Implementation Typology

Our proposed implementation typology has to take a business perspective and address both the generic and specific issues of interoperability (Sect. 3) like process aspects, semantics of the data shared, and the technical representation of open standards as machine-readable documents.

- **Business perspective—value exchange**. Instead of replacing paper business documents with electronic ones, this paper proposes to electronically support value exchange between any two organizations [24]. Logistics services like 'transport,' 'transshipment,' and 'storage' are the basis for business transactions and have properties like duration, locations, and goods with particular characteristics like reefer or dangerous goods. These properties have data requirements like goods properties, conditions for cleaning tankers, and handling instructions of for instance reefer or dangerous cargo. IGs are not required anymore, as long as a shared semantic model and choreography can be implemented by all stakeholders.
- **Choreography—process aspects**. From a modeling perspective, a choreography of related business interactions has to be specified as business process model (BPM) and supported by internal business processes [26]. The choreography should support the following functionality for business transactions:
  - **Search and find**—by posting a goal, matches with published logistics services can be made.
  - **Booking and ordering**—prices and delivery conditions are negotiated leading to an order and a final plan of a service provider to execute the order of individual legs in a chain.
  - **Visibility**—providing progress of the execution of an order and potentially detecting any delays that may cause redesign of logistics chains or chain coordination.

This functionality has been developed and validated in various business settings [25]. The interactions, e.g., a booking or a transport order, have to contain rules for minimal data requirements formulated on a semantic model.

- **Data semantics of supply and logistics**. Semantics in supply and logistics represents the actual physical objects and relevant concepts for logistics services like (packaged or bulk) cargo (e.g., boxes or pallets with products like flowers, pharmaceutical and electronics, sand, oil, and grain), assets used by LSPs and carriers (containers, trucks, vessels), and geographical (locations, regions) and time-related concepts (duration, time windows). A business transaction between any two organizations is about sharing the expected, the planned, and the actual situation (time and location), and agreeing on conditions and rates for value exchange, the cargo, and the assets utilized. Figure 3 shows the high-level semantic model for logistics.
- **Implementation of open standards by an organization**. Each organization will have its internal IT system with a particular data structure and business processes. These have to support the choreography and the shared semantics, based on the logistics services of that organization. This external business view needs to be transformed into database views of the internal IT systems of an organization.
- **Representation of open standards**. The final issue is that of technical representation of these models. OWL is in our view the best candidate for technical representation of a semantic model:

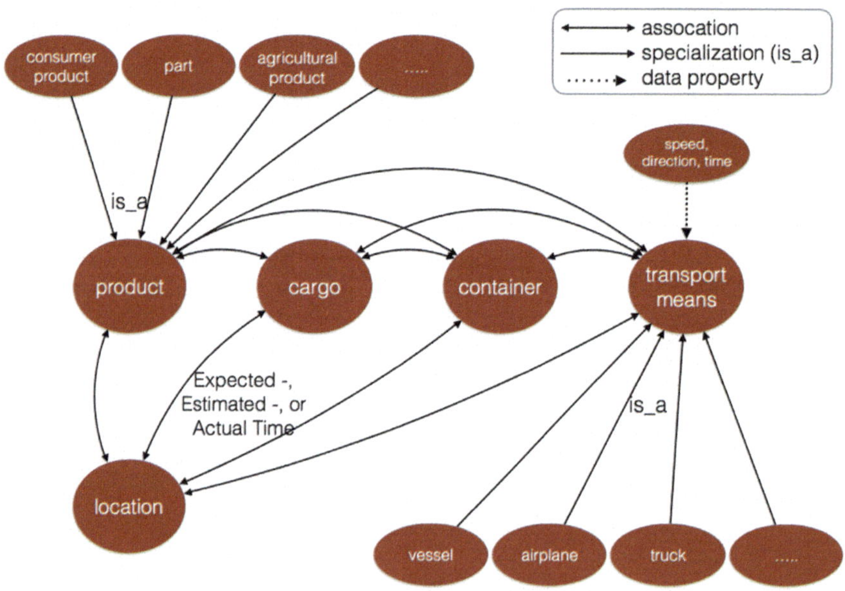

Fig. 3 High-level semantic model

- **Hierarchies**—an OWL model can import another and extend the functionality of another model. The high-level semantic model (Fig. 3) can function as upper ontology, with mode specific models importing and specializing the upper ontology.
- **View**—a view is an OWL model that is a subset of another one. Each organization can create its view to support their logistics services and use it to interface with their IT systems.
- **Extendable**—by creating a new view, a semantic model can be extended with new functionality to model not yet foreseen data requirements.
- **Meta-model**—OWL has a defined meta-model for sharing various OWL models between open source or COTS (Commercial Off-The-Shelve) tools.

Ontologies for interoperability are to our knowledge not yet applied in supply and logistics. Besides representing interactions by an ontology, they can also be represented by a rule set in a standard rule language like RuleML.

5 Conclusions and Recommendations

This paper has presented the state of the art of implementation of open standards for interoperability in the supply and logistics domain for B2B and B2G. These implementations lead to closed systems, which prevent logistics innovations like synchro-modal planning, agility, and resilience. A solution to address this issue is proposed by introducing a new implementation typology complementing the requirements of the EIF [8] and Industrial Data Space [3]. The proposed implementation typology is based on value exchange [24] for business transactions [25] with a choreography [26] and the Ontology Web Language for technical representation of semantic models. The proposed solution does not require any additional investments in IGs.

Constructing these solutions still requires a number of challenges for future research, like governance and standardization, adoption and implementation strategies considering existing investments and available functionality, trust in participating organizations and validating their implementation, and the role of innovative technology like blockchain technology and/or micro-services to construct a system of systems in a fully distributed manner. Since there are already quite a large number of platforms and solutions, each with their particular services, wide-scale adoption and implementation will probably still take some time. Adoption can be fastened by stimulating new entrants of platform providers to implement the proposed approach.

Acknowledgements This paper is realized by the CORE (Consistently Optimized Resilient Secure Global Supply-Chains) project funded by the EU (funding scheme: FP7-SEC-2013.2.4-1).

References

1. Biggs, P., Johnson, T., Lozanova, Y., & Sundberg, N. (2012). Emerging issues for a hyperconnected world. *The global information technology report*, pp. 47–56.
2. Montreuil, B., Meller, R. D., & Ballot, E. (2013). Physical internet foundations. In: *Service orientation in holonic and multi agent manufacturing robots* (pp. 151–166). Heidelberg: Springer.
3. Otto, B., Jürjens, J., Auer, S., Menz, N., Schon, J., Wenzel, S., & Cirullies J. (2016) Industrial data space—digital sovereignity over data. Fraunhofer-Gezsellschaft zur Forderung der angewandten Forschung e.V., München.
4. McFarlane, D., Giannikas, V., & Lu, W. (2016). Intelligent logistics: Involving the customer. *Computers in Industry, 81,* 105–115.
5. Endsley, M. R. (1995). Toward a theory of situation awareness in dynamic systems. *Human Factors: The Journal of the Human Factors and Ergonomics Society, 37*(1), 32–64.
6. Wang, W., Tolk, A., & Wang, W. (2009). The levels of conceptual interoperability model: Applying systems engineering principles to M&S. In *Spring Simulation Multiconference*.
7. The Digital Transport and Logistics Forum (DTLF). (2017). An outline for a generic concept for an innovative approach to interoperability in supply and logistics chains, Brussels.
8. European Union. (2017). New European Interoperability Framework—Promoting seamless services and data flows for European Public Administrations. Publications Office of the European Union, Luxembourg.
9. Erl, T. (2005). *Service oriented architecture—Concepts, technology and design*. Prentice-Hall.
10. Heath, T., & Bizer, C. (2011). Linked data—Evolving the web into a global data space. In: *Synthesis lectures on the semantic web: Theory and technology*. Morgan & Claypool Publishers.
11. Barros, A., & Oberle, D. (2012). *Handbook of service description—USDL and its methods*. Berlin: Springer.
12. International Organisation for Standardisation. (2017). ISO 9735:1988—Electronic data interchange for administration, commerce and transport (EDIFACT)—Application level syntax rules, 1988. [Online]. Retrieved October 27, 2017 from iso.org.
13. United Nations Economic Commission for Europe (UNECE), 2005. [Online]. Retrieved October 27, 2017 from www.unece.org.
14. World Wide Web Consortium (W3C). (2017). Extensible Markup Language (XML) 1.0 (5th ed.), 26 11 2008. [Online]. Retrieved October 27, 2017 from www.w3c.org.
15. World Wide Web Consortium (W3C). (2017). XML Schema Current Status, 5 4 2012. [Online]. Retrieved October 27, 2017 from www.w3c.org.
16. Object Management Group (OMG). (2017). About the Unified Modeling Language Specification version 2.5, 5 2015. [Online]. Retrieved October 27, 2017 from www.omg.org.
17. World Wide Web Consortium (W3C). (2017). Web Ontology Language OWL, 11 12 2012. [Online]. Retrieved October 27, 2017 from www.w3.org.
18. Object Role Modeling—The official site for conceptual data modeling. Object Role Modeling. [Online]. Retrieved October 27, 2017 from www.orm.net.
19. Pedersen, J. T. (2012). One Common Framework for information and communication systems in transport and logistics: Facilitating interoperability. In P. Glinska & M. Hajdul (Eds.), *Sustainable transport* (pp. 165–196). Berlin: Springer.
20. Dutch Organisation for Applied Science (TNO). (2017). Logistics Core Ontology. [Online]. Retrieved October 27, 2017 from ontology.tno.nl.
21. World Customs Organisation (WCO). (2017). WCO Data model version 3.6.0. [Online]. Retrieved October 27, 2017 from www.wcoomd.org.
22. Fensel, D., Kerrigan, M., & Zaremba, M. (2008). *Implementing semantic web services—The SESA framework*. Berlin: Springer.
23. Choudry, V. (1997). Strategic choices in the development of interorganizational information systems. *Information Systems Research, 8*(1).
24. Spohrer, J. K. S. (2009). Service Science, Management, Engineering, and Design (SSMED)—An emerging discipline—Outline and References. *International Journal on Information Systems in the Service Sector*.

25. Dietz, J. (2006). *Enterprise ontology, theory and methodology*. Berlin: Springer.
26. Schonberger, A., Wilms, C., & Wirtz, G. (2009). *A requirements analysis of Business-to-Business integration*. Bamberg: Fakultat Wirschaftsinformatik und angewandte Informatik Otto-Friedrich-Universitat.

Part VIII
Interoperability Application Scenarios

Integrating Business Process Interoperability into an Inter-enterprise Performance Management System

María-José Verdecho, Juan-José Alfaro-Saiz and Raúl Rodríguez-Rodríguez

Abstract Supply chain performance mainly depends on how well are defined and managed their inter-enterprise business processes. For that purpose, performance management systems should consider inter-organisational processes in their structures as well as be defined at two levels: intra-enterprise and inter-enterprise. In the literature, several performance management systems already developed consider to some extent business processes management. However, the introduction of inter-organisational business process interoperability measurement is still an issue that needs to be researched in more detail. To solve this gap, this paper presents a performance management system for inter-organisational contexts that integrates business process interoperability measurement. In addition, an application of this system is shown for the agri-food supply chain.

Keywords Business process interoperability · Performance management system · Supply chain

1 Introduction

In global markets, collaboration among the enterprises is becoming more a necessity than an option to achieve and maintain competitiveness [1, 2]. In this situation, the business processes of collaborative enterprises should work together to reach common objectives [2].

Supply chain performance mainly depends on how well are defined and managed their inter-enterprise business processes. Also, the performance of the intra-enterprise business process of the individual enterprises impacts on the performance of the supply chain. For that reason, it is important to manage business processes from both levels: an inter-enterprise perspective (global) and an intra-enterprise perspective (individual enterprise).

M.-J. Verdecho (✉) · J.-J. Alfaro-Saiz · R. Rodríguez-Rodríguez
Department of Business Organisation, CIGIP-Research Centre on Production Management and Engineering, Universitat Politècnica de València, 46022 Valencia, Spain
e-mail: mverdecho@cigip.upv.es

K. Popplewell et al. (eds.), *Enterprise Interoperability VIII*, Proceedings of the I-ESA Conferences 9, https://doi.org/10.1007/978-3-030-13693-2_22

In the literature, there are different performance management systems (PMSs) for inter-enterprise contexts. One of the most important performance measurement frameworks is the balanced scorecard (BSC) by Kaplan and Norton [3]. The BSC has been modified for inter-enterprise performance management, for example the works developed by Brewer and Speh [4], Bititci et al. [5], Folan and Browne [6], and Alfaro et al. [7]. These performance frameworks present in their structure different performance elements (levels, perspectives, objectives, performance indicators, etc.). Some of them include in their structures the deployment of performance measurement from the strategic to the operational level what facilitates consistency in performance definition, management and evolution. However, it has to be noted that not all PMSs can manage the interoperability of business processes.

One of the definitions in Interop [8] characterises process interoperability as the 'ability of different processes to work together and exchange information, data, control information, etc'. Therefore, solving the problem of interoperability becomes essential, as it will favour the evolution of organisations towards interoperable environments, allowing efficient coordination of flows among enterprises.

In a PMS, there are four important characteristics to manage interoperability of business processes. First, it has to be managed at both levels: strategic and process. If business process interoperability is managed only at the strategic level, it may overlook interoperability of the specific business processes where real operation takes part. If business process interoperability is managed only at the business process level, the strategic dimension of pursuing collaboration with low effort may be disconnected from the business process operation avoiding a solid deployment of performance measurement. Second, as previously stated, it is important to manage both the strategic and business process levels at both contexts: intra-enterprise and inter-enterprise scopes. Third, it is important to connect in an integrated PMS structure the two previous characteristics. This means that the PMS should link in a coherent manner the four blocks: intra-inter-strategic performance measurement; intra–inter-process performance measurement; inter-strategic and process measurement; and intra-strategy and process measurement. Finally, the fourth characteristic includes the definition of performance elements to increase interoperability of business processes. This implies that the system should define, where necessary, specific objectives and KPIs to increase the efficiency of collaboration regarding clarity, visibility, alignment, coordination, synchronisation integration, flexibility and monitoring of collaborative business processes [9].

The PMSs that consider business processes management, in a higher or lower depth, are Supply Chain BSC Framework [4]; Gunasekaran Framework [10]; Process-Based Framework [11]; Supply Chain Performance Metrics Framework [12]; Extended Enterprise Balanced Scorecard [6]; Performance Measurement framework for Collaborative Supply Chain [13]; A Framework to Analyse Collaborative Performance [14]. However, none of these works complies with the four characteristics exposed to build a solid PMS.

The main aim of this paper is to present an integrated PMS that directly includes in its structure the measurement and management of business process as well as the management of interoperability by including the four characteristics explained. The structure of this paper is as follows. First, the business process interoperability concept is presented. Second, a performance management system integrating business process interoperability is developed. Third, an application of this system to an agri-food supply chain is shown. Finally, conclusions are exposed.

2 Business Process Interoperability

Inter-enterprise business process is 'a process where two or more enterprises participate, independently of the degree of cooperation/collaboration existing between them' [15]. Figure 1 presents an inter-enterprise business process. In the left side of the figure, the collaboration of three enterprises (A, B and C) working together through four business processes (lines connecting the enterprises) is shown. Enterprises A and B work together within two business processes (two lines connecting the

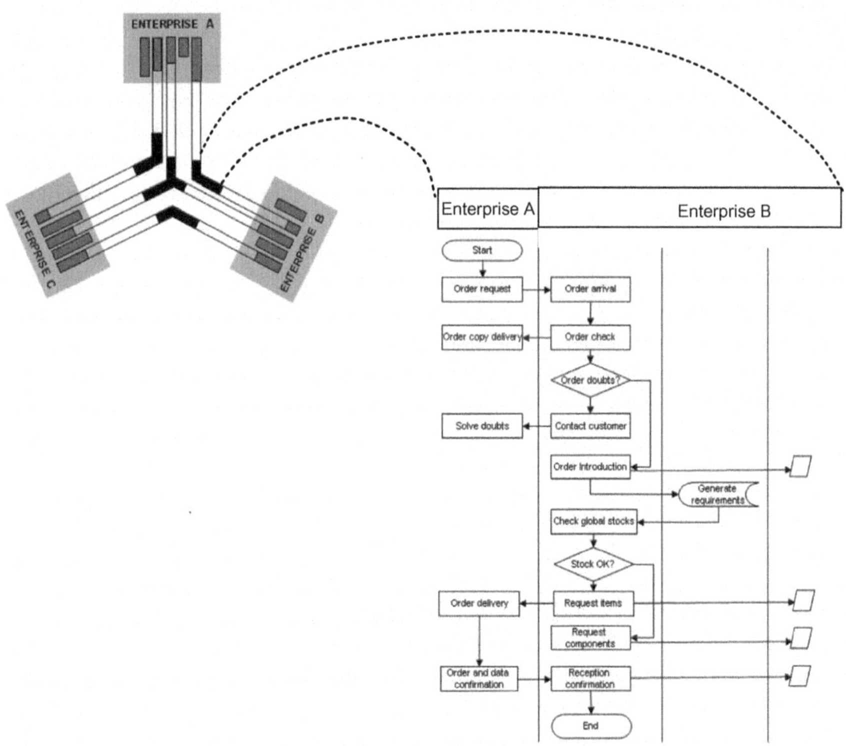

Fig. 1 Inter-enterprise business process. *Source* Alfaro-Saiz et al. [15]

A and B enterprises). Enterprises A and C work together in two business processes and, enterprises B and C work together in two business processes as well. In addition, one of the business processes is performed by all three enterprises (the business process that connects all three enterprises). In the four inter-enterprise business processes (lines connecting the enterprises), there is a part of the business process that is performed solely by an enterprise (white area) and another part that is performed jointly by different enterprises (black area) and, therefore, it is in this black area where interoperability of the business process occurs. In the right side of the figure, we represent the activities and information flow of the business process interoperability performed by enterprises A and B.

In order to manage business process interoperability, the next section presents PMSs that integrates into its structure the management of business process by dealing with the four characteristics defined in Sect. 1.

3 A Performance Management System Integrating Business Process Interoperability

As stated before, companies need tools that support managing business process interoperability for inter-organisational scopes following a solid structure. The performance management system presented in this section fulfils these characteristics by using the performance management system [7] based on three phases: (1) definition of the strategic framework, (2) definition of the process framework and (3) monitoring. The PMS [7] has been selected as it includes in its structure three out of four characteristics defined in Sect. 1. In addition, it can accommodate the fourth characteristic as will be presented in this section.

The characteristics of a performance management system for managing the business process interoperability of inter-enterprise contexts are the requirements that are to be accomplished by the system to be consistent. This implies that the system might comply with all the characteristics to manage the inter-organisational scope. The first characteristic is to consider both the strategic and process levels. In addition, that system should aid the decision-making of the companies that collaborate. For that reason, the second characteristic is to consider two scopes: inter-enterprise and individual enterprise. Both scopes and levels must be aligned in order to maintain traceability between performance elements. For that reason, the third characteristic is that the PMS should link in a coherent manner the four blocks: intra-inter-strategy performance measurement; intra-inter-process performance measurement; inter-strategy and process measurement; and intra-strategy and process measurement. Finally, the fourth characteristic deals with the definition of performance elements to increase interoperability of business processes and the efficiency of collaboration regarding information synchronisation, alignment, etc.

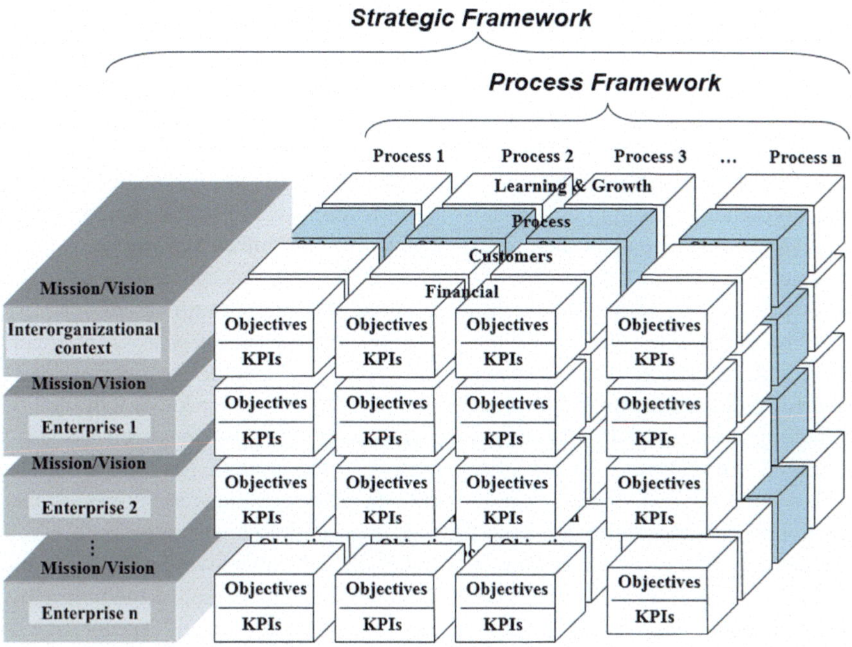

Fig. 2 Performance management system for integrating business process interoperability. *Source* Adapted from Verdecho et al. [16]

In the PMS [7], at the enterprise level, the system starts from the vision and vision of each individual enterprise defining the most important strategic points of the business. Therefore, the system starts by defining the strategic framework (phase 1).

Figure 2 represents the structure of the PMS which distinguishes between two types of sub-frameworks: strategic and process framework.

The development of the strategic framework consists of defining: mission and vision; objectives and key performance indicators (KPIs) at the strategic level. The elements are proposed for the four performance perspectives of the balanced score-card: financial, customer, process, and learning and growth. These perspectives aid to structure performance measurement following cause–effect relationships. These perspectives are the cubes represented in the figure where each cube contains the objectives and KPIs associated with one perspective, scope (intra-enterprise or inter-enterprise) and level (strategic or process). In the strategic framework definition, it is important to verify the consistency among the performance elements defined. This implies to analyse the correspondence and alignment between the strategic objectives at the intra-enterprise and the inter-enterprise scopes.

After the strategic framework, the process framework composed of the performance elements (objectives and KPIs) of the key business processes associated with the inter-organisational environment (Phase 2) is defined. The key processes are the ones directly linked to a common product/service produced by the partners as well as processes that support the fulfilment of those products/services. In the process framework definition, it is also important to verify the consistency among the performance elements defined. This implies to analyse the correspondence and alignment between: the strategic objectives and the process objectives at the intra-enterprise scope; the strategic objectives and the process objectives at the inter-enterprise scope; and the process objectives at both scopes: intra-enterprise and inter-enterprise.

In Fig. 2, the blue cubes indicate the location where business process interoperability is to be measured in the system. The PMS has a specific perspective designated for process measurement (process perspective); this perspective will include all the objectives and KPIs for managing business process. Business processes are monitored at both levels: the process perspective at the strategic framework and the process perspective of all the processes within the process framework. It is important to maintain this traceability between the objectives at both levels in order to verify the coherence of the system. In these blue cubes then, the objectives and KPIs for the business processes are defined, linked and monitored. The extent of the objectives within the blue cubes should contain the key objectives in terms of aspects such as time, efficiency and quality of the products (that are considered classical process objectives) as well as aspects related to interoperability such as degree of development and formalisation of business process models, alignment of processes. All these aspects should assure the seamless coordination and efficiency of business processes to decrease interoperability issues in the time towards an interoperable environment within the intra- and inter-enterprise contexts.

4 Application to an Agri-food Supply Chain

In this section, the performance management system is defined for an agri-food supply chain. Table 1 presents the objectives and KPIs of the strategic framework at the supply chain level. The strategic framework comprises 23 objectives and 28 KPIs. As can be observed in the table, for each objective, there is at least one KPI to assess its evolution. The process perspective integrates the business process interoperability objectives to be achieved and KPIs to be monitored. On the one hand, the process perspective includes the time and quality objectives and KPIs (lead time, shelf life, product safety). On the other hand, this perspective includes the business process representation, formalisation, semantics, IT connections objectives and KPIs (in italics).

Table 1 Strategic framework for the inter-enterprise level. Agri-food supply chain

Perspectives	Objectives	KPIs
Financial	Reduce production costs	KPI1 = % variation of production cost
	Increase profitability	KPI2 = % ROI variation ROI
		KPI3 = % ROA variation
	Reduce inventory cost	KPI4 = %variation of inventory cost
	Reduce transaction cost	KPI5 = % variation of transaction cost
Customer	Increase customer satisfaction	KPI6 = % customer satisfied/total customers
	Decrease customer complaints	KPI7 = % variation of customer complaints
	Backorders	KPI8 = % number of backorders to the total number of orders
	Lost sales	KPI9 = % number of lost sales to the total number of sales
Process	Reduce lead time	KPI10 = % variation of lead time
	Increase shelf life	KPI11 = % variation of shelf life
	Increase taste properties	KPI12 = % variation of brix value
	Increase product safety	KPI13 = % accomplishment of laboratory checks and monitoring processes according to certification schemes
	Improvement of storage and distribution conditions	KPI14 = % accomplishment of relative humidity and temperature complying with standard regulations
	Reduce environmental impacts	KPI15 = % variation of consumption of water
		KPI16 = % variation of consumption of energy
		KPI17 = % variation of recycling and re-use
	Increase on-time delivery orders	KPI18 = % on-time deliveries/total deliveries
	Increase traceability	KPI19 = % of materials with quality certification of origin
	Increase formalisation of the collaborative processes	*KPI20 = % involvement of key people for defining and reviewing standard process models*
		KPI21 = % key people satisfied with the model/total people
		KPI22 = % collaborative processes modelled

(continued)

Table 1 (continued)

Perspectives	Objectives	KPIs
	Improve IT interoperable connections	*KPI23 = % non-time efficient external connections/total external connections*
	Increase semantic interoperability	*KPI24 = % accomplishment of a semantic ontology definition*
Learning and growth	Increase innovation capability	KPI25 = Number of personnel suggestions
		KPI26 = % of personnel suggestions implemented
	Increase training	KPI27 = % training hours done/training hours planned
	Increase compromise of personnel	KPI28 = Number of policies and incentives developed
	Improve knowledge management	KPI29 = implementation of a knowledge management system

5 Conclusions

Enterprise collaboration among enterprises is today a necessity to achieve and maintain competitiveness. In inter-organisational contexts, the business processes of the enterprises that collaborate must work together to reach common objectives. For that reason, managing and monitoring the interoperability of business processes is a key issue. Performance management systems can aid to collect and structure the necessary information to compete in these environments, but proper performance structures of measurement must be defined and used. This paper has conceptualised the concept of business process interoperability and introduced its measurement into a performance management system for inter-organisational contexts. The future lines comprise its extension in other contexts to receive feedback that aids to redefine the work.

Acknowledgements Authors of this publication acknowledge the contribution of Project 691249, RUC-APS: 'Enhancing and implementing Knowledge based ICT solutions within high Risk and Uncertain Conditions for Agriculture Production Systems (www.ruc-aps.eu)' funded by the European Union under their funding scheme H2020-MSCA-RISE-2015 and the Project GV/2017/065 'Development of a decision support tool for the management and improvement of sustainability in supply chains' funded by the Regional Government of Valencia.

References

1. Matopoulos, A., Vlachopoulou, M., Manthou, V., & Manos, B. (2007). A conceptual framework for supply chain collaboration: Empirical evidence from the agrifood industry. *Supply Chain Management: An International Journal, 12*(3), 177–186.
2. Verdecho, M. J., Alfaro Saiz, J. J., Rodríguez Rodríguez, R., & Ortiz Bas, A. (2012). A multi-criteria approach for managing inter-enterprise collaborative relationships. *OMEGA, 40,* 249–263.
3. Kaplan, R. S., & Norton, D. P. (1992). The balanced scorecard—Measures that drive performance. *Harvard Business Review, 70*(1), 71–79.
4. Brewer, P. C., Speh, T. W. (2000). Using the balanced scorecard to measure supply chain performance. *Journal of Business Logistics, 21*(1), 75–93.
5. Bititci, U. S., Mendibil, K., Martinez, V., & Albores, P. (2005). Measuring and managing performance in extended enterprises. *International Journal of Operations & Production Management, 25*(4), 333–353.
6. Folan, P., & Browne, J. (2005). Development of an extended enterprise performance measurement system. *Production Planning and Control, 16*(6), 531–544.
7. Alfaro, J. J., Ortiz, A., & Rodríguez, R. (2007). Performance measurement system for enterprise networks. *International Journal of Productivity and Performance Management, 56*(4), 305–334.
8. Interop. (2008). *Interoperability ontology* [online]. Interop-vlab platform. Retrieved October 28, 2008 from http://interop-vlab.eu/.
9. Cabral, I., Grilo, A., Gonçalves-Coelho, A., & Mourão, A. (2016). An agent-based model for analyzing the impact of business interoperability on the performance of cooperative industrial networks. *Data & Knowledge Engineering, 105*, 107–129.
10. Gunasekaran, A., Patel, C., & Tirtiroglu, E. (2001). Performance measures and metrics in a supply chain environment. *International Journal of Operations & Production Management, 21*(1–2), 71–87.
11. Chan, F. T. S., & Qi, H. J. (2003). Feasibility of performance measurement system for supply chain: A process-based approach and measures. *Integrated Manufacturing System, 14*(3), 179–190.
12. Gunasekaran, A., Patel, C., & McGaughey, R. E. (2004). A framework for supply chain performance measurement. *International Journal of Production and Economics, 87*(3), 333–347.
13. Angerhofer, B. J., & Angelides, M. C. (2006). A model and a performance measurement system for collaborative supply chains. *Decision Support Systems, 42*(1), 283–301.
14. Gruat La Forme, F. A., Botta Genoulaz, V., & Campagne, J. P. (2007). A framework to analyse collaborative performance. *Computers in Industry, 58*(7), 687–697.
15. Alfaro-Saiz, J. J., Rodriguez-Rodriguez, R., Verdecho, M. J., & Ortiz, A. (2009). Business process interoperability and collaborative performance measurement. *International Journal of Computer Integrated Manufacturing, 22*, 877–889.
16. Verdecho, M. J., Alfaro-Saiz, J. J., & Rodriguez-Rodriguez, R. (2012). Prioritization and management of inter-enterprise collaborative performance. *Decision Support Systems, 53,* 142–153.

Interoperability Challenges in Building Information Modelling (BIM)

Justine Flore Tchouanguem Djuedja, Mohamed Hedi Karray, Bernard Kamsu Foguem, Camille Magniont and Fonbeyin Henry Abanda

Abstract The issue of interoperability is even more present in the building sector as the building is a complex object. Building Information Modelling (BIM) is a shared knowledge resource for information about a facility as a building. The diversity of actors involved and the very long life cycle of the building decrease the ability to solve identified problems. When tackling the importance of interoperability in BIM, overviewing challenges and listing existing solutions is an important step. This work highlights the need of interoperability in BIM. A review of the state of art on encountered interoperability challenges in BIM also helps to enumerate existing solutions. Because of various factors, some of the solutions proposed need improvements. One improvement could be the enhancing of a widely used BIM-based ontology. This identification work is part of a larger project, namely MINDOC. It is a foundation for future proposals in the area of BIM for sustainable construction.

Keywords Interoperability · BIM · Challenges

1 Introduction

Building Information Modelling (BIM) is a shared knowledge resource for information about a facility forming a reliable basis for decisions during its life cycle [1]. BIM enhances decision-making from design to construction and demolition of a construction project [2]. The information produced by any given BIM authoring software can be used by other software for different applications or purposes such as feasibility study, sunlight and thermal simulation, cost estimation and material management [2, 3]. In collaboration with some software vendors, the former International

J. F. Tchouanguem Djuedja (✉) · M. H. Karray · B. K. Foguem
Université Fédérale de Toulouse Midi-Pyrénées, INP-ENIT, Tarbes, France
e-mail: justine-flore.tchouanguem-djuedja@enit.fr

C. Magniont
LMDC, Université de Toulouse, INSA, UPS, Toulouse, France

F. H. Abanda
Oxford Brookes University, Oxford, UK

K. Popplewell et al. (eds.), *Enterprise Interoperability VIII*, Proceedings of the I-ESA Conferences 9, https://doi.org/10.1007/978-3-030-13693-2_23

Alliance for Interoperability (BuildingSmart) introduced OpenBIM, using the open BuildingSmart data model. OpenBIM is an approach for the design, realization and operation of buildings based on open standards and workflows. OpenBIM includes Industry Foundation Classes (IFC), an object-oriented standardized terminology and a common data model, which aims to facilitate data exchanges between BIM tools in building industry. It also comprises four other methodologies standards, namely Information Delivery Manual (IDM), BIM Collaboration Format (BCF), International Framework for Dictionaries (IFD) and Model View Definitions (MVD). IDM is in charge of processes description; it aims to provide the integrated reference for process and data required by BIM by identifying the discrete processes undertaken within building construction, the information required for their execution and the results of that activity. IDM will specify, for example, where a process fits and why it is relevant. BCF ensures coordination changes whilst IFD maps items and MVD translates processes into technical requirements. OpenBIM helps to avoid different interpretations and mistakes. The global aim is to work together without errors.

To ensure building sustainability, the BIM model should be in a format that can easily be exchanged and shared with other actors in a construction project. Those actors could be an energy simulation tool, a material database, a thermal or sunlight simulation tool, the project owner, a structural engineer or an architecture engineer.

To improve the use of BIM, so many other tools tap into it depending on the stages of a project, hence leading to a need for a good data exchange between the relevant tools. The paradigm of information exchange is enshrined in a concept known as interoperability.

This paper tackles the importance of interoperability in BIM and provides an overview of both challenges and existing solutions. The first part presents the need for interoperability in BIM. Next, a state of the art on encountered challenges is introduced. It is followed by the presentation of existing solutions to deal with issues.

2 The Need of Interoperability in BIM

Interoperability is the ability of diverse systems, organizations and/or individuals to work together, using the parts or equipment of each other, to achieve a common goal, regardless of their divergences. Ide and Pustejovsky [4] define it as a measure of the degree of that ability. The need for interoperability faces many obstacles, namely interoperability barriers. Archimède and Vallespir [5] distinguish three kinds of interoperability barriers: conceptual, technological and organizational barriers. Conceptual barriers concern only information problems like their representation at a high level of abstraction or the level of programming. They mainly concern syntactic and semantic incompatibility of the information to be exchanged. That is the difference of data formats, the ambiguity of meaning or understanding. Technological barriers concern IT problems, which is related to the use of computers. Organizational barriers concern human problems like responsibility, authority, or organizational structure and management. By considering these definitions, we conclude that Architecture, Engineering and Construction (AEC) domain is faced with conceptual and technological interoperability barriers. Syntactic and semantic incompatibility hinder interoperability at a very high level since it corrupts the ability of two or more systems or elements to exchange information and to use the information that has been exchanged. Resolve this will procure advantage of the structuration of the data exchange and the codification of the data including vocabulary, so that the receiving systems will be able to interpret it. When BIM systems are interoperable, different construction stakeholders working in the same office or on different sites can share information about the different phases of a project. A simple case would be when, in the same office, the structural engineer and the architect work separately on a 3D model of the same building. The two models must be then put together to plan the project in a tool such as Navisworks. Also, all updates by each actor must be report to the global 3D model along the life cycle of the building. To be done, this work needs an appropriate interoperability at all levels. This paper focuses on highlighting interoperability challenges in BIM, particularly when used for sustainable construction. Interoperability in BIM is an important need for it serves for rules checking [6] and energy performance assessment (EPA) [7]. The latter demonstrates how interoperability can improve BIM-based EPAs. Choi et al. [7] develops for this aim a material library and an openBIM-based energy analysis software, validated by a case study. Interoperability here takes advantage of the fact that more than 70% of the information needed for the building energy analysis is already contained in BIM data. In the construction industry, EPA should be boosted through a perfect interoperability between BIM data and energy simulation models. Interoperability is also an important issue in complex area such as architectural precast facades [8]. The erection of that kind of facade requires close collaboration amongst various actors: architects, precast fabricators, structural engineers and general contractors; it is a good candidate for the usage of interoperability in AEC domain. BIM's interoperability can also be useful to check the compatibility of buildings with *Haute Qualité Environnementale* (HQE) and Building Research Establishment Environmental Assessment Method (BREEAM) standards, or with the E+/C− (Energie+/Carbone−) label. For rules checking in AEC domain, interoperability can improve communication between BIM's software and rule-checking environments.

3 Review of the State of Art on Interoperability Challenges in BIM

Despite the benefits of interoperability, there exists some main factors hindering its full potential in sharing construction project information.

Steel et al. [9] has mentioned four levels of interoperability in BIM whilst focusing on IFC-based interoperability:

- File level is the ability of two tools to successfully exchange files
- Syntax level is the ability of two tools to successfully parse those files without errors
- Visualization level is the ability of two tools to faithfully visualize model being exchanged
- Semantic level is the ability of two tools to come to a common understanding of the meaning of a model being exchanged.

Considered those levels, Steel et al. [9] enumerates lot of interoperability issues:

Issue 1: Very large size of the models being used. It results in failures when generating 2D drawings or rendering in 3D, and the inability to load models because of number of objects or memory consumption restrictions.
Issue 2: The use and reuse of geometries which results in inappropriate position of objects when tools are changed.
Issue 3: Alternative Visualizations which modify model appearance in different tools, depending on the objectives in play.
Issue 4: A loose approach to the use of object identifiers which causes difficulties of versioning in case of the merging of models from different actors for example.
Issue 5: Coverage of a BIM-based language by implementing tools, or coverage of the domain by the intended language.
Issue 6: Variation of levels of parameterisation support by different tools.

In summary:

- Firstly, there is a challenge in capturing and translating knowledge from experts into a BIM software. The perspective of an architecture engineer who enters information is not the same as the perspective of an energy analyst. For example, a beam could be viewed as a volume of concrete and a mass of steel reinforcing bars by the architecture side, whilst it would be a thermal bridge for the energy designer.
- Secondly, sometimes, in one specialty, different kinds of information from BIM must be considering. For instance, in construction: facility management, architectural, manufacturing and geometric BIM information are needed. This involves various software systems to undertake the construction of a building [2].
- Thirdly, the fact that data evolve following the different phases of a construction project is a challenge. Data present in building models evolve as the project progresses. Thus, design and construction models are rarely the same, especially in very complex projects. At the beginning of the building design, the model contains only an assembly of 3D objects. For example, for a concrete wall, all starts with a 3D drawing. Then, little by little other information are added: its cost, its role, the time dimension, etc. These additions take place one after the other throughout the life cycle of the building and will allow at certain moments to evaluate the environmental impact of this wall and then to ideally make the necessary adjustments before the construction. The problem is that the information needed for an efficient assessment is not necessarily available at the right moment. Furthermore, if the used information change, the assessment becomes incorrect.

The key question is 'what are solutions to interoperability challenges with regard to BIM for sustainable construction?' To answer this question, it is imperative to learn from previous studies. The next section tries to give some answers.

4 Existing Solutions for Interoperability Issues in BIM

Many solutions have been proposed to solve interoperability issues in building domain. Grilo and Jardim-Goncalves [10] cited some of them. Amongst others, there are international/regional or national standards, labels, ontologies and so on. Afsari et al. [11] proposes to use the JavaSript Object Notation (JSON) format to introduce ifcJSON schema and its data content. ifcJSON is the first implementation of IFC data model based on JSON data exchange format, and it would be an alternative to ifcXML. From a standardized JSON schema, valid ifcJSON documents produced are to be used for Web-based data transfer and to improve interoperability of cloud-based BIM applications. Unfortunately, there are currently no tools capable of previewing geometric data contained in ifcJSON. Hu et al. [12] addressed interoperability challenge between architectural and structural models and amongst multiple structural analysis models. They have proposed the prototyping of IFC-based Unified Information Model (UIM) and various algorithms in two software system architectures: Client/Server (Unitive-BIM) and Browser/Server (Web-BIM) platforms. IFC-based UIM is a data model implemented as a central data server. For model display, the Web-BIM platform is based on WebGL (GL means Graphic Library), whilst Unitive-BIM is based on OpenGL. Pauwels et al. [13] bring AEC and semantic web technologies together by proposing an OWL (Web Ontology Language) ontology for IFC file format. IFC is as a matter of fact, the main file format used in building industry nowadays. Establishing that the semantic Web technologies were likely to overcome the interoperability problems in building field, they took advantage of the production of ifcOWL ontology by the BuildingSmart's LDWG team, for their proposal. The BuildingSmart's LDWG team has relied on earlier work [14] to 'convert the IFC schema into an OWL ontology and to convert IFC STEP Physical Files (SPF) into Resource Description Framework (RDF) graphs that follow the ifcOWL ontology'. Törmä [15] has introduced a way to combine Web of Data (WoD) and IFC technologies. The need for interoperability increases as BIM model data evolves. Within BIM, Torma [15] distinguishes type-level interoperability from instance-level interoperability. Type-level interoperability concerns the common interpretation that different tools share for the same object, whilst instance-level interoperability concerns the types of entities representing different aspects of the same real-world object. The Web of building data consists of three spheres corresponding to the degree of complexity needed for a proper exchange of data between actors. The deepest sphere corresponds to the part of WoD represented according to the IFC ontology: it is the Web-based BIM. Introduced by de Farias et al. [16], FOWLA means Federate Architecture for OWL Ontology and aims to improve interoperability of BIM at data level. This rule-based federated architecture aims to leverage semantic Web technologies for interoperability between the AEC/FM (Facility Management) and other ontologies, to solve data structure heterogeneity issues.

BuildingSmart [1] describes some International/National or Regional standards as an attempt to solve interoperability issues in BIM. They are constituted of:

- TC184/SC4 by ISO within WG3 (Product modelling), the T22: Building construction group.
- ISO10303-STEP, part 225 called: 'Application Protocol (AP): Building Elements Using Explicit Shape Representation'.
- IFC developed by Industry Alliance for Interoperability (IAI) to improve interoperability of applications from different software vendors. The latter has adopted ISO EXPRESS language to describe its models.

In fact, a lot of classes in IFC have been defined according to ISO10303 standard and its derivatives: ISO10303-46, ISO/CD 10303-46:1992, ISO10303-42, ISO/CD 10303-42:1992. IFC also facilitates exchange between already used format in BIM, such as rvt or rfa, through translation processes.

Another attempt cited in [10] is the BIM standardization to link GIS to AEC by two means: by linking IAI/IFC to GIS or by linking GIS-BIM-CAD.

For energy simulations and for predicting current energy demand and carbon emissions, Arayici et al. [17] have proposed an interoperability specification to promote early collaboration.

The schema in Fig. 1 summarizes the content of this document.

5 Concluding Remarks

Translation in order to achieve its aim in building domain, BIM needs a reliable environment to enable flexible information exchange between all project actors. The

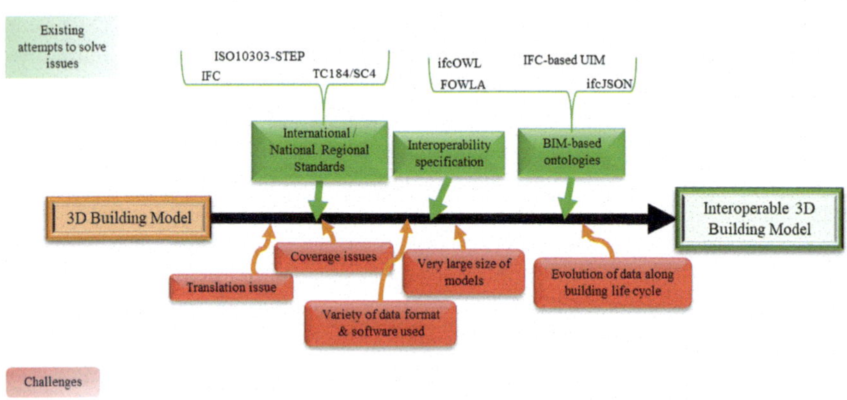

Fig. 1 Interoperability in BIM

so-called interoperability is a key to ensure a sustainable construction of buildings. Encouraged by the complexity of building as an object, interoperability challenges are more and more present in BIM as it wants to evolve, trying to respond to increasing needs of various actors. The major interoperability problems identified in the BIM are: the translation or coverage issues, the variety of tools dealing with different kinds of information and the very large-size models. There is also the evolution of data all along the life cycle of the building and the alternative visualization issue. International standards such as TC184/SC4, ISO10303-STEP or IFC were amongst first attempts to reach interoperability in BIM. They were completed by interoperability specification, and then the construction of many BIM-based ontologies or data model implementation like ifcOWl, FOWLA, IFC-based UIM or ifcJSON. Considering the state of the art review, it is sure that all these efforts are significant. Yet, they are insufficient to ensure interoperability in the field of AEC and particularly for sustainable construction. Because of the lack of coverage of the domain or because of the quality of ontology construction, some of solutions proposed need improvements. One improvement could be the enhancing of a widely used BIM-based ontology.

Acknowledgements This work was done in the scope of MINDOC project funded by the region Occitanie.

References

1. 'About', *buildingSMART*.
2. Kensek, K. (2015). *Manuel BIM: Théorie et applications*. Eyrolles.
3. Abanda, F. H., Vidalakis, C., Oti, A. H., & Tah, J. H. M. (2015). A critical analysis of Building Information Modelling systems used in construction projects. *Advanced Engineering Software, 90,* 183–201.
4. Ide, N., & Pustejovsky, J. What does interoperability mean, anyway? Toward an operational definition of interoperability for language technology, p. 8.
5. Archimède, B., & Vallespir, B. (2017). *Enterprise interoperability: INTEROP-PGSO vision*. Wiley, Hoboken.
6. Pauwels, P., et al. (2011). A semantic rule checking environment for building performance checking. *Automation in Construction, 20*(5), 506–518.
7. Choi, J., Shin, J., Kim, M., & Kim, I. (2016). Development of openBIM-based energy analysis software to improve the interoperability of energy performance assessment. *Automation in Construction, 72,* 52–64.
8. Sacks, R., Kaner, I., Eastman, C. M., & Jeong, Y.-S. (2010). The Rosewood experiment—Building information modeling and interoperability for architectural precast facades. *Automation in Construction, 19*(4), 419–432.
9. Steel, J., Drogemuller, R., & Toth, B. (2012). Model interoperability in building information modelling. *Software & Systems Modeling, 11*(1), 99–109.
10. Grilo, A., & Jardim-Goncalves, R. (2010). Value proposition on interoperability of BIM and collaborative working environments. *Automation in Construction, 19*(5), 522–530.
11. Afsari, K., Eastman, C. M., & Castro-Lacouture, D. (2017). JavaScript Object Notation (JSON) data serialization for IFC schema in web-based BIM data exchange. *Automation in Construction, 77,* 24–51.

12. Hu, Z.-Z., Zhang, X.-Y., Wang, H.-W., & Kassem, M. (2016). Improving interoperability between architectural and structural design models: An industry foundation classes-based approach with web-based tools. *Automation in Construction, 66,* 29–42.
13. Pauwels, P., Zhang, S., & Lee, Y.-C. (2017). Semantic web technologies in AEC industry: A literature overview. *Automation in Construction, 73,* 145–165.
14. Pauwels, P., Terkaj, W., Krijnen, T., & Beetz, J. (2015). *Coping with lists in the ifcOWL ontology*, p. 10.
15. Törmä, S. (2014). Web of building data—Integrating IFC with the Web of Data. In A. Mahdavi, B. Martens, & R. Scherer (Eds.), *eWork and eBusiness in architecture, engineering and construction* (pp. 141–147). CRC Press.
16. de Farias, T. M., Roxin, A., & Nicolle, C. (2016). A federated approach for interoperating AEC/FM ontologies. In *LDAC2016—4th Linked Data in Architecture and Construction Workshop*, Madrid, Spain.
17. Arayici, Y., Fernando, T., Munoz, V., & Bassanino, M. (2018). Interoperability specification development for integrated BIM use in performance based design. *Automation in Construction, 85,* 167–181.

An Application of Managing Deviations Within a Supply Chain Monitoring Framework

Zheng Jiang, Jacques Lamothe, Julien Lesbegueries, Frederick Benaben and Frederic Hauser

Abstract Cloud Collaborative Manufacturing Networks (C2NET) project is a European H2020 project started in January 2015. The main focus of this project is to develop a cloud platform that supports the services for the collaborative planning processes management among partners within supply chain networks. This paper presents an application of the generation and visualization of supply chain deviations. This application is based on a monitoring framework, which provides the following main functionalities: collecting data from legacy ERP systems, generating automatically models of supply chain planed and current status, detecting deviations between two statuses, and suggesting adaption processes associated to deviations and visualization of deviation assessment.

Keywords Supply chain monitoring · Supply chain modeling · Supply chain visibility

1 Introduction

1.1 Background

During the last decades, European Small and Medium Enterprises (SMEs) have faced huge challenges in accessing to advanced information management systems and collaborative tools, as a result of the restricted resources. As a consequence, the ongoing project, i.e., C2NET [1], focuses on creating a cloud platform, to better manage the supply chain plans and data associated. The platform provides three main functionalities: a data collection framework (DCF), which gathers data from legacy ERP systems and IoT (Internet of Things) devices; optimization service (OPT),

Z. Jiang (✉) · J. Lamothe · J. Lesbegueries · F. Benaben
Centre Génie Industriel, IMT Mines Albi Carmaux, Toulouse University Albi, Toulouse, France
e-mail: zheng.jiang@mines-albi.fr

F. Hauser
Supply Chain Direction, Pierre Fabre Dermo-Cosmetics Lavaur, Lavaur, France

K. Popplewell et al. (eds.), *Enterprise Interoperability VIII*, Proceedings of the I-ESA Conferences 9, https://doi.org/10.1007/978-3-030-13693-2_24

which estimates optimization algorithms for optimizing a (production or delivery) plan; and collaboration tools, which provides applications assisting collaboration including monitoring plan execution (MPE) component and orchestration planning process (OPP) component.

The aim of this paper is mainly focused on introducing a supply chain deviation management application within the MPE component of which could effectively support the agility of a supply chain network of enterprises.

Several issues are associated to supply chain agility [2]: an issue of visualization for the purpose of utilizing huge amount of data to generate models of supply chain situations; an issue of detecting of the disturbances; an issue of decision to quickly propose adaptation solutions; and an issue of deployment in order to coordinate the response in practice. The present paper is focusing on the visibility issues.

1.2 Literature Review

Odette developed the definition of supply chain monitoring (SCMo) [3] as "the effort of the actors in a supply chain to manage and control visibility of information regarding flows of products and services in different levels and directions." Afterward, the SCMo is applied to the field of industries as "the permanent checking that current inventory is synchronized with the respective demand."

It is well proven that supply chain visibility is a significant contributor regarding agility [2], risk [4–6] or responsiveness [7] management, with the aim to prevent wrongly made decisions, excessive inventories, bullwhip effect, and risk to profit [8]. Realizing these facts, supply chain visibility [5] owns the feasibility of sharing customer demand data, inventory information, transportation costs, and other factors of a supply chain. It thus could facilitate identification and react of the risks of supply chain for the enterprises [6].

In last years, researches have focused on the usage of the data by supply chain visibility. Due to the large amount of collected data and visibility objectives, some big data issues are concerned in C2NET project:

- Quality: to obtain accurate, complete, and consistent data at the level of aggregation [9].
- Modeling: to automatically generating models from the interpretations of the transformed data. It can be achieved by using a metamodel [10].
- Detection: identify events according to patterns from a flow of data. It can be achieved by a complex event processing (CEP) engine [11].

2 Monitoring Plan Execution Framework

In Fig. 1, it is presented the five interacting components in MPE framework [12]. The Pilot system contains the legacy ERP systems of a company. The data associated to status (planned and current) of supply chain status are provided. Besides, IoT devices can also provide data that focus in the field (current status).

These data can be input of the modeling service depending on what kind of data source is supported and interested in the pilot side. Due to privacy politics of some companies, the direct access to legacy systems is not supported. Concerning this kind of interoperability problems, it is supported that the actor extracts excel files from legacy systems and uploads manually as the input. As explained above, this paper is supposed to focus on the MPE framework. Consequently, DCF features are bypassed.

Data from pilot systems comes to the modeling service and transformed into a wireless sensor networks (WSN) event format. Interpretation of the data is made by a complex event processing (CEP) engine that subscribes to dedicated events. The interpretation is generating instances of a reference metamodel [13], called as R-model (reference model) that contains the planned status of a supply chain network.

R-model is passed to the detection service and stored in a graph database (Neo4j). The detection service subscribes to it and duplicates the R-model when there is a new one. The R-model duplication is used as S-model (situation-model). At the same time, the detection service also subscribes to the CEP engine for new data arrivals. The S-model will be updated according to these new data. After the updating, a comparison is made to detect deviations between R-model and S-model.

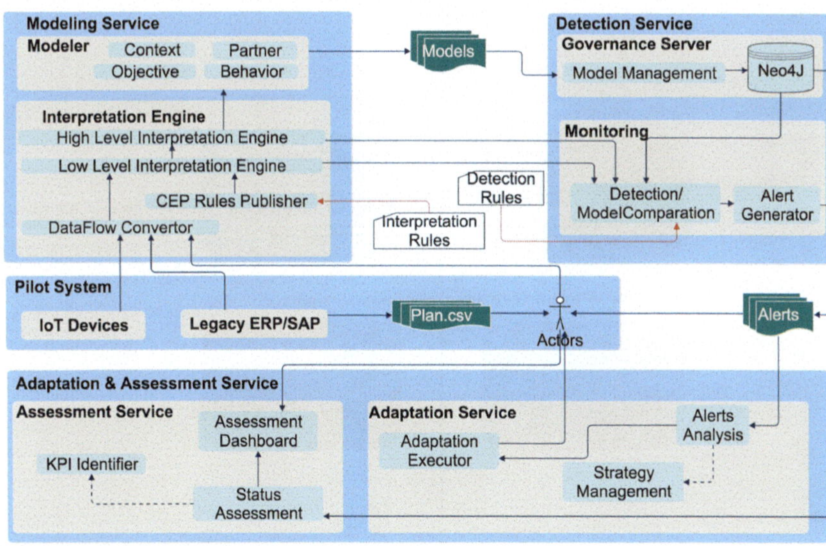

Fig. 1 Monitoring plan execution framework

Once deviation(s) has been detected, it will generate an alert to notify the associated actors. Deviation(s) is also sent to the adaptation service for further analysis.

The adaptation service obtains detected deviations and makes analysis to determine whether they are significant. The deviation by default is absolute deviation, but there can be also a threshold if needed. The adaptation service will categorize the deviations and associate pre-defined processes as a solution. Actors will be notified with the solution. The execution of the adaptation processes will be carried on when associated actors validate.

The assessment service is subscribed to the deviations that are detected in order to visualize them through a dashboard. The dashboard provides different views of assessment of the deviations.

3 The Deviation Assessment Dashboard

3.1 The Dashboard Template

As it has been introduced in Sect. 2, the services of MPE are mostly backend services and hidden from the users. Due to visibility issues caused by the amount of the data (instances in a model), it is more interesting to display the detected deviations as results instead of the huge original models. In order to achieve this objective, here we propose a deviation assessment dashboard template (see Fig. 2).

In the dashboard (template), there are four panels that contain the associated information of detected deviations.

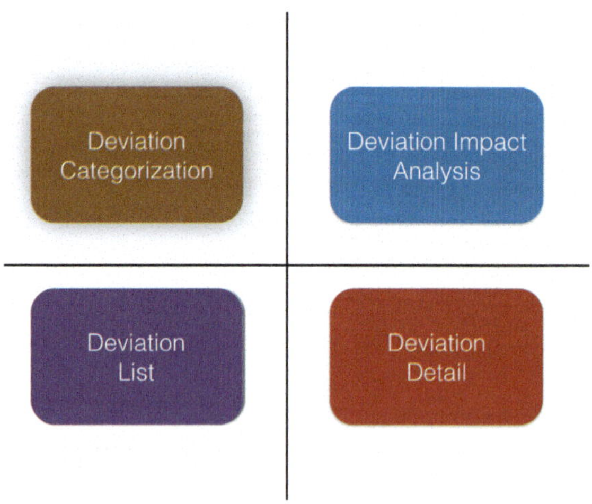

Fig. 2 Deviation assessment dashboard template

Table 1 Template of a criteria matrix

	Factor 1	Factor 2	Factor 3
Criticality A	Case 1	Case 1	Case 2
Criticality B	Case 2	Case 3	Case 3
Criticality C	Case 3	Case 3	Case 4

Deviation Categorization In the deviation categorization panel, the deviations are categorized according to a criteria matrix. In the criteria matrix, there are configurable dimensions used to identify the position of a detected deviation. Each element of the matrix represents one situation that the deviation is facing. Within the element also a series of adaptation processes are pre-defined in terms of unraveling the deviation.

For the dimensions of the matrix, one of the regular dimensions is the criticality of a supply chain concept. It could be a product priority, order priority, customer criticality, etc. The other dimensions can be different factors the supply chain manager is interested in, for example, it could be stock level, horizon, etc. In Table 1, a template of a criteria matrix is given.

This matrix can be adapted according to different requirements of the supply chain manager. It can be used as more than a two-dimensional matrix but also multi-dimensional matrix if there are more than one factor to be considered. Finally, the matrix should converge into a matrix, which represents the mapping between category of deviations and series of adaptation processes. In Fig. 3, an evolution of the criteria matrix is shown.

Deviation Impact Analysis It is important to not only consider the deviation itself, but also in a context. In practice, there can be deviations that are detected, but not causing any problem, or leading to a serious disruption. These cases need to be distinguished thanks to the impact analysis.

Deviation List A list that contains the deviations, with associated information in order to filter/sort.

Deviation Detail The detailed information of one deviation.

The proposed deviation assessment dashboard template is a skeleton for defining a usable dashboard. Due to the variety of supply chain context, it is hardly probable to define a dashboard that displays all the interested information for all the contexts. This template can be customized to a certain use case that contains the concerned information of a potential deviation of a company or a supply chain network.

In the following Sect. 3.2, an instance of the template is given. This instance is a customization for a scenario of Pierre Fabre Dermo Cosmetic (PFDC) supply chain.

	Factor 1	Factor 2	Factor 3
Criticality A	Case 1	Case 1	Case 2
Criticality B	Case 2	Case 3	Case 3
Criticality C	Case 3	Case 3	Case 4

*

	Factor A	Factor B	Factor C
Criticality A	Case 1	Case 1	Case 2
Criticality B	Case 2	Case 3	Case 3
Criticality C	Case 3	Case 3	Case 4

	Factor 1*A	Factor 1*B	Factor 1*C	Factor 2*A	Factor 2*B	Factor 2*C	Factor 3*A	Factor 3*B	Factor 3*C
Criticality A	Case 1	Case 1	Case 2	Case 1	Case 1	Case 2	Case 2	Case 2	Case 3
Criticality B	Case 3	Case 4	Case 5	Case 3	Case 6	Case 6	Case 5	Case 6	Case 6
Criticality C	Case 6	Case 6	Case 7	Case 6	Case 6	Case 7	Case 7	Case 7	Case 8

Deviation Case	Adaptation Process
Case 1	Process 1 and 2
Case 2	Process 2
Case 3	Process 2 and 3
Case 4	Process 3 and 4
Case 5	Process 5
Case 6	Process 5
Case 7	Process 5
Case 8	Process 5/None

Fig. 3 Evolution from two-dimensional matrix to final process mapping

3.2 PFDC Dashboard

For this scenario, the dashboard has been customized to meet the interest of the informational concentration of the deviations of PFDC. An implemented user interface is given in the following Fig. 4 [14].

Deviation Categorization
For the criteria matrix in the deviation categorization panel, the first dimension of the criteria matrix, it is defined by four product priorities, A+, A, B+, and B. Table 2 shows this definition (Fig. 5).

For the factor dimensions, in this case, there are two dimensions of factor: impact and horizon. For the impact factor, it is going to be introduced in the next subsection.

For the horizon factor, there are three different horizons are defined in this scenario: short-term (0–2 weeks), mid-term (2–8 weeks), and long-term (8–12 weeks) (Fig. 6).

Deviation Impact Analysis
In this scenario, the impact analysis of a deviation is done thanks to counting a series

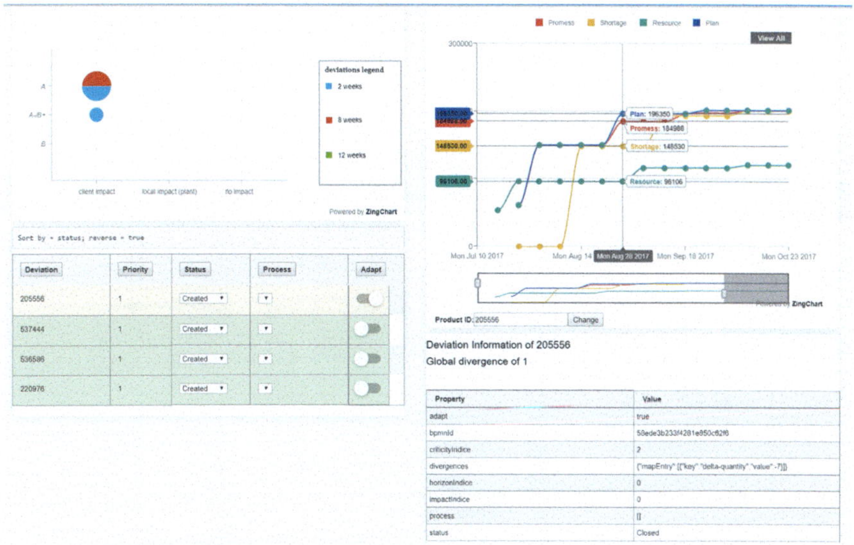

Fig. 4 Implanted customized PFDC dashboard

Table 2 PFDC scenario product criticality definition

Criticality	Product characteristic
A+	Products in central business and critical in local context
A	Products in central business
B+	Products not in central business but critical in local context
B	Products not in central business

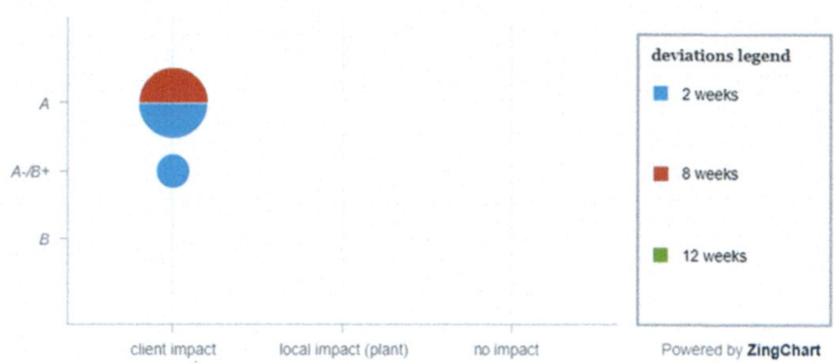

Fig. 5 Deviation categorization panel

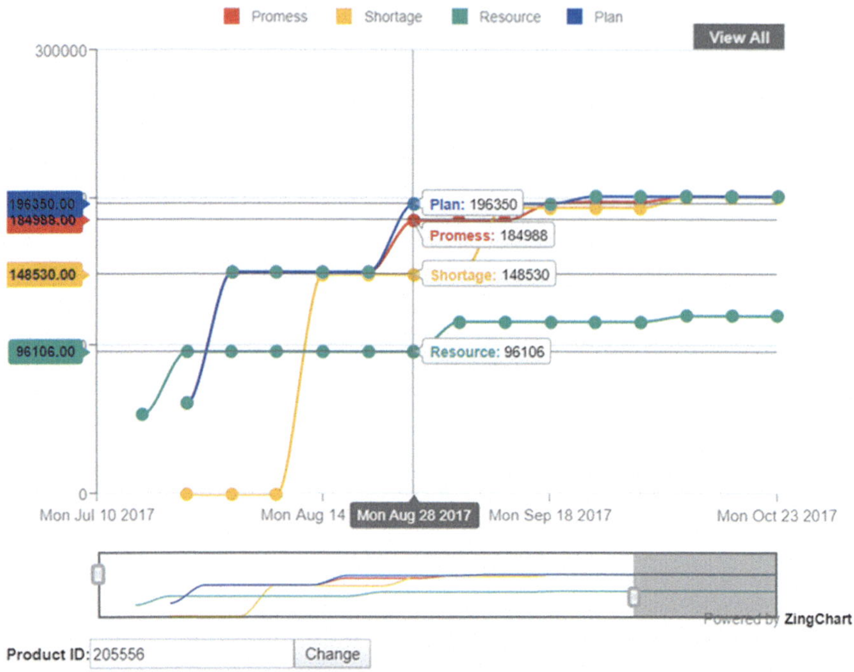

Fig. 6 Impact analysis panel

Table 3 Comparison among the planning factors

Condition	Impact
C > D	There will be a series shortage caused
B > D	There is a promise breakdown but not a shortage
A > D	There is a consumption of safety stock but not a shortage
D > A or B or C	There is no impact nor a consumption of safety stock

of factors of the planning and the storage that are computed from the legacy DRP systems (local DRP of the subsidiaries and central DRP of central warehouse). By comparing the variables, the impact of a deviation can be defined.

The factors are:

- Plan (A): the quantity of a product that is planned to reach.
- Promise (B): the quantity that is promised to deliver to the subsidiaries.
- Shortage (C): the quantity that will cause a shortage to the market.
- Resource (D): the available stock level of the product.

These factors are calculated as accumulated from the beginning of a DRP plan till a "target" date. By the following comparison among the factors (Table 3), the impact of a deviation can be identified:

Sort by = status; reverse = true

Deviation	Priority	Status	Process	Adapt
205556	1	Created ▾	▾	
537444	1	Created ▾	▾	
536586	1	Created ▾	▾	
220976	1	Created ▾	▾	

Fig. 7 Deviation list panel

Deviation Information of 205556

Global divergence of 1

Property	Value
adapt	true
bpmnId	58ede3b233f4281e850c62f6
criticityIndice	2
divergences	{"mapEntry":[{"key":"delta-quantity","value":-7}]}
horizonIndice	0
impactIndice	0
process	[]
status	Closed

Fig. 8 Deviation detail panel

By defining this, the impact dimension in the above section can be defined as: client impact (C > D), local impact (B or A > D), and no impact (D > A, B and C).

Deviation List

In the deviation list panel, the detected deviations are listed here, and it can be sorted by different properties. Here in the list, an adaptation process can be selected for resolving a deviation (Fig. 7).

Deviation Detail

In the deviation detail panel, the detailed information associated to one deviation are listed. User can view it to understand the deviation (Fig. 8).

4 Conclusion

This paper introduced a prototype a deviation assessment dashboard in order to visualize the result (deviations) of the MPE framework. It has been integrated as a part of the C2NET platform. This dashboard is composed of four panels: deviation categorization, deviation impact analysis, deviation list, and deviation detail. User can visualize deviations in the supply chain plan execution. At this moment, the interface is customized according to PFDC requirements. In case of other interest of this monitoring feature, the dashboard is able to be adapted to meet other requirements.

Acknowledgements The authors wish to acknowledge the European Commission for their support of funding the project C2NET (Cloud Collaborative Manufacturing Networks).

References

1. C2Net|D6.6—White Paper of C2NET platform/openness and portability—Deliverables [online], 2016. Retrieved October 24, 2016 from http://c2net-project.eu/deliverables/-/blogs/d6-6-white-paper-of-c2net-platform-openness-and-portability.
2. Charles, A., Lauras, M., & Van Wassenhove, L. (2010). A model to define and assess the agility of supply chains: Building on humanitarian experience. *International Journal of Physical Distribution & Logistics Management, 40*(8–9), 722–741. Author, F., Author, S., & Author, T. (1999). *Book title* (2nd ed.). Location: Publisher.
3. Merle, O. (2003). Best-practice concept for supply chain monitoring. In P. P. Sachsenmeier & P. M. Schottenloher (Eds.), *Challenges between competition and collaboration* (pp. 277–296). Berlin, Heidelberg: Springer.
4. Christopher, L., & Lee, H. (2004). Mitigating supply chain risk through improved confidence. *International Journal of Physical Distribution & Logistics Management, 34*(5), 388–396.
5. Nooraie, S. V., & Parast, M. M. (2015). A multi-objective approach to supply chain risk management: Integrating visibility with supply and demand risk. *International Journal of Production Economics, 161,* 192–200.
6. Messina, D., Santos, C., Soares, A. L., & Barros, A. C. (2016). Risk and visibility in supply chains: An information management perspective. In *Handbook of research on information management for effective logistics and supply chains* (vol. 34).
7. Gunasekaran, A., Lai, K., & Edwin Cheng, T. C. (2008). Responsive supply chain: A competitive strategy in a networked economy. *Omega, 36*(4), 549–564.
8. Li, G., Fan, H., Lee, P. K., & Cheng, T. C. E. (2015). Joint supply chain risk management: An agency and collaboration perspective. *International Journal of Production Economics, 164,* 83–94.
9. Hazen, B. T., Boone, C. A., Ezell, J. D., & Jones-Farmer, L. A. (2014). Data quality for data science, predictive analytics, and big data in supply chain management: An introduction to the problem and suggestions for research and applications. *International Journal of Production Economics, 154,* 72–80.
10. Wang, T., Truptil, S., & Benaben, F. (2016). An automatic model-to-model mapping and transformation methodology to serve model-based systems engineering. *Information Systems and e-Business Management*, 1–54.
11. Luckham, D. (2008). The power of events: An introduction to complex event processing in distributed enterprise systems. In: *International Workshop on Rules and Rule Markup Languages for the Semantic Web* (pp. 3–3). Springer, Berlin.

12. Jiang, Z., Lamothe, J., & Bénaben, F. (2017). A monitoring framework of collaborative supply chain. *IFAC-PapersOnline, 50*(1), 13072–13077.
13. Jiang, Z., Lamothe, J., & Bénaben, F. (2016). Meta-modeling of collaborative supply chain. In *Enterprise Interoperability VII* (pp. 307–320). Springer International Publishing.
14. Zingchart. https://www.zingchart.com/. Last accessed 2017/09/01.

Part IX
Interoperability in Manufacturing and Repair and Operation (MRO)

Toward Information System Architecture to Support Predictive Maintenance Approach

Alexandre Sarazin, Sébastien Truptil, Aurélie Montarnal and Jacques Lamothe

Abstract The prognostic and health management (PHM) approach aims at supporting maintenance operations in order to ensure the functionality of a system. In order to achieve this objective, a PHM approach is composed of a prognostic component, able to send a prognostic of failure, and a component able to give the health status of the system. Nowadays, this approach suffers from a lack of exploitation of the emerging technologies. This article presents a novel architecture for PHM approach able to extract added value from data. This lambda architecture embeds two layers: a speed layer and a storage layer. Thanks to the storage layer, maintenance rules can be applied as well as the result of machine learning algorithms to the speed layer in order to realize the prognostic aspect of the PHM. In addition, the system has to deal with heterogeneous data, which comes with the necessity to handle the big data issues as well as making it interoperable. This is achieved thanks to a service-oriented architecture approach and the use of complex event processing.

Keywords Prognostic and health management · Maintenance · Lambda architecture

1 Introduction

The development of new technologies or concepts such as the Industry 4.0 [1], the Internet Of Things, the Internet of Services, the Smart factory will bring new requests from industries in order to increase their competitiveness. As a consequence, indus-

A. Sarazin (✉) · S. Truptil · A. Montarnal · J. Lamothe
Centre de Génie Industriel, Université de Toulouse—IMT Mines Albi, Albi, France
e-mail: alexandre-m.sarazin@mines-albi.fr

S. Truptil
e-mail: sebastien.truptil@mines-albi.fr

A. Montarnal
e-mail: aurelie.montarnal@mines-albi.fr

J. Lamothe
e-mail: jacques.lamothe@mines-albi.fr

K. Popplewell et al. (eds.), *Enterprise Interoperability VIII*, Proceedings of the I-ESA Conferences 9, https://doi.org/10.1007/978-3-030-13693-2_25

tries are asking for more and more functionalities, which in turn result in increasingly complex systems.

Despite this complexity, these systems must meet quality and reliability requirements [2, 3]. Indeed, in [3], the authors remind us that the consequences of failures can generate a financial loss (for example due to delay) but also unacceptable damages to human as well as to material, and eventually degrade the image of a company.

In order to limit these effects and the related economical risks, many efforts have been made over the past two decades by researchers and industries to find methods for predicting and preventing malfunctions and the resulting disasters. These methods have been developed in several fields of application, such as electronics [4, 5], public health [6] and the aeronautical field [7, 8], and can be grouped under the term prognostic and health management (PHM).

Although PHM approach has been developed for more than 10 years, in [9] the authors explain that these approaches are focused on remaining useful life (RUL), which is one specific aspect of PHM but, still, other sides have to be considered, such as the No Fault Found Problems for example. The No Fault Found Problems characterize failure situations that are not detectable or reproducible by experiments [10]. In [10], the authors explain that 90% of the costs of maintenance in the aeronautical field is due to activities related to the No Fault Found Problems. In [11], the authors report that the cost of these activities for the US Department of Defense is between $2 billion and $10 billion per year.

In parallel, first the cloud computing explosion and then the big data emergence have brought with them new possibilities. In particular, in [12], based on a survey on the topic of cloud-enabled prognosis, the authors highlight that applying these new technologies to real-time maintenance prognosis enabled by the cloud technologies and the Internet of Things (IoT) at a wider level, results in three main gaps that can be summarized by data transmission, storage and analysis.

This paper presents a novel approach, which aims at monitoring units thanks to the new technologies (Industry 4.0, IoT, etc.) in order to predict failures, and diagnose and improve the maintenance activities with the objective of completing the PHM approach by using the new identification and tracking capabilities (i.e., datasets related to the lifecycle of the units). Regarding the number of units per industry, a huge quantity of data will be generated. These data about different events linked to device have to be treated automatically to permit us to extract and understand the device state.

The first part of this article describes the main characteristics of a prognostic and health management (PHM) approach. Then the objectives of a PHM able to exploit the features of the new technologies are listed. Based on these objectives, the second part of this article aims at proposing an architecture able to reach them. Finally, the third part insists on the importance of the knowledge base for which an architecture of metamodel is proposed.

2 Prognostic and Health Management (PHM) Approach

2.1 Definition and Gaps of PHM

In [5], the author describes the prognostic and health management approach as follows:

- **Prognostic**: is the process of predicting the reliability of a product based on the deviation of its function from its "nominal" state.
- **Health Management**: represents the process of measuring, saving and monitoring the data in order to detect failures of the product in real time.

Therefore, a PHM approach aims at offering (i) a model of the health status of a product, (ii) a system that allows the detection of problems and (iii) plans maintenance activities.

In [9], the authors underline the fact that existing PHM approaches have the following gaps:

- They are dedicated to specific equipment.
- They are applied to a specific problem without explanation of the chosen methods.
- Visualization tools are missing as well as the possibility to share information. Thus, there is a lack of support for decision.

2.2 Objectives of the PHM Approach

In order to build a PHM approach, all the data extracted about the unit's state have to be processed in order to understand the health status of a system. The main reason for this is that they all bring different information that either "consolidate" each other (e.g., enhance the reliability of the data) or complete each other (e.g., necessity of data of different nature to infer a failure). In [13], the authors highlight the importance of being able to provide a real-time health status of a machine, and propose a system able to gather data from machines in real time, analyze them through a RUL prediction service to finally share the machine's status. To go a bit further and in an attempt to address the No Fault Found Problem too, as proposed by [14], it is planned to use the Lambda architecture to exploit new technologies for two purposes:

Real-time and a reactive analysis: This module aims to obtain the health status of the studied system and issue alerts in case of failure. On the one hand, it is based on expert rules, including thresholds to trigger the alerts. On the other hand, two models are generated: (i) an expected model of the health status of the unit based on its lifecycle and (ii) a model of the current state of the unit. These two models are then used by a comparison algorithm to detect unexpected anomalies, in which case the system will ultimately trigger alerts.

Such model comparison approach was studied and applied in the works described in [15]; however, it was then applied to the crisis management field and now needs to be adapted to the industrial maintenance.

While it helps users making more efficient and objective decisions thanks to accurate resulting of well measured and evaluated data, the visualization of the models will bring a more intuitive maintenance information system to industries.

Capitalized knowledge: This module aims to store and take advantage of previous situations in order to improve the prediction and predictability of the alerts. The main objective of this module is to capitalize knowledge from experts and also from previous cases, especially in order to treat the problem of *No Fault Found Problem* situation.

New technologies have brought new challenges. Therefore, these two modules have to handle several aspects of these new technologies in order to provide the good information, at the right person, at the right time. This aim could be seen as the definition of interoperability [16] paired with the characteristics of big data. Indeed, big data embeds a set of techniques and technologies that require new forms of integration to uncover large hidden values from large datasets that are heterogeneous, complex, and of a massive scale [17]. Big Data is usually characterized through its famous 4 Vs:

- **Volume**: concerns the massive amount of data that needs now to be stored and processed.
- **Variety**: it is necessary to manage the diversity of types or formats.
- **Velocity**: refers to the frequency with which data is issued and should then be processed.
- **Value**: refers to the added value of the information.

These PHM approaches have to manage all these particularities in order to exploiting all the interest of new technologies. Table 1 aims at summarizing the objective of each modules regarding each aspect of the Big Data.

Table 1 Objectives of the PHM approach

<table>
<tr><th>Objective</th><th>Real-time and a reactive analysis</th><th>Capitalized knowledge</th></tr>
<tr><td>Volume</td><td>Need to treat the data with no storage in order to reduce the treatment time</td><td rowspan="2">Need large space of storage to store all the data. The lifecycle of the data has to be defined in order to reduce the needed space</td></tr>
<tr><td>Velocity</td><td>Need to treat the data in real time (thus with no storage)</td></tr>
<tr><td>Variety</td><td colspan="2">Need to transform in an automatic way the data in order to aggregate it or to build the model of health status in real time</td></tr>
<tr><td>Value</td><td>Create alerts based on the data or transform it to the health status model</td><td>Assess the usefulness of the data for further predictions and keep it if it sounds useful</td></tr>
</table>

3 Architecture to Support Prognostic and Health Management (PHM) Approach

In this section, the components of the architecture are described. Figure 1 represents the overview of the interactions between the components. All components are based on a service-oriented architecture and a publish/subscribe approach in order to be the most flexible possible.

- **Homogeneity layer**: this service aims at providing a common shared input for data. Its objectives are (i) to provide pub/sub approaches in order to be able to connect to every kind of solutions as inputs and (ii) to absorb the diversity of the incoming data. About the second point, a unified approach for ensuring the interoperability of the data can be used, just as described by Wang et al. [18], who describes an interesting approach.
- **Knowledge base**: this element is the most important of the architecture. Its objective is to gather knowledge about previous experiences and knowledge from experts in order to be able to predict the needs of maintenance. This prediction could be based on business rules (e.g., the use of a threshold on a specific value) and executed by the **rules service** or on the comparison of models (i.e., between the expected model of the unit and its actual health status model).
- **Storage Layer**: this service aims at storing all the data that can be in turn analyzed as a whole dataset. The Hadoop solution from Apache is selected for this service.

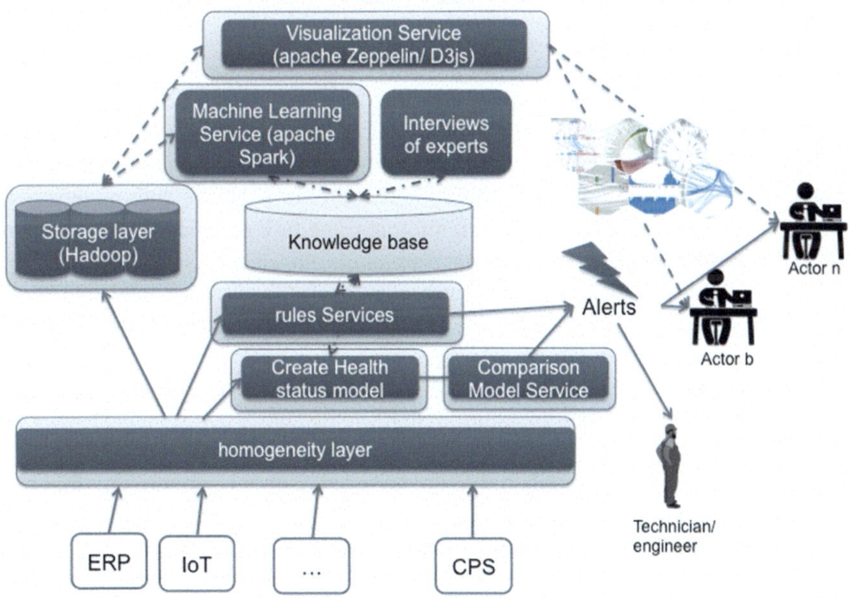

Fig. 1 Overview of the targeted PHM approach architecture

- **Visualization service**: this service aims at showing past data to user. This service aims at filling the gap of visualization tools from existing PHM approach [10]. This service is based on Zeppelin solution from Apache and on specific development.
- **Machine Learning Service**: this service aims at executing machine learning algorithms in order to deduce the needs of maintenance, based on previous experiences. The main objective of this service is to fill the existing gap of No Fault Found Problem. Spark solution from Apache is selected for this service.
- **Create Health status model**: this service aims at creating the health status model of a system based on the collected data and the knowledge base. The first version of this service could be built on similar existing works initially applied to the crisis management [19].
- **Comparison Model Service**: this service is used in order to calculate the divergence between two models. Based on the result, alerts could be generated. This service is based on the work described in [15].

The speed layer is a real-time and reactive analysis module composed of the following elements: homogeneity layer, rules services, create health status model service and comparison model service. These services are essentially based on Complex Event Processing (CEP) that allows the treatment of data with the least storage possible and thus with the highest speed possible, unlike the capitalized knowledge module used to deduce the configuration of the previous services.

4 Metamodel for the Knowledge Base

As explained in the previous section, the knowledge base is a key element of the approach. Indeed, it aims at helping managing rules and, thus, health status model. Therefore, thanks to the knowledge base, it will be possible to infer new statements from a set of given assumptions. According to [20], this approach could be seen as a knowledge-based system due to this functionality. In [21], the authors describe three categories of knowledge: domain, inference and task knowledge. The domain knowledge describes the concepts, properties and instances for a particular domain, the inference is about deducing new knowledge from the current knowledge itself and the task knowledge concerns the objectives to be achieved, consisting of properly using the inferences for this.

In our approach, the knowledge base is mainly the domain knowledge that is taken into account and stored. In addition, in [22], the author defines an ontology as a formal explicit specification of a shared conceptualization for a domain of interest. This definition allows us to explain that the knowledge base is an ontology that has to be structured, which can be achieved through the use of a metamodel [23].

This metamodel has to describe the key information related to maintenance activities. Maintenance activities could be seen as a crisis situation. Indeed, in a crisis situation people aim to prevent risks or reduce effect of the crisis. In the maintenance

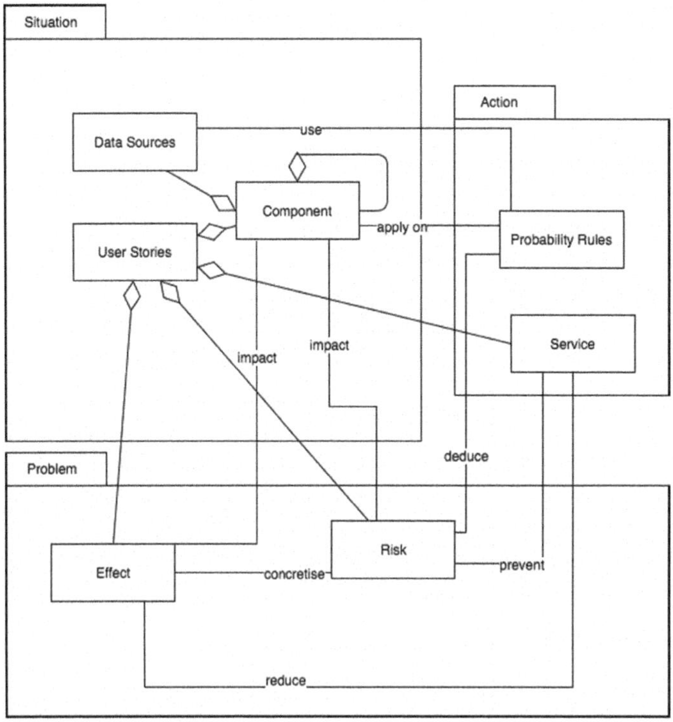

Fig. 2 Metamodel of the knowledge base

domain, the objectives are equivalent. Thus, we build the metamodel in a similar way that the work presented in [24].

The metamodel, represented by Fig. 2, is divided into three packages:

- Situation: composed of the **components** of the system. A component is a part of a system and thus could be decomposed into sub-components. **Data Sources** are linked to the components because it represents the sensors or other solutions that could send data related to this component's state. **User stories** correspond to past events. User stories will be used as the input for the machine learning algorithms.
- Problem: this package describes the **risks** and **effects** of components of the system. An effect is the realization of a risk when specific trigger appends.
- Action: This package is composed of business **services**, which represent the maintenance activities that prevent risk and reduce effects, and **probability rules**. This concept encompasses the definition of expert rules as well as results of machine learning algorithms.

5 Conclusion and Future Work

The increasing number of the available data must allow building a proactive maintenance approach. Indeed, once a failure occurs, the achievement of a diagnostics must prevent the appearance of the same type of failure. This prediction, called prognostics, could not be based only on diagnostics because of unknown reason of failure situations. The existence of this kind of situations, called No Fault Found Problems, implies that the use of machine learning algorithms is a promising approach.

Therefore, this paper proposes a new architecture able to connect to new technology and to predict failure whatever the situation is. This architecture aims at fulfilling the gaps that have been highlighted for the current PHM approaches, namely:

- *They are dedicated to specific equipment*:

The use of a metamodel allows to implement different abstract level to consider the system to be maintained. As such, two dimensions are considered: (i) the auto-aggregation of the concept Component allows to study systems of systems, i.e., any "sub-equipment of equipment" and (ii) the concepts can be generalized to any kind of equipment, based on its nomenclature.

- *They are applied to a specific problem without explanation of the chosen methods*:

First, the use of both a knowledge base consolidated brings a history of all maintenance past cases, and allows to capitalize the experience. Consolidated with a rules service, specific methods, i.e., rules are pre-integrated according to experts' knowledge, to determine the occurrence of the most common failures. Finally, in turn, the machine learning service goes further in attempting to provide a prognostic for Not Fault Found Problems.

- *Visualization tools are missing as well as the possibility to share information. Thus, there is a lack of support for decision.*

The visualization service aims to address this gap by providing the end-users with past data visualization.

In the next steps, this architecture will be applied for the maintenance of aircrafts. In particular several steps will be implemented in a short term: (i) the validation of the metamodel and possibly its extension or refinement according to the experts point of view, (ii) collecting the dataset on real past cases from aeronautical companies and feed the knowledge base, (iii) interview experts of the domain to set up a set of business rules for the rules service. At a slightly longer term, developing and refining on several iterations the machine learning service will concern a large contribution of these works as well.

References

1. Hermann, M., Pentek, T., & Otto, B. (2016, January). Design principles for industrie 4.0 scenarios. In *2016 49th Hawaii International Conference on System Sciences (HICSS)* (pp. 3928–3937). IEEE.
2. Wang, Z., Zarader, J.-L., & Argentieri, S. (2012). A novel aircraft engine fault diagnostic and prognostic system based on SVM. In *Proceedings of International Conference on CMD*.
3. Tsui, K. L., Chen, N., Zhou, Q., Hai, Y., & Wang, W. (2015). Prognostics and health management: A review on data driven approaches. *Mathematical Problems in Engineering*.
4. Bowles, J. B. (1992). A survey of reliability-prediction procedures for microelectronic devices. *IEEE Transactions on Reliability, 41*(1), 2–12.
5. Pecht, M., & Jaai, R. (2010). A prognostics and health management roadmap for information and electronics-rich systems. *Microelectronics Reliability, 50*(3), 317–323.
6. Tsui, K. L., Chiu, W., Gierlich, P., Goldsman, D., Liu, X., & Maschek, T. (2008). A review of healthcare, public health, and syndromic surveillance. *Quality Engineering, 20*(4), 435–450.
7. Xu, J., Wang, Y., & Xu, L. (2014). PHM-oriented integrated fusion prognostics for aircraft engines based on sensor data. *IEEE Sensors Journal, 14*(4), 1124–1132.
8. Orsagh, R. F., Brown, D. W., Kalgren, P. W., Byington, C. S., Hess, A. J., & Dabney, T. (2006). Prognostic health management for avionic systems. In *Proceedings of IEEE Aerospace Conference*.
9. Lee, J., Wu, F., Zhao, W., Ghaffari, M., Liao, L., & Siegel, D. (2014). Prognostics and health management design for rotary machinery systems—Reviews, methodology and applications. *Mechanical Systems and Signal Processing, 42*(1), 314–334.
10. Williams, R., Banner, J., Knowles, I., Dube, M., Natishan, M., & Pecht, M. (1998). An investigation of 'cannot duplicate' failures. *Quality and Reliability Engineering International, 14*(5), 331–337.
11. Anderson, K. (2012). Intermittent fault detection & isolation system. In *Proceedings of the Maintenance Symposium and Exhibition*. Department of Defense, November 2012.
12. Gao, R., Wang, L., Teti, R., Dornfeld, D., Kumara, S., Mori, M., et al. (2015). Cloud-enabled prognosis for manufacturing. *CIRP Annals-Manufacturing Technology, 64*(2), 749–772.
13. Lee, J., Lapira, E., Bagheri, B., & Kao, H. A. (2013). Recent advances and trends in predictive manufacturing systems in big data environment. *Manufacturing Letters, 1*(1), 38–41.
14. Villari, M., Celesti, A., Fazio, M., & Puliafito, A. (2014, November). Alljoyn lambda: An architecture for the management of smart environments in iot. In *2014 International Conference on Smart Computing Workshops (SMARTCOMP Workshops)* (pp. 9–14). IEEE.
15. Barthe-Delanoë, A. M., Truptil, S., Bénaben, F., & Pingaud, H. (2014). Event-driven agility of interoperability during the Run-time of collaborative processes. *Decision Support Systems, 59*, 171–179.
16. Konstantas, D., Bourrières, J.-P., Léonard, M., & Boudjlida, N. (2006). *Interoperability of enterprise software and applications*. Springer Science & Business Media.
17. Hashem, I. A. T., Yaqoob, I., Anuar, N. B., Mokhtar, S., Gani, A., & Khan, S. U. (2015). The rise of "big data" on cloud computing: Review and open research issues. *Information Systems, 47*, 98–115.
18. Wang, T., Truptil, S., & Benaben, F. (2017). An automatic model-to-model mapping and transformation methodology to serve model-based systems engineering. *Information Systems and e-Business Management, 15*(2), 323–376.
19. Fertier, A., Montarnal, A., Barthe-Delanoë, A. M., Truptil, S., & Bénaben, F. (2016). Adoption of big data in crisis management toward a better support in decision-making. In *Proceedings of Conference on Information System for Crisis Response and Management*.
20. Keller, U., & Feier, C. (2005). State of the art and requirements on reasoning with semantic web services. *RW2 Project Deliverable D, 1*.
21. Godoy, C. P. (2005). *Knowledge-based reasoning over the web* (Thèse de doctorat. Ph.D. thesis). Universidad del pais vasco.

22. Gruber, T. R. (1993). A translation approach to portable ontology specifications. *Knowledge Acquisition, 5*(2), 199–220.
23. MOF, O. (2002). OMG Meta Object Facility (MOF) Specification v1. 4.
24. Lauras, M., Truptil, S., & Bénaben, F. (2015). Towards a better management of complex emergencies through crisis management meta-modelling. *Disasters, 39*(4), 687–714.

A Unified Architecture for Proactive Maintenance in Manufacturing Enterprises

Alexandros Bousdekis, Gregoris Mentzas, Karl Hribernik, Marco Lewandowski, Moritz von Stietencron and Klaus-Dieter Thoben

Abstract Since industrial maintenance is a key operation, modern manufacturing firms need to minimize maintenance losses and to improve their overall performance. In addition, emerging information technologies such as the Internet of things (IoT), cyber-physical systems, proactive computing and big data analysis in the context of Industry 4.0 are able to enhance maintenance management with the aim to implement a new maintenance strategy: proactive maintenance. To this end, we propose a unified conceptual architecture for proactive maintenance in a sensor-based industrial environment. Furthermore, we describe how we aim to implement it with the use of existing services and tools, the integration of which will result in the UPTIME information system. Finally, we present our plans for its evaluation in three industrial cases: a white goods/home appliances industry, a steel industry and an aviation industry.

Keywords Proactive maintenance · E-maintenance · Industry 4.0 · IoT

A. Bousdekis (✉) · G. Mentzas
Information Management Unit (IMU), Institute of Communication and Computer Systems (ICCS), National Technical University of Athens (NTUA), Athens, Greece
e-mail: albous@mail.ntua.gr

G. Mentzas
e-mail: gmentzas@mail.ntua.gr

K. Hribernik · M. Lewandowski · M. von Stietencron
BIBA—Bremer Institut für Produktion und Logistik GmbH, University of Bremen, Hochschulring 20, 28359 Bremen, Germany
e-mail: hri@biba.uni-bremen.de

M. Lewandowski
e-mail: lew@biba.uni-bremen.de

M. von Stietencron
e-mail: sti@biba.uni-bremen.de

K.-D. Thoben
Faculty of Production Engineering, University of Bremen, Badgasteinerstr. 1, 28359 Bremen, Germany
e-mail: tho@biba.uni-bremen.de

K. Popplewell et al. (eds.), *Enterprise Interoperability VIII*, Proceedings of the I-ESA Conferences 9, https://doi.org/10.1007/978-3-030-13693-2_26

1 Introduction

Maintenance is a crucial manufacturing operation since failure of critical assets has been rated as the most significant risk to operational performance, because of how it affects the overall business performance [1]. Maintenance strategies are categorized into: (i) breakdown maintenance; (ii) time-based maintenance; (iii) predictive maintenance [2]. Manufacturers increasingly see maintenance as a strategic business function for maintenance costs, downtime reduction and asset lifecycle increase. The development of Internet of things (IoT), the advancements in big data processing, the proactive computing theory as well as the Industry 4.0 concept has led to a novel lever for maintenance, i.e. proactive maintenance [2].

Due to the emergence of the new technologies and computing paradigms, several approaches, frameworks and architectures for intelligent maintenance have appeared in both academia and industry [3–6]. However, currently, there is still a lack of services and tools capable of processing real-time big data from heterogeneous sources in an efficient way, implementing complex algorithms and provide meaningful insights about potential problems along with a continuous self-improvement approach [2, 7]. Moreover, predictive maintenance strategy and related information systems have several challenges during their deployment in manufacturing companies due to their complexity and their implementation lifecycle [8]. In the current paper, we present a unified architecture for proactive maintenance which forms the basis for the development of a unified information system capable of covering the whole prognostic lifecycle and linking maintenance with other industrial operations. The proposed architecture is generic in order to be applicable in manufacturing companies of different processes, machines and products. We also present the H2020 UPTIME project which will lead to the unification of five research results (services and tools) in the frame of proactive maintenance.

The current paper is organized in the following way: Sect. 2 describes the new technological enablers and the emerging concepts that can contribute to the evolution of maintenance management. Section 3 presents the proposed conceptual architecture, and Sect. 4 describes how the architecture will be implemented. Section 5 presents an industrial scenario for adopting the proposed architecture, while Sect. 6 provides the conclusions and the plans for future work.

2 Technological Enablers and Emerging Concepts

Nowadays, most of the manufacturing enterprises have not incorporated a complete predictive maintenance strategy and suitable real-time information systems for sensor data processing in order to exploit its advantages [9]. However, predictive maintenance is becoming essential as products have become more and more complex. Therefore, quality and reliability of equipment have become important aspects. To this end, the costs of time-based maintenance have increased and predictive maintenance has arisen as a new, novel maintenance strategy [10]. Apart from the manufacturing companies, technology providers also need to get involved with the development of innovative solutions. These solutions should be applicable to both existing and new equipment. Moreover, the technology providers should be able to demonstrate successful use cases, while the technology adopters should be willing to change their maintenance strategy. To do this, the latest technological advancements should be exploited and the new ecosystem partners should be identified [11, 12].

The emergence of **IoT** enhances the monitoring capabilities of enterprises with sensors enabling them to decide and act ahead of time [13], i.e. to resolve problems before they appear. The Industrial IoT (IIoT) is considered to be the IoT concept with the highest overall potential, although it has not gathered yet the interest that smart homes or wearables have gathered, due to the high investments required and the long periods of implementation needed. The IIoT can take advantage of the Industry 4.0 concept. **Industry 4.0** indicates production systems functioning according to the cyber-physical systems paradigm that bring together production planning and logistics management, but also social needs in order to establish global value creation networks [14].

Today's manufacturing enterprises have started monitoring and detecting signals that indicate abnormal situations so that they know when the equipment is degraded and a breakdown might occur. However, currently, manufacturing enterprises can have more information about manufacturing systems' health state if they increase their data analytics maturity. There are four levels of **data analytics maturity**, each one building on the previous one: monitor, diagnose and control, manage, optimize [15]. Taking advantage of the **big data** generated from the large amount of sensors within the IIoT needs the implementation of information systems capable of processing sensor-generated, real-time data in the complicated, uncertain and dynamic industrial environment. However, the availability of big data poses challenges to the real-time processing and storage in an efficient and scalable way with the aim to predict undesirable situations and enable to decide and act ahead of time.

Building manufacturing value-driven solutions requires automated services that are suitable for predictive maintenance [1, 7]. To this end, the e-maintenance paradigm has evolved. **E-maintenance** is referred to the utilization of information and communication technologies in the context of information systems that are able to take into account the constraints existing in the manufacturing environment in order to support decision-making in a proactive way [1]. Therefore, e-maintenance solutions can take advantage of proactive computing. **Proactive computing** is referred to the use of information systems for mitigating the impact of a future undesired event, or to take advantage of a potential future opportunity. Since prediction is an indispensable element of proactivity, decision-making can be conducted ahead of time according to the "Detect-Predict-Decide-Act" proactivity principle [13]. The proactivity principle can be applied to every operation of manufacturing companies (e.g. logistics management, production planning, etc.), and therefore, the interactions among the various operations (e.g. production tasks driven by predictive maintenance) can be facilitated. Proactive decision-making can be seen as the event-driven variation of prescriptive analytics [13]. Based on the aforementioned technologies and concepts, a new maintenance strategy indicating the condition-based predictive maintenance in Industry 4.0 has arisen, **proactive maintenance** [2]. However, there is the need for a conceptual architecture supporting its implementation in a unified information system capable of exploiting the aforementioned technologies and concepts.

3 The Proposed Unified Conceptual Architecture for Proactive Maintenance

To exploit the capabilities that IIoT, big data processing technologies and proactive computing provide in the context of Industry 4.0, there is the need for a unified architecture that will be able to facilitate the implementation of proactive maintenance in an IoT-based industrial environment. The proposed architecture combines and extends existing condition-based predictive maintenance approaches, architectures and international standards [8, 9, 16, 17] to provide the basis for the development of a unified information system capable of covering the whole prognostic lifecycle and linking maintenance with other industrial operations, i.e. production, logistics, quality along with a continuous improvement loop. Unification takes place when bringing together approaches and algorithms, but also tools and services for addressing the various phases of the proactive maintenance architecture. Its implementation needs to support the management layers of the company, i.e. operational (e.g. maintenance engineers), management (e.g. factory manager), strategic (e.g. board of directors) by aggregating and interpreting data captured from the production system and effectively sharing the massive amount of information horizontally and vertically in the manufacturing company. For example, a maintenance engineer is interested in the real-time visualization of results (diagnostics, prognostics, recommendations, etc.) for conducting their day-to-day duties; the factory manager is interested in an aggregation of information per week to study the number and type of failures, the implemented actions, etc., while the board of directors require more aggregated form of information, e.g. for studying (e.g. in the form of reports) the production process performance at a monthly level.

The proposed architecture is generic in order to be applicable to any production process. It takes advantage of predictive maintenance management, IIoT and big data, as well as proactive computing and the e-maintenance concept. Thus, it aims to reframe the way maintenance engineers perform their work in manufacturing companies. Furthermore, it aims to have a wider implication in manufacturing companies, since it will enable the transition of maintenance strategy to proactive by utilizing the most recent advancements in maintenance management and computer science.

The proposed conceptual architecture is depicted in Fig. 1. It consists of three layers which are further described below. The ***User Interaction Layer*** deals with an integrated GUI for configuration by the user as well as visualization and real-time monitoring of the generated information. The ***real-time processing layer*** has to do with real-time processing of sensor-generated data by applying the proactivity

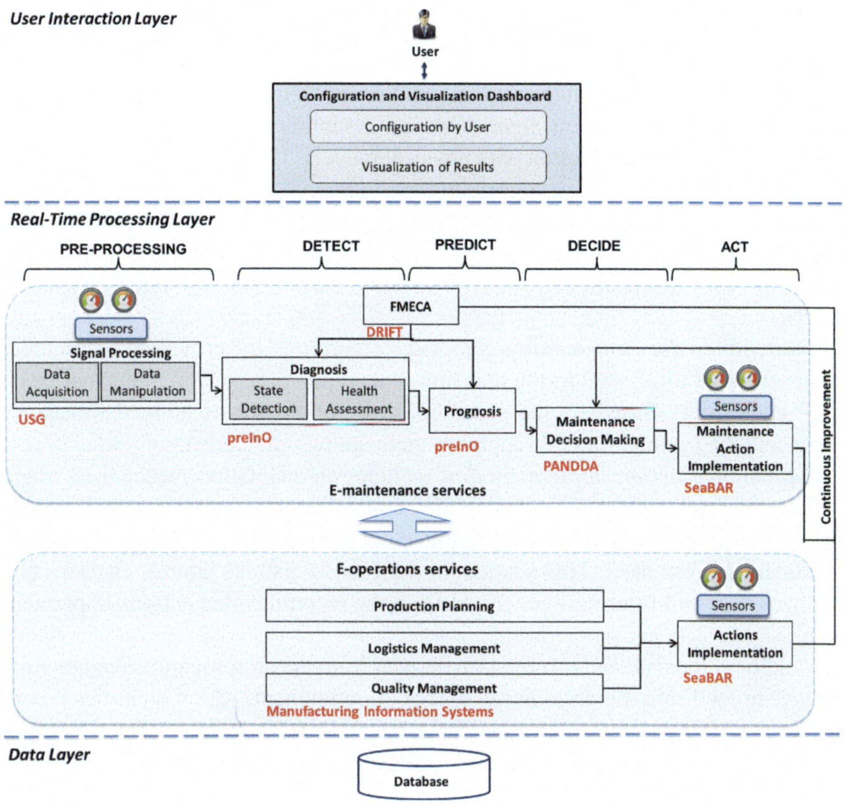

Fig. 1 The proposed conceptual architecture and how it will be implemented

principle of "Detect-Predict-Decide-Act". The ***Data Layer*** deals with storage and incorporates a relational and a NoSQL database where the information required by the other two layers is stored. For the raw sensor data itself, the storage concept is enhanced by a database for time series to ensure efficient and reliable storage of this data.

The following explains the scope of each conceptual block of the real-time processing layer.

Signal processing includes data acquisition and manipulation by exploiting modern technologies with the aim to connect, acquire and integrate data from heterogeneous data sources.

Diagnosis encompasses real-time statistical/machine learning algorithms embedded in appropriate and continuously processing software in order to identify unusual and potentially dangerous states of manufacturing systems. This is done with comparison to models for the normal behaviour of the equipment which has been modelled from domain experts or learned from real-time data observations. The diagnostic models continuously learn from the actual equipment behaviour by updating and improving the incorporated diagnostic models.

Prognosis deals with state prediction of manufacturing systems. The analysis is carried out by different algorithms and essentially by combination of data-driven algorithms in conjunction with physical models. Therefore, appropriate prognostic and/or other predictive algorithms are implemented (e.g. predictions about production-related issues, spare parts inventory and quality management) considering the data and information gathered from actions implementation (e.g. failure modes) in order to update and improve continuously the incorporated prognostic models.

Maintenance decision-making comprises algorithms for providing recommendations ahead of time considering real-time state predictions and maintenance engineers' knowledge. Based on the predictions, the optimal mitigating maintenance actions and the optimal time for applying them are recommended.

Maintenance action implementation includes an adaptation mechanism which provides feedback and learning capabilities to the diagnosis, prognosis and maintenance decision-making phases of the architecture with the aim to update and improve the models that are used. This adaptation mechanism gathers, stores, analyses and visualizes data and information derived from the recommended actions implementation.

In addition, the data and information derived from the aforementioned adaptation mechanism feed into the ***data-driven FMECA*** component, which identifies potentially critical failure modes. On this basis, it conducts analysis of the criticalities that might arise. The resulting failure modes and other associated results feed into diagnosis and prognosis.

Industrial operations management consists of information about production, logistics and quality. This information is stored in manufacturing company's legacy systems and is utilized so that other operations affected by maintenance are also taken into account. In this way, not only a single manufacturing operation (i.e. maintenance), but also the overall business performance can be optimized. Therefore, maintenance activities can be scheduled together with production, logistics and quality activities.

4 The UPTIME Information System

We aim to develop the UPTIME system based on the conceptual architecture described in Sect. 3. The system will derive from the unification of extended versions of existing e-maintenance services and tools. These will provide the required support for: (i) configuration at design time based on expert knowledge; (ii) real-time information processing through all the steps of the proposed architecture; and, (iii) visualization of useful information (either historical or real-time) through web-based dashboards. To provide effective guidance, support and information sharing to maintenance engineers as well as to other enterprise management levels, the

UPTIME e-maintenance services will address the various steps of the conceptual architecture as shown in Fig. 1. USG (Universal Sensor Gateway) will implement the signal processing phase; preInO will address the diagnosis and the prognosis phases; PANDDA (ProActive seNsing enterprise Decision configurator DAshboard) will deal with maintenance decision-making; SeaBAR (Search-Based Application Repository) will address the maintenance and operational actions implementation; and DRIFT (Data-Driven FMECA Tool) will deal with data-driven failures modes, effects, criticality analysis (FMECA). The aforementioned tools will also process and interact with operations-related information from the manufacturing information systems (e.g. ERP, MES). The novel services and tools that will provide the baseline research inputs to the system are explained below.

USG was developed, implemented and evolved throughout several national (German) and EU-funded research projects: BOMA project (FP7), ThroughLife (FP7), CyProS, Fortissimo HighSeas Experiment, PreInO. USG serves as a modular data acquisition, manipulation and integration device to the Product Lifecycle Management ecosystem of a product. It can connect to both analogue and digital data sources via numerous protocols, acquire data and integrate the heterogeneous data sources towards a configurable data set. The USG can store and intelligently handle and filter the data acquired and provide it to other stakeholders. Due to its modular structure, it is both sensor and platform agnostic and can be easily adapted to any use case. The USG takes data from any sensor or other data system as input.

The **preInO** processing engine was developed within the national (German) preInO project. It can detect and predict the state of a manufacturing system such as wind turbines, special-purpose vehicles, production machinery, etc. For this, context-dependent time series from the past can be analysed and evaluated, while the data can come from different sources. The analysis is carried out by combinations of different detection and prediction algorithms for condition estimation and remaining life prediction of the analysed component. A library with typical algorithms for data analytics and predictions was implemented, but it is also possible to add new algorithms to the system. Based on the uniform access on time-related data, the transformation of data into information will be enabled by several methods, algorithms and application-oriented calculation workflows.

The **PANDDA** system was developed in the context of the EU ProaSensE (FP7) project. At design time, through a GUI, PANDDA enables decision-makers to add knowledge such as a list of alternative actions, the cost of the failure, cost functions of actions, etc. At the real-time processing layer, PANDDA is triggered by predictions about future failures and generates recommendations of optimal actions that mitigate or eliminate their impact along with the optimal time for their application. Moreover, PANDDA incorporates a continuous improvement functionality addressed by a sensor-enabled feedback (SEF) mechanism, the role of which is twofold: (i) to expose to the user the estimations of costs in real-time, during the application of the recommended actions, and (ii) to update the estimated cost functions of the maintenance actions with the aim to improve the generated proactive recommendations.

SeaBAR was first developed in the frame of a commercial R&D project, while it was extended in the frame of the national (German) projects: "Knowledge Cloud" and "Vertumnus—Early diagnosis, measurement, evaluation and design of change in value creation networks". SeaBAR is a modular software platform built on Big Data and Enterprise Search technology. The SeaBAR platform supports end-users by means of data aggregation, data analysis and visualization. Its core engine collects, normalizes and indexes data coming from heterogeneous unstructured, semi- and full-structured data sources such as sensor data, product lifecycle information and issue histories offering flexible, web-based dashboards and visualization tools. SeaBAR includes pre-defined context-sensitive and interactive dashboards and many visualization options.

DRIFT will be partially built on the knowledge gathered by the EU-funded HIPR (FP7) project and partially on the EU-funded ThroughLife (FP7) project. DRIFT is a tool that, on the basis of the information gathered in other modules, uses them to identify what are the failure modes, effects and criticalities of the components and system. The resulting failures modes and other associated results feed into diagnosis and prognosis. Tool's inputs are: failures, effects recommended actions, etc. The outcome of the FMECA will be the calculation of a Risk Priority Index and the criticality for each potential failure mode.

5 Industrial Scenarios

The UPTIME solution will be deployed and evaluated in three industrial use cases: a white goods/home appliances industry, a steel industry and an aviation industry case. In this section, we present a practical application of the proposed architecture in the aforementioned industries. The implementation of a unified information system implementing proactive maintenance will lead to lower maintenance costs and to improved overall business performance compared to breakdown maintenance, time-based maintenance and the existing implementations of predictive maintenance strategies, as outlined at the following business cases.

The **white goods/home appliances business case** deals with an automatic production line which produces drums for dryer. The drum is a carbon steel cylinder for holding and rotating the clothes during drying. The manufacturing processes several steps and requires the synchronization of mechanical, electrical, hydraulic and pneumatic tools and moving parts. Currently, only preventive and reactive maintenance are implemented. Typical examples of failures are: wearing or mechanical tools, pneumatic failures of moving parts, misplaced parts and lubrication absence. Data collection is based on a proprietary tool to gather main outputs from equipment PLC, while SAP PM is governing the spare parts. In the future vision, sensors will be used to generate real-time data to monitor the component wear status (welding seaming wheels, hemming heads) and punching areas on the product (e.g. deviation on the punched profiles indicates wear or damages to punchers), temperature sensors on spinning parts indicate lubrication problems or wear, pressure sensors on hydraulic units (punchers and embossing) with the aim to conduct proactive maintenance.

The **steel industry business case** involves cold-rolling mill stands for the production of rolling products with the closest possible thickness tolerances and an excellent surface finish. Cold rolling is a process that aims to shape a metal part through the deformation that is caused by two metal rolls rotating in opposite direction. In the future vision, maintenance activities will not be performed at frequent predetermined intervals, which is based on generic performance data or previous experience. The machines will have sensing equipment installed which will generate large amounts of data. In this way, the UPTIME platform will process all these data in order to provide information about the current health state of the machines, predictions about future failures and proactive recommendations. The incorporated models will be able to self-assess, and thus, decision-making will be better supported. This fact will raise the equipment maintenance to more than a low-level topic of study and can affect positively the products' entire life cycle. Consequently, the company will reduce its maintenance costs and will improve its performance.

The **aviation industry business case** involves the maintenance of manufacturing equipment in the aviation domain. Its Hoisting Equipment comprises fittings, to be mounted at aircraft components for assembly or transportation, and lifting gears, to be applied to lift during the assembly of an airframe. Fittings and lifting gears are maintained and recertified on a regular basis. By collecting, analysing and processing product instance-specific data (e.g. indicating the condition of lifting gears and fittings and their quantity and delivery time as well as the optimal maintenance actions to be applied), it will be possible to provide the required resources (spare parts, manpower, etc.). Information that could be extracted through smart technologies (i.e. RFID, sensors) deals with specific damages that tend to occur at fittings or lifting gears coming from a certain assembly line or from a certain production facility, reasons for these damages, parts typically missing or broken, higher or lower amounts of incoming equipment for maintenance during specific periods of time, indication of specific fittings or lifting gears which need costly repair measures more often than other types.

6 Conclusions and Future Work

In this paper, we propose a unified conceptual architecture for proactive maintenance aiming to improve maintainability and increase operating life of production systems for high-quality products with optimized losses. The proposed architecture takes advantage of IoT, the advancements in big data processing and the proactive computing theory. We described each conceptual block of the architecture, and we explained how we will develop an information system (called UPTIME) based on the aforementioned architecture in the context of the H2020 UPTIME project. The system will be developed based on existing services and tools that are able to address each conceptual block. These will be extended in order to be integrated and fulfil more manufacturing needs and requirements. Consequently, there is the need for interoperable interfaces among the services, the sensors and the protocols in the context of the aforementioned architecture in order to overcome the interoperability challenges of data inconsistency, scalability and accommodating the scope of data [18]. The unification of the novel e-maintenance services and tools in the proposed architecture will help overcome existing commercial software and research prototype limitations and will conclude in a novel proactive maintenance solution. The system will be applicable at the level of component, machine and production system, depending on the placement of sensors throughout the production lifecycle and the data availability in the manufacturing company's systems (e.g. ERP, MES). The evaluation will be conducted in three real industrial cases: a white goods/home appliances, a steel and an aviation industry case. Moreover, we aim to demonstrate and diffuse the UPTIME solution and its evaluation results to more manufacturing companies as future potential users to generate business flows and to maximize the impact on the industry.

Acknowledgements This work is partly funded by the European Commission project H2020 UPTIME "Unified Predictive Maintenance System" (768634).

References

1. Aboelmaged, M. G. S. (2015). E-maintenance research: A multifaceted perspective. *Journal of Manufacturing Technology Management, 26*(5), 606–631.
2. Bousdekis, A., & Mentzas, G. (2017). Condition-based predictive maintenance in the frame of Industry 4.0. In *IFIP International Conference on Advances in Production Management Systems* (pp. 399–406). Springer, Cham.
3. Pistofidis, P., Emmanouilidis, C., Koulamas, C., Karampatzakis, D., & Papathanassiou, N. (2012). A layered e-maintenance architecture powered by smart wireless monitoring components. In *2012 IEEE International Conference on Industrial Technology (ICIT)* (pp. 390–395). IEEE.
4. Bousdekis, A., Papageorgiou, N., Magoutas, B., Apostolou, D., & Mentzas, G. (2015). A real-time architecture for proactive decision making in manufacturing enterprises. In *OTM Confederated International Conferences on the Move to Meaningful Internet Systems* (pp. 137–146). Springer, Cham.
5. Macchi, M., Martínez, L. B., Márquez, A. C., Fumagalli, L., & Granados, M. H. (2018). Value assessment of e-maintenance platforms. In *Advanced maintenance modelling for asset management* (pp. 371–385). Springer, Cham.
6. Fumagalli, L., & Macchi, M. (2015). Integrating maintenance within the production process through a flexible E-maintenance platform. *IFAC-PapersOnLine, 48*(3), 1457–1462.
7. Camarinha-Matos, L. M., Goes, J., Gomes, L., & Martins, J. (2013). Contributing to the Internet of Things. In *Doctoral conference on computing, electrical and industrial systems* (pp. 3–12). Springer, Berlin.
8. Guillén, A. J., Crespo, A., Gómez, J. F., & Sanz, M. D. (2016). A framework for effective management of condition based maintenance programs in the context of industrial development of E-Maintenance strategies. *Computers in Industry, 82,* 170–185.
9. Bousdekis, A., Magoutas, B., Apostolou, D., & Mentzas, G. (2015). A proactive decision making framework for condition-based maintenance. *Industrial Management & Data Systems, 115*(7), 1225–1250.
10. Jardine, A. K., Lin, D., & Banjevic, D. (2006). A review on machinery diagnostics and prognostics implementing condition-based maintenance. *Mechanical Systems and Signal Processing, 20*(7), 1483–1510.

11. Gartner, https://www.gartner.com/doc/3065317/using-advanced-analytics-predict-equipment. Last accessed 2017/09/10.
12. PwC, http://www.pwc.com/gx/en/industries/industry-4.0.html. Last accessed 2017/09/10.
13. Engel, Y., Etzion, O., & Feldman, Z. (2012). A basic model for proactive event-driven computing. In *Proceedings of the 6th ACM International Conference on Distributed Event-Based Systems* (pp. 107–118). ACM.
14. Wang, S., Wan, J., Zhang, D., Li, D., & Zhang, C. (2016). Towards smart factory for industry 4.0: A self-organized multi-agent system with big data based feedback and coordination. *Computer Networks, 101,* 158–168.
15. Gartner, https://www.gartner.com/doc/2826118/industrial-analytics-revolutionizes-big-data. Last accessed 2017/09/10.
16. Voisin, A., Levrat, E., Cocheteux, P., & Iung, B. (2010). Generic prognosis model for proactive maintenance decision support: application to pre-industrial e-maintenance test bed. *Journal of Intelligent Manufacturing, 21*(2), 177–193.
17. Choudhary, R., Perinpanayagam, S., & Butans, E. (2016). Design and analysis of communication model for implementation of CBM systems based on OSA-CBM framework. In *Aerospace Conference, 2016 IEEE* (pp. 1–7). IEEE.
18. Wan, J., Tang, S., Shu, Z., Li, D., Wang, S., Imran, M., et al. (2016). Software-defined industrial internet of things in the context of industry 4.0. *IEEE Sensors Journal, 16*(20), 7373–7380.

Toward Predictive Maintenance of Walls in Hard Rock Underground Facilities: IoT-Enabled Rock Bolts

John Lindström, Jens Eliasson, Petter Kyösti and Ulf Andersson

Abstract The paper addresses the first one-and-a-half cycles, out of four planned, in an action research effort concerned with predictive maintenance of walls and ceilings in tunnels of hard rock underground facilities by using Internet-of-Things-enabled Rock Bolts (IoTeRB). The IoTeRB concept is developed together with a consortium of companies ranging from rock bolt manufacturers, sensor specialists, researchers, and cloud-service providers to data analysts. Thus, the action research effort is a multi-disciplinary endeavor. The result of the paper is an action plan for the second cycle concerning technology and business development which, according to the design criterion, will move the IoTeRB toward commercialization.

Keywords Availability · Efficiency · Intelligent · IoT · Mining · Predictive maintenance · Productivity · Rock bolts · Sustainable · Smart rock reinforcement

1 Introduction

The paper concerns the first one-and-a-half cycles out of four planned, of an action research effort at a micro-SME company, Thingwave AB, which together with partners develops Internet-of-Things-enabled Rock Bolts (IoTeRB) for hard rock underground facilities (such as mines, tunnels, and deep storage). Rock bolts help to stabilize the rock in underground facilities and are therefore a must in order to be able to have humans and expensive machinery and equipment in deep cavities or tunnels. In larger underground mines, for instance, there are millions of rock bolts of various types installed in the rock wall and commonly applied together with concrete and metal nets, building a stabilizing layer on top of the bare rock wall. There are various types of rock bolts, and they have different features and functionality in order to stabilize the rock under various conditions such as seismic movements and blasting

---

J. Lindström (✉) · P. Kyösti · U. Andersson
ProcessIT Innovations R&D Centre, Luleå University of Technology, 971 87 Luleå, Sweden
e-mail: john.lindstrom@ltu.se

J. Eliasson
ThingWave AB, 954 32 Gammelstad, Sweden

K. Popplewell et al. (eds.), *Enterprise Interoperability VIII*, Proceedings of the I-ESA Conferences 9, https://doi.org/10.1007/978-3-030-13693-2_27

Fig. 1 IoT-enabled rock bolt installed in mine tunnel

deep underground with increasing pressure. By IoT-enabling the rock bolts, a lot more information about the rock wall, the rock bolts themselves, as well as the environment surrounding the rock bolts, can be collected, combined, analyzed, and used to support decision-making concerning, for instance, rock stability, safety, predictive maintenance and asset management. Thus, besides improving safety and monitoring of underground facilities, the aim is also to increase the efficiency, availability, and productivity of the underground operations (e.g., in mines).

As the IoTeRB are more expensive than "normal/plain" rock bolts, how many IoT-enabled to use must be considered. Commonly, the rock bolts are applied in patterns with approximately one meter in between the bolts. To be able to predict problems in the rock walls, depending on the rock movements and seismic activity, between 1 and 10% of the rock bolts have to be IoT-enabled. As per standard, the IoTeRB are equipped with a strain/elongation sensor applied on the bolt's dynamic zone in order to detect if the bolt is no longer attached to the wall, is broken or has been stretched too much and needs to be replaced. Additional sensors can be attached to the head of the bolt after it has been applied to the rock wall, and the IoT-attachment is screwed to the head of the bolt (see Fig. 1). Additional sensors may include: smoke and temperature sensors (detect fires), gas sensors (detect dangerous levels of gases like CO/CO_2 or blast gases), or vibration sensors (detect movements in the rock and seismic activities). The IoT-attachment has a small electronic circuit board including an antenna and can securely communicate with a gateway, which then securely relays the data collected by the IoTeRB to a central monitoring/data collection/analytic platform (local or in the cloud). The data are then analyzed and decision-support information is made available according to the settings. The IoTeRB can be configured to communicate at different rates; however, communications consume battery power, and they commonly report every hour or day unless some value exceeds the normal parameter ranges configured—which then triggers immediate data transfer and notifications if wanted. Further, lifecycle management features such as self-registration, self-configuration, remote re-configurations or software updates, and asset management functionality such as: sensors function OK, battery level, rock bolt intact, and attached to the wall are also available.

The strategy to develop the IoTeRB is a four-cycle action research effort involving researchers, software developers and partners, (Big) data analysts, vibration/seismic analysts/researchers, monitoring/data collection/analytics platform partners, electronics partners, sensor developers and providers, rock bolt developers/manufacturers, as well as potential customers. The four cycles are:

1. Initial technology development (monitoring/data collection platform, sensor communication/integration) based on the Arrowhead framework [1, 2]—TRL levels 4–6.
2. Intermediary technology development—TRL levels 6–8, including; multi-year low-power operation, and lifecycle management function supporting, for instance, easy-deployment/self-registration and self/re-configuration of large-scale installations. Further, basic data modeling, analytics, and visualization will be covered.
3. Information analytics/fusion—data modeling/making additional sense and creating value out of the sensor data, and novel machine-learning approaches.
4. Integration to other systems—enabling integration to automation systems and lifecycle management systems such as ABB 800XA, Siemens, Honeywell, Rockwell, Mobilaris Mining Intelligence, and GeoScience Integrator.

The current status of the IoTeRB evolution is at mid-stage of cycle two, and thus there is a need to complete some additional intermediary technology development and also start to address modeling, analytics/visualization of data. The use of the Arrowhead framework enables interoperability through the Service-Oriented Architecture (SOA), and allows for efficient integration of new types of sensors, protocols, cloud services as well as to external information- or lifecycle management systems, etc. [1]. There are many protocols and standards that are used within the IoT world, but not existing in the industry. Interoperability with standards from IIC, IPSO, OMA, IETF, and others, are important and need to get a strong industrial acceptance.

The initial design criterion was to get the rock bolt, the electronics, sensors, embedded software, secure communications, and data collection/analytics platform as well as very basic analytics and notifications to work together. In addition, the IoTeRB should be integrable with information and automation systems as well as having efficient lifecycle management functionality. Efficient lifecycle management functionality is a key in order to keep the total cost during the life cycle as low as possible as well as to be able to trust the data from the IoTeRB. After meeting customers and making test installations in mines during the first one-and-a-half cycles, an additional design criterion has been added—making additional sense and creating maximum value out of the sensor data collected. Thus, the third cycle was added to the planning, as this was understood early during the second cycle.

This paper addresses the question as to how to design IoTeRB to enable predictive maintenance of hard rock underground facilities. Further, the purpose is to make senior management teams and R&D managers at companies with underground facilities aware of the potential of this new approach and how it can be used. In addition, besides providing IoTeRB, ThingWave AB and its partners are interested in providing "soft parts" such as knowledge, analytics, services, and lifecycle management of the rock bolts as integrated product-service offerings or as Product-Service Systems/Industrial Product-Service Systems (PSS/IPS[2]) [3, 4] or Functional Products (FP) [5, 6]. This paper concerns an important aspect of achieving such offers.

The paper is organized such that the research approach follows the introduction and related work. This is followed by the results section, including an analysis and, finally, the discussion and conclusions section.

2 Related Work

The use of IoT in industrial applications, i.e., in Industry 4.0 or Industrial Internet, has gained a lot of momentum in the last five years. Originating in Wireless Sensor Networks [7], but nowadays with a protocol convergence, we are starting to see true Industrial IoT systems being deployed. Factories and production lines can now be equipped with wireless (or wired) sensors and actuators, all communicating using the Internet Protocol (IP) and modern standardized application layers such as CoAP, HTTP(S), MQTT, or XMPP. This approach enables systems to be deployed rapidly without time-consuming and thus costly wire installations. Large European R&D projects, such as Arrowhead [1, 2], EMC^2 [8] and Productive 4.0 [9], are all pushing the frontier for the use of IoT and Service-Oriented Architecture (SOA) in the industrial domain.

Rock bolts have been used by the mining industry for over a hundred years. They come in different types, where the main categories are (1) grouted, (2) non-grouted, and (3) cable bolts. Rock bolts are used to reinforce walls, tunnels, and other cavities, and they perform a vital task in order to keep mines and tunnels structurally intact and thus safe. However, rock bolts can be damaged by seismic activity and stresses in the rock masses [10].

If rock bolts are elongated beyond their elastic limit they lose load-bearing capability. If that happens, ceilings or walls can come down, crushing everything (and everyone) in the vicinity. A ceiling or wall collapse can be very expensive in terms of economic costs due to damaged machinery, production downtime, injuries, and even loss of human lives. By equipping rock bolts with IoT technologies, real-time monitoring is made possible [11]. ThingWave AB has been working on their IoTeRB concept for several years and now offers systems for large-scale rock bolt monitoring [12]. A conceptual image of an installed IoTeRB is shown in Fig. 2.

In order to fully utilize the capabilities of online monitoring, it is vital that all data collected can be processed in an efficient manner. Therefore, it is highly beneficial if modeling tools can be used to predict rock mass stress [13], and that the rock bolt sensors can be used to validate the models (possibly in real time). Typical physical parameters that must be monitored include strain/elongation, force, bolt breakage, seismicity, as well as other types of sensors to detect gases, temperature changes.

By viewing IoTeRB as a source of maintenance information, sensor data and inferred status information can be used to operate reinforcement in a predictive maintenance [14, 15] approach. All IoTeRB transmit information at regular inter-

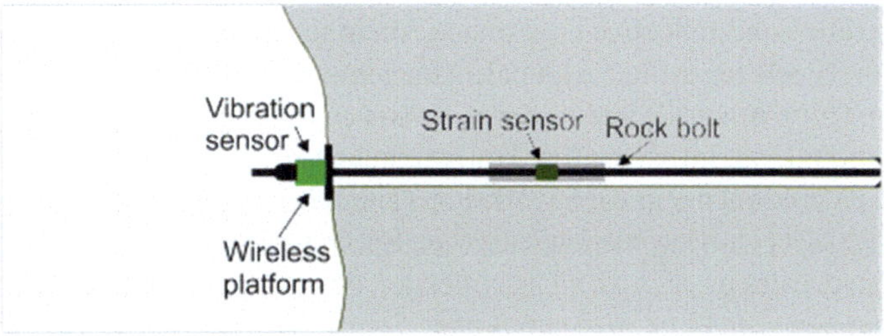

Fig. 2 Schematic of IoT-enabled rock bolt in mine tunnel

vals about their status, health, and measured parameters to a monitoring platform. The monitoring platform can thereby provide operators with information about, for example, regions of a mine such as certain tunnels where large rock stress is detected, and the appropriate measures can be taken to prevent collapses.

3 Research Approach

The research approach employed in this study has been based on an in-depth qualitative study using action research at a micro-SME, ThingWave AB, and its partners located in northern Sweden. ThingWave AB intends to increase its revenue by addition of soft parts to the hardware and local software of the IoTeRB through, e.g., analytics of sensor data, cloud services (data collection and modeling), knowledge, and application know-how. The business model has been initially developed from a product and, currently, to a product with integrated services and will later be developed into something similar to PSS/IPS[2] or FP provided by a consortium.

The design criteria for the action research effort were described above, at the end of the introduction. The research targeted in this paper is the first one-and-a-half cycles (and is currently moving from the planning phase toward action taking) of an action research [16] effort. The researchers have had the roles of external expert/advisor, developer, and project manager. A literature review has been part of the first phase in both cycles, and these results were used as input to the following diagnoses. Action research has been conducted as defined in [16]. The characteristics of action research are: (1) that action researchers act in the studied situations, and (2) that action research involves two goals. The goals pertain to solving the problem (the role of the consultant) and making a contribution to knowledge (the role of the researcher); further, that action research requires interaction and cooperation between researchers and the client personnel and, finally, that action research can include all types of data gathering methods [17]. In accordance with [18], the action research approach encompasses four phases: diagnosing, planning action, taking action, and evaluating the action in relation to a certain context and with a specific purpose. Until now, six phases have been completed in a planned four-cycle effort (in total 16 phases) using an iterative and reflective case management methodology. The data collected during the first cycle regarding technical, management, and organizational needs were collected at ThingWave AB and its partners as well as at potential customers during three workshops [19] (involving four key respondents from the R&D, production and maintenance departments at each workshop) and three semi-structured open-ended interviews [20, 21] with key respondents (i.e., R&D manager, development engineers and maintenance engineers). The data collection during the second cycle was done in a similar manner, with three workshops and 15 semi-structured open interviews with key respondents (i.e., R&D managers and development engineers). The workshops and interviews were conducted from 2014 until 2017. The respondents were well aware of and knowledgeable regarding underground facilities (mines and tunnels), production systems/equipment, IT, monitoring, predictive maintenance, automation systems, and lifecycle management issues.

To collect additional data, after the workshops, semi-structured interviews with open-ended questions [20, 21] were used. This allowed the respondents to provide answers and extra details as well as add complementary information where deemed necessary [22]. The duration of the interviews was approximately two hours. In order to reduce response bias, the respondents were selected from various parts of the organization as well as different levels. In order to strengthen the validity of the study, data were continuously displayed using a projector during the interviews. This enabled the respondents to immediately read, provide feedback, and accept the collected data. Further, the collected data were displayed and analyzed using matrices (cf. [23]), and the outcome of the planning action efforts was summarized into prioritized matrices comprising the diagnoses in terms of technical and business development requirements (see Tables 1 and 2) as well as a high-level plan for action taking concerning the second cycle.

4 Results—Toward Intelligent Predictive Maintenance in Hard Rock Underground Facilities

The IoTeRB have been developed with the intention of enabling multi-purpose or general usability of the IoT parts. The IoT parts have been tested and verified outside of this research effort, where they were proven to work with IoT-enabling composite band rollers in conveyor belts measuring the rolling speed and direction, temperature and smoke (to avoid overheating and detect potential fires), and vibrations (need to maintain or replace roller). However, the focus of this research effort has been on hard rock underground facilities. The technical requirements found for the second cycle can be found in Table 1. The requirements in both Tables 1 and 2 are prioritized from 1 (high) to 10 (low).

Further, the business development requirements found for cycle 2 can be found in Table 2. The business development requirements will take some time to manage/mitigate and will thus spill over to cycles 3 and 4, planning-wise.

Table 1 Technical requirements for the second cycle

#	Technical requirement	Prio.
1	Find suitable/develop necessary sensors to measure the rock bolt and surrounding parameters: rock bolt intact (strain/elongation), rock bolt load and fastened (force), vibrations/seismicity, gases, temperature, etc.	1
2	Develop the IoT parts, electronics, embedded software, cloud service for data collection/analytics and monitoring	1
3	Develop basic modeling and analytics for the data collected concerning predictive maintenance of walls/tunnels in underground facilities. Which triggers/notifications/alarms do the customers want?	2
4	Develop basic visualization of analytic results for a variety of modalities such as mobile phones, tablets, or PC	2
5	Plan for integration toward other information-, automation-, or lifecycle management systems	3
6	Test with additional types of rock bolts	4
7	Verify the IoTeRB overall system's functionality with at least two installations and at least two different types of rock bolts	2
8	Evaluate system security, robustness, power/battery consumption and lifecycle management functionality	2

Table 2 Business development requirements for the second cycle

#	Business development requirement	Prio.
1	Start to build a consortium of technology and business partners with long-term interest	2
2	Define the customer value—i.e., the value proposition—and differentiation toward the competition	1
3	Advance the business modeling from a product with services toward PSS/IPS[2] or FP	3
4	Investigate customer/legal requirements for certification in different regions of the world	2

In order to prioritize the requirements, it was decided that all within the range 1–4 should be addressed and 5–10 should not be addressed. Thus, according to the prioritization, all current requirements in Tables 1 and 2 were selected. ThingWave AB is concerned with almost all technical requirements, whereas the further specialized consortium partners are mainly concerned with technical requirements #1, 2, 6, 7 and 8. The customers will get involved mainly in technical requirements #5–8. Further, regarding the business development requirements, ThingWave AB and its consortium partners will be involved in all requirements, whereas the customer will be concerned primarily with #1–3.

The plan for action taking during the second cycle, based on Tables 1 and 2, is outlined in Fig. 3. To provide an overall understanding for the IoT-enabled rock bolt concept, Fig. 3 outlines how the hardware, software, electronics, cloud services and

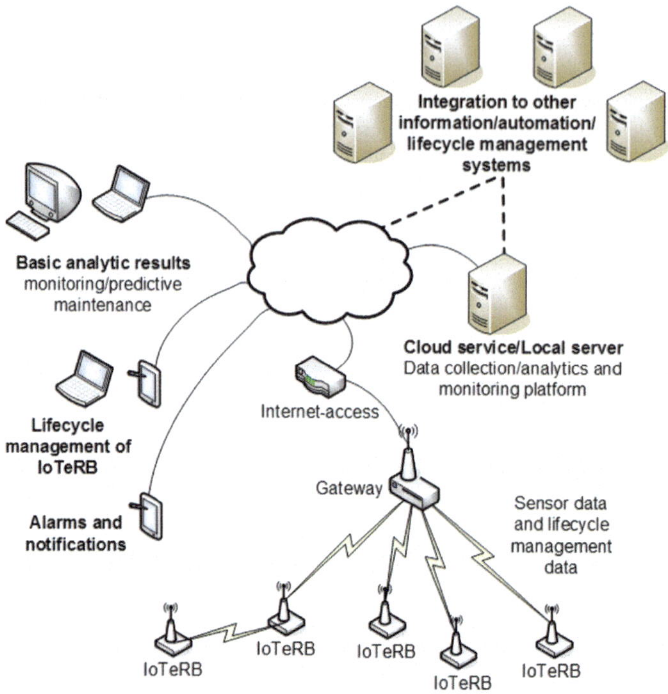

Fig. 3 Action plan for development during the second cycle—to achieve an overall IoT-enabled rock bolt concept

knowledge and know-how interoperate and work together. Further, the integrability to other information-, automation-, or lifecycle management systems is also highlighted (although that will be thoroughly addressed in later cycles) as it affects the design of the other parts and components in the IoTeRB.

Regarding the business development requirements, #1 is addressed in Fig. 3 by using technology from a number of business partners (sensors, electronics, communications, rock bolts, cloud service, etc.). Further, #2–4 will be initiated during the second cycle and iterated and refined during cycles three and four. However, as these requirements have a large impact on the technology, they need to be considered as early as possible.

The plan is to use a mining case, using at least two installations (with different conditions) and at least two different types of rock bolts (to make sure the functionality is generic), to test and verify the design criteria and functionality of the IoTeRB.

5 Analysis

The technical requirements go hand in hand with the business development ones, and they both affect each other. A technology that is too limited will not allow for advanced business models, whereas a simple business model will likely not require

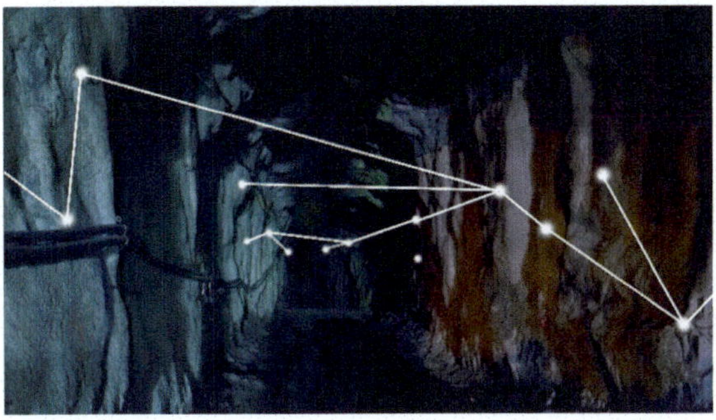

Fig. 4 Toward intelligent underground rock reinforcement—facilitated by the IoTeRB concept visualization

sophisticated technology. The action plan aims to achieve an overall IoTeRB system with all basic functionality tested and verified. Further, advanced data modeling/analytics/visualization and integration toward additional systems will be covered in the following cycles 3 and 4. These are all steps toward what can be depicted as "intelligent" walls in underground hard rock facilities. The vision of intelligent underground facilities is outlined in Fig. 4.

Regarding the design criteria, they are still the same after the action planning, as stated in the introduction. The business development requirements are harder and take longer to manage compared to most of the technical ones. Partly, this is due to the fact that ThingWave AB and the consortium of partners are more accustomed to engineering matters than business development. In addition, more than one business model may need to be tried before a suitable one can be found.

It is expected that possibilities for new business and general data mining/analysis, based on collected and streaming data, will be explored by the partner consortium (as well as together with customers) in order to increase efficiency and operational safety in underground facilities such as mines or tunnels. This will enable the consortium partners to develop new services/products and further advanced offers based on analytics of data combined with advanced maintenance knowledge.

6 Discussion and Conclusions

The paper makes a contribution to literature by outlining a concept of IoTeRB to be used for walls in underground hard rock facilities such as open storage spaces, tunnels, and mines. In addition, the paper highlights that thinking about system security, a security baseline for IoT device/networks, and lifecycle management functions are required from an early stage. The paper contributes to practice as the IoTeRB may lead to "intelligent" walls in, for instance, mines, which can provide information for decision-making regarding predictive maintenance of the walls as well as on the (changing) state of the rock. The managerial contribution of the paper is that the IoTeRB can lead to improved safety, productivity, and availability in, for instance, road tunnels or mining operations. Further, managers need to make sure customer business value is defined and that business models are considered from early on as that affects the technology design and competence needs.

The research so far, comprising the first one-and-a-half cycles, has produced an action plan (see Tables 1 and 2 as well as Fig. 3) concerning technical and business development requirements. During the next cycles, additional such requirements are expected to emerge as well as more organizational-, marketing-, and consortium-related ones as the concept approaches commercialization. Further, the design criteria may also need to be changed or updated when learning more from installations and pilots with different requirements.

Examples of potential future additions to the IoTeRB concept are data stream mining/analytics and artificial intelligence (AI) that can be used to get closer to real time and improve the depth of the data mining and thus further improve the support for decisions. However, real time requires a higher battery capacity and carefully considered computing schemes. In addition, notification/warnings, improved visualization, and further intelligent decision-making support will be of interest when continuing the action research effort. The decision support will benefit management, rock mechanics, tunnel builders, and those who operates the IoTeRB and integrated systems.

Finally, the technology developed will be generalizable after some additions and can be applied in other contexts for monitoring and predictive maintenance purposes.

References

1. Delsing, J. (Ed.). (2017). *IoT automation—Arrowhead framework*. Boca Raton, FL, USA: CRC Press.
2. Delsing, J., Eliasson, J., de Venter, J., Derhamy, H., & Varga, P. (2016). Enabling IoT automation using local clouds. In *Proceedings of IEEE World Forum on Internet of Things* (pp. 501–507), Reston, VA, USA, December 12–14, 2016.
3. Mont, O. (2001). Introducing and developing a Product-Service System (PSS) concept in Sweden. In *The International Institute for Industrial Environmental Economics (IIIEE)*. Sweden: Lund University.
4. Meier, H., Roy, R., & Seliger, G. (2008). Industrial Product-Service Systems—IPS2, CIRP Annals Manufacturing Technology 2008 (pp 1–24).
5. Alonso-Rasgado, T., Thompson, G., & Elfstrom, B.-O. (2004). The design of functional (total care) products. *Journal of Engineering Design, 15*(6), 515–540.
6. Lindström, J., Sas, D., Lideskog, H., Löfstrand, M., & Karlsson, L. (2015). Defining 'Functional Products' through their constituents. *International Journal of Product Development, 20*(1), 1–24.
7. Akyildiz, I. F., Weilian, S., Sankarasubramaniam, Y., & Cayirci, E. (2002). A survey on sensor networks. *IEEE Communications Magazine, 40*(8), 102–114.
8. Weber, W., Hoess, A., Oppenheimer, F., Koppenhöefer, B., Vissers, B., & Nordmoen, B. (2015). EMC2 a platform project on embedded microcontrollers in applications of mobility, industry and the internet of things. In *Euromicro Conference on Digital System Design* (pp. 125–130), Funchal, Portugal, August 2015.

9. Productive 4.0, https://www.artemis-emc2.eu/fileadmin/user_upload/Publications/2017_HiPEAC/HiPEAC_2017_EMC2_03_Productive4.0_Knut_Hufeld.pdf. Last accessed on 22-Apr-2017.
10. Thompson, A.G., Villaescusa, E., & Windsor, C.R. (1999). *Rock support and reinforcement practice in mining*. CRC Press. ISBN 9789058090454.
11. Eliasson, J., Punal, P., Mäkitaavola, H., & Delsing, J. (2014). A feasibility study of SOA-enabled networked rock bolts. In *Proceedings of 2014 IEEE 19th International Conference on Emerging Technologies & Factory Automation (ETFA 2014)*, Barcelona, Spain, October 2014.
12. ThingWave, https://www.thingwave.eu/. Last accessed 22-Jul-2017.
13. Han, Z., Weatherley, D., & Puscasu, R. (2015). Application of discrete element method to model crack propagation. In *13th International Congress of Rock Mechanics*, Montreal, Canada, May 2015.
14. Sule, S. (2016). Predictive maintenance, its implementation and latest trends. *Proceedings of the Institution of Mechanical Engineers, Part B: Journal of Engineering Manufacture*.
15. Gross, J. M. (2002). *Fundamentals of preventive maintenance*. New York: AMACOM.
16. Reason, P., & Bradbury, H. (Eds.). (2001). *Handbook of action research: Participative inquiry and practice*. London, UK: Sage Publications.
17. Gummesson, E. (2000). *Qualitative methods in management research* (2nd ed.). Thousand Oaks, CA, USA: Sage Publications.
18. Coghlan, D., Coughlan, P., & Brennan, L. (2004). Organizing for research and action: Implementing action research networks. *Systemic Practice and Action Research, 17*(1), 37–49.
19. Remenyi, D. (2013). *Field methods for academic research: Interviews, focus groups & questionnaires in business and management studies* (3rd ed.). Reading, UK: Academic Conferences and Publishing International Limited.
20. Patton, M. Q. (1990). *Qualitative evaluation and research methods*. London, UK: Sage Publications.
21. Kvale, S., & Brinkmann, S. (2009). *InterViews: Learning the craft of qualitative research interviewing*. LA, USA: Sage Publications.
22. Fontana, A., & Frey, J. (1994). Interviewing. In N. Denzin & Y. Lincoln (Eds.), *Handbook of qualitative research*. Thousand Oaks, CA, USA: Sage Publications.
23. Miles, M., & Huberman, M. (1994). *An expanded sourcebook—Qualitative data analysis* (2nd ed.). Thousand Oaks, CA, USA: Sage Publications.

Part X
Modelling and Frameworks

A Performance Measurement Extension for BPMN

One Step Further Quantifying Interoperability in Process Model

Xabier Heguy, Gregory Zacharewicz, Yves Ducq, Said Tazi and Bruno Vallespir

Abstract Business process model and notation (BPMN) is becoming the most used modeling language for business process. One of the important upgrades of BPMN 2.0 is the fact that data objects are now handling semantic elements. Nevertheless, BPMN does not enable the representation of performance measurement in the case of interoperability problems in the exchange of data objects, which remains a limitation when using BPMN to express interoperability issues in enterprise processes. We propose to extend the metamodel of BPMN in order to fill this gap. The extension, named performanceMeasurement, is defined using the BPMN extension mechanism. This new element will allow to represent performance measurement in the case of interoperability problems as well as interoperability concerns which have been solved. We illustrate the data interoperability capabilities with an example from a real industrial case.

Keywords Performance measurement · Interoperability · BPMN · Modeling · CBP

1 Introduction

The global economic context requires enterprises to acquire and maintain an efficient information system. An adapted and well-defined ERP is today a sine qua non condition for the success of a company. In addition, exchanges of information

X. Heguy · Y. Ducq (✉) · B. Vallespir
Univ. Bordeaux, CNRS, IMS UMR 5218, 33405 Talence, France
e-mail: yves.ducq@u-bordeaux.fr

X. Heguy
ONETIK, D252, 64240 Macaye, France

S. Tazi
University of Pau and Pays de l'Adour, Pau, France

G. Zacharewicz
IMT - Mines Ales, 30100 Ales, France

K. Popplewell et al. (eds.), *Enterprise Interoperability VIII*, Proceedings of the I-ESA Conferences 9, https://doi.org/10.1007/978-3-030-13693-2_28

between various information systems are increasingly necessary: In particular, information exchanges are growing with customers, suppliers, subcontractors or when the enterprise is bought out and integrated in another company, but also within an internal department. Also, assuming one ERP covering all sectors of the company is not always possible, which entails grafting many heterogeneous ERP or at least modules. The crucial problem of interoperability then arises.

Cross-organizational business process (CBP) modeling aims to describe the interactions between different organizations [1] but also between different systems within a single organization. Process modeling at the highest level should enable the various partners to understand the articulation of the different processes in a simple and intuitive way. However, defining user needs is not necessarily collaborative. Users rarely understand the differences between inter-organization processes and internal processes.

BPMN 2.0 [2] is a standard for the modeling of enterprise business processes. BPMN is currently the most used language among process modelers (64% of industry penetration according to P. Harmon's study) [3]. It provides a metamodel and notation for defining and visualizing them. BPMN 2.0 allows the modeling of CBP by clearly identifying independent resources pools and collaborative swim lanes within the same resource group. However, the representation of data in BPMN 2.0 does not reflect the concept of interoperability. Whether data are exchanged within the same information system or between two entities, the representation of the data remains the same. This is a weakness because this model hides the challenge and the difficulty of solving interoperability problems, in particular in the eyes of decision-makers. This lack of visibility can lead to incomprehension between IT technicians and managers and may lead them to underestimate both the impact of the lack of interoperability in the representation of the existing system and to make difficult the evaluation of costs and delays to produce the system to set up. In an earlier work, we introduced an extension to represent data exchanges presenting an interoperability issue as well as interoperability concerns which have been solved [4].

But BPMN does not either permit to represent performance measurement. Yet, making interoperability issues visible for all involved participants is not enough. An interoperability issue causes a data treatment in order to solve this problem. And this data treatment needs to commit resources, which is a waste of time and money. Then, in order to make the consequences of an interoperability issue evident, it is necessary to measure the performance of the actual process and to make visible the results of the measurement. In this way, all involved participants can easily make a comparison of these results and deduce that solving interoperability issues may result beneficial.

The technical structuring of BPMN is based on the concept of extensible layers around a core set of basic elements. This extensibility makes it possible to define an overlay of elements in order to better represent concepts inherent in the targeted sector of activity. Each new layer is constructed in extension of a lower layer. BPMN extensions are most often used to represent the specific needs of an industry [4]. But they can also be used to fill a general gap. This is the proposition of this paper: prolonging the BPMN model specification by adding an extension which permits to display the results of performance measurement in that case of interoperability.

2 Background

2.1 *Interoperability*

The most common definition of interoperability tells us that it is: "the ability of two (or more) systems or components to exchange information and use it" [5]. The InterOp NoE (network of excellence in interoperability) defined interoperability as "the ability of a system to work with another system without effort on the part of the user" [6]. To complement these definitions, we can say that interoperability is "the ability of systems, natively independent, to interact in order to build harmonious and intentional collaborative behaviors without deeply modifying their individual structure or behavior" [7]. These definitions demonstrate that interoperability is rapidly becoming complex and that its success depends on the resolution of a number of barriers [8].

There are three categories of barriers: conceptual (syntactic and semantic incompatibilities), technological (incompatibility of IT architecture and platforms, infrastructure, operating system, etc.) and organizational (incompatibilities of organization structure and management techniques) [9].

Three interoperability concerns are identified: interoperability of data, interoperability of service and interoperability of business.

According to EIF (enterprise interoperability framework) [9], we can notice three approaches of interoperability: integrated, unified and federated.

2.2 *Performance Measurement*

A problem of interoperability induces a processing of conformity of the data so that they can be exploited in the continuation of the process (most often reentry). This processing is a non-value-added task (NVA). Resolving the interoperability problem leads to the disappearance of this task. In order to highlight the need for this disappearance, it is necessary to display the figures represented by the resources used in this processing using performance indicators.

We have defined the following four performance domains: cost, quality, reliability and time. Cost, quality and time are the three most used indicators. Reliability is a part of quality. Indeed, if a datum does not comply with reliability, then it does not comply with quality. But in the case that interests us, it seemed that it should be interesting to measure it, because it is one of the most important sources of the non-quality of data. Indeed, most of the business software integrates control mechanisms. These mechanisms can force to insert relevant data (the data type is as it is expected), complete (all the fields have to be filled) and consistent (by limiting the values of the inserted data). But they will never be able to prevent the insertion of a bad value (non-reliable).

3 Related Works

Several works treat the problems of performance measurement visibility in BPMN models. Pavlovski and Zou [10] propose an extension that permits to identify the non-functional requirements. They introduce two new artifacts: The first, operating condition, indicates that a constraint is related to a flow object. The second, control case, specifies the business controls which will be set up to deal with the risk of the mode of operation. But these extensions do not permit to display any performance measurement.

Lodhi et al. [11] propose to extend BPMN. Their goal is to represent the impact of business objects in execution. They use different colors that indicate the performance level of activities and the use of different swim lanes for each dimension (for instance cost or time) and three lanes to show if the performance is low, medium or high. Other extensions are proposed that are out of the scope of this paper. But the proposed extension does not display the values of the performance measurement, and we think that these values have to be shown in order to be adequately appreciated.

The contribution of Saeedi et al. [12] adopts an annotation approach to make visible performance measurement. The proposed extension is based on the use of tables in which the results of performance measurement of a task or a process are displayed. In order to calculate the performance measurement of a process, they use reduction rules. They consider cost, time and reliability requirements. Then, this work is not interoperability oriented, and it does not consider data quality. Besides, their work has been very helpful for us.

All mentioned works cover the problems of performance measurement. But they do not cover the specific issue that is interoperability in data interchange between different information systems. This issue is traditionally reserved for IT technicians. This is surely useful, but it does not offer a possibility to make these problems visible for all the collaborating business partners (and not only IT technicians). With the extension we define in our work, we offer the possibility of that shared understanding, which is one of the main goals of BPMN.

4 Extension

Our aim is to create a sustainable extension of BPMN, which allows displaying performance measurement values of cost, quality, reliability and time in the framework of process diagrams, in order to solve interoperability issues in data exchanges. With dataInteroperabilityBarrier and dataInteroperabilitySolute we defined in our previous work [4], this new extension can be used in very different business contexts, for example in inter-company data exchanges, in Web applications, etc. It will make possible to show practically the inconveniences caused by an interoperability problem. The goal of this extension is not to solve interoperability problems (there is no automatic resolution of the problem), but to reveal it. It will permit to make evident the interoperability problems in the AS IS model and show in the TO BE the benefits of solving them.

We propose the performanceMeasurement extension. It allows displaying in any task or task group the values of cost, quality, reliability and time indicators.

4.1 *BPMN Extension Mechanisms*

BPMN has been designed to be extensible. The technical structuring of BPMN is based on the concept of extensible layers around a core of simple elements. Extensibility is used to define an overlay of elements to better represent concepts inherent in the targeted industry. BPMN provides generic extension elements in the metamodel.

BPMN has extension elements in order to add additional elements and attributes to the existing BPMN elements. These extension elements are: ExtensionDefinition, ExtensionAttributeDefinition, ExtensionAttributeValue and Extension. The Extension element links an extension to the BPMN model. The structure of this element is defined by the ExtensionDefinition element, which adds extra attributes. Each attribute is defined by a name and a type in the ExtensionAttributeValue element, and this value is set by the ExtensionAttributeDefinition element.

This extension permits the inclusion of the table definition that makes possible to display the values of performance measurement of cost, quality, reliability and time. The structure of the proposed element is defined through ExtensionDefinition and ExtensionAttributeDefinition elements.

The structure of BPMN is described through two representations:

- The Meta-Object Facility (MOF) metamodel in which the concepts are characterized;
- The XML Schema Definition (XSD) in which the format for interchange is settled [13].

The class diagram of the MOF metamodel is presented in the specification manual. It is divided in different sections. We will only represent the section that concerns our study (see Fig. 5).

We have defined the following ExtensionDefinition elements: cost, quality, time and reliability. The corresponding ExtensionAttributeDefinition elements are: implementationCost and executionCost for the cost, qualityValue for the quality, executionTime for the time and reliabilityValue for the reliability.

We can see in Fig. 6 how the new extension is linked to the Definitions class (Figs. 1 and 2).

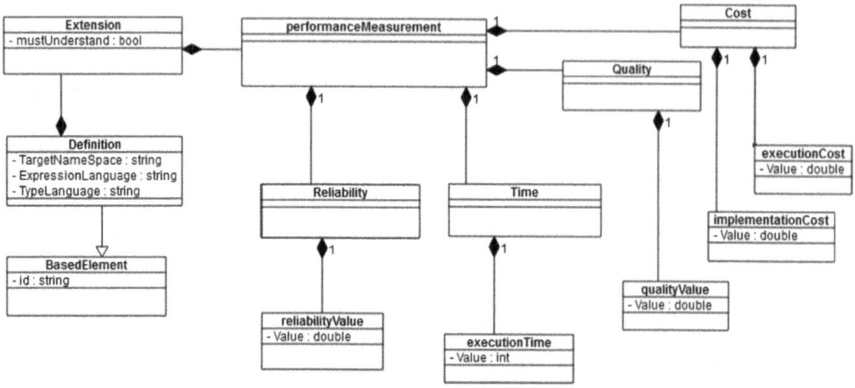

Fig. 1 BPMN class extension

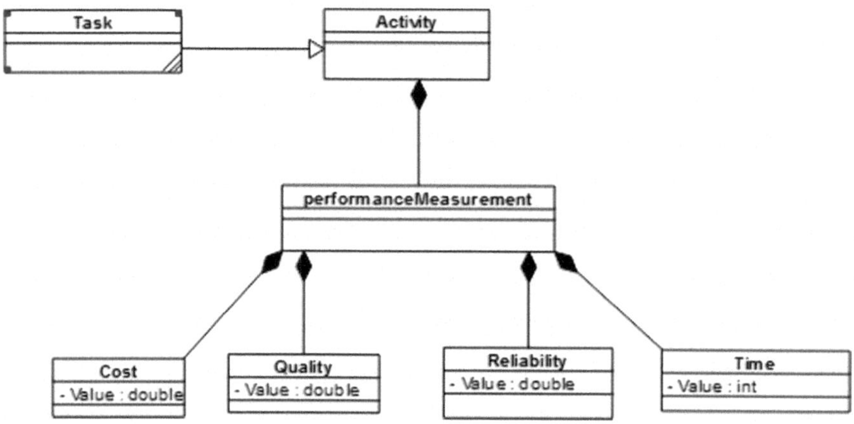

Fig. 2 Link with the BPMN metamodel

4.2 Performance Measurement Aggregation

As explained previously, an interoperability issue causes the necessity or a (or more) non-value-added task to address this issue. The final goal being to solve the interoperability issues, it is necessary to compare the performances of the existing system (AS IS) with those of the future system (TO BE) in order to appreciate the benefits of this solving. But when the interoperability issue will be solved, the non-value-added task will disappear. Then, it will not appear in the TO BE model. We will then compare the performance of a unique task (in the TO BE model) with those of a set of tasks (in the AS IS model). We have then to be able to aggregate the performance measurements of two or more tasks.

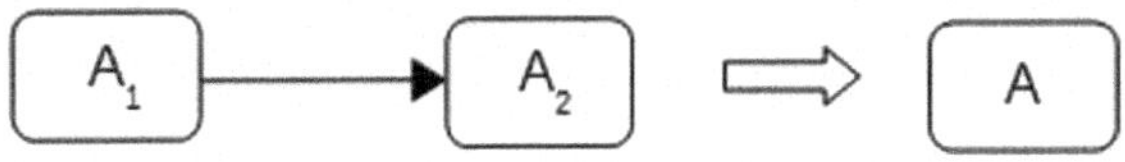

Fig. 3 Sequential process reduction

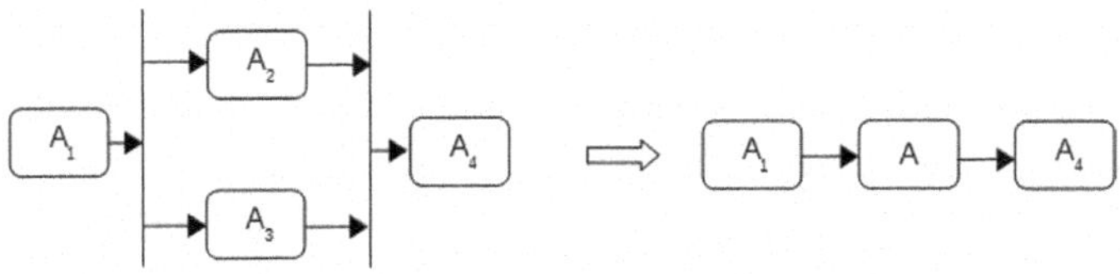

Fig. 4 OR process reduction

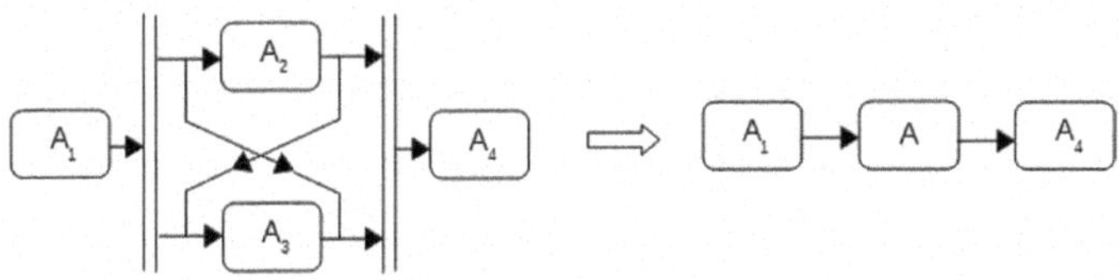

Fig. 5 AND process reduction

In this section, we propose a performance measurement aggregation model based on Ducq's works to determine the different aggregation typologies in a data interoperability scope [9].

(1) Processes reduction

We propose to use the stochastic workflow reduction (SWR) algorithm [14] to reduce the tasks. This algorithm uses six reduction rules: sequence, parallel, conditional, fault tolerant, loop and network. We consider that only the first three rules are relevant in the scope of data interoperability. Indeed, fault tolerant and loop rules are more IT oriented. And a network process block represents a sub-process. In the case we are discussing, there are most of the time two tasks: the data reception and the data processing (the non-value-added task) which is not enough to constitute a sub-process.

The reduction rules are successively used to a process. Following, a unique task remains. The performance measurement values of that task will then match with those of the entire business process. We reduce the A_i tasks in the A task.

Sequential reduction: Here, all tasks are executed consecutively (Fig. 3).

OR reduction: In this aggregation type, we can equally execute task A_2 or A_3, having different performance characteristics. See Fig. 4.

AND reduction: In this case, we have to execute both tasks (A_2 and A_3), having different performance characteristics. See Fig. 5.

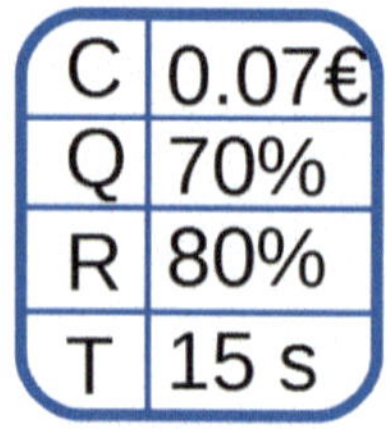

Fig. 6 performanceMeasurement representation

(2) Performance measurement aggregation

Once we solved the process reduction problem, we will define the formulas we will use in order to calculate the values of the aggregated tasks performances.

Sequential aggregation: Cost and time are equal to the sum of costs and times, respectively. Quality and reliability are equal to the product of qualities and reliabilities.

Cost:

$$C(A) = C(A_1) + C(A_2)$$

Quality:

$$Q(A) = Q(A_1) * Q(A_2)$$

Reliability:

$$R(A) = R(A_1) * R(A_2)$$

Time:

$$T(A) = T(A_1) + T(A_2).$$

OR aggregation: The values of cost and time are equal to the maximum of costs and times, respectively. The values of quality and reliability are equal to the minimum.

Cost:

$$C(A) = \max(C(A_1), C(A_2))$$

Quality:

$$Q(A) = \min(Q(A_1), Q(A_2))$$

Reliability:

$$R(A) = \min(R(A_1), R(A_2))$$

Time:

$$T(A) = \max(T(A_1), T(A_2)).$$

AND aggregation: The value of cost is equal to the sum of the costs. The value of the time is the value of the maximum time. The values of quality and reliability are equal to the products of qualities and reliabilities, respectively.

Cost:

$$C(A) = C(A_1) + C(A_2)$$

Quality:

$$Q(A) = Q(A_1) * Q(A_2)$$

Reliability:

$$R(A) = R(A_1) * R(A_2)$$

Time:

$$\max(T(A_1), T(A_2)).$$

The aggregation based on multiplicative operator has a drawback: It can amplify the errors. But it makes them more visible, which is a good thing in our work's scope.

5 Graphical Representation

We propose to provide the graphical representations for the performanceMeasurement extensions visible in Fig. 7. This extension represents a table where the values of cost, quality, reliability and time are displayed. This table is coupled with every task or the task group from which we want to display the performance.

6 Use Case

To illustrate the interest of performanceMeasurement extension, we will study the case of Onetik SME. This company of the Basque Country manufactures and markets cheeses. It uses the Nodhos ERP. It is an SME, but it is a part of a supply chain, having many providers and clients. We can therefore think that it is a relevant use case.

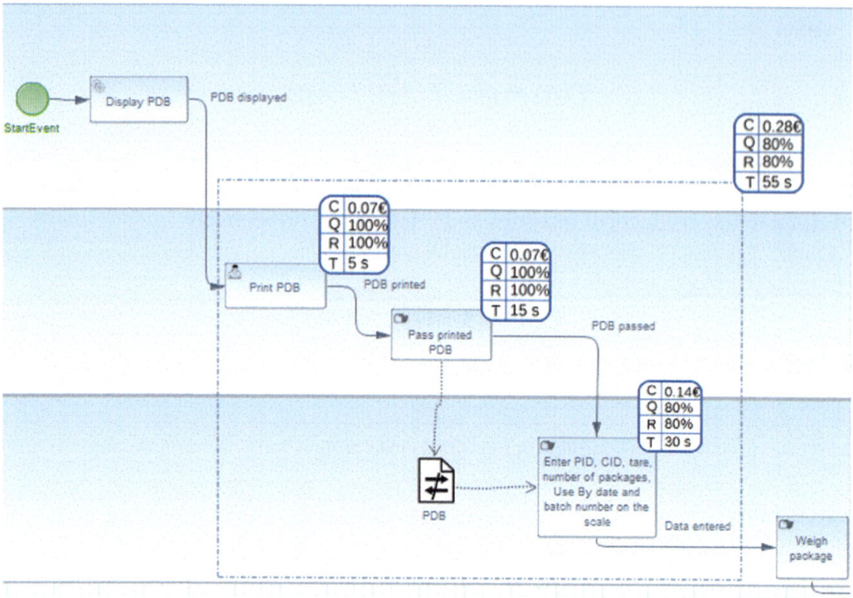

Fig. 7 AS IS model

The shipment management module of this ERP does not give it complete satisfaction, and its use is source of errors and therefore, among other things, disputes with customers that are costly for the company. The replacement of the ERP is not envisaged in the short term mainly for financial reasons. The management of Onetik has then decided to graft the shipment management module of another ERP (InteGraal Agro).

In the framework of this project, two models were carried out using BPMN: One is corresponding to the existing system (AS IS) and the other to the desired system (TO BE). Each model consisted of several models of the various processes. We will represent the (simplified) process of order preparation.

This AS IS model (Fig. 7) represents (a part of) real case proposed by the current organization of the information system. This is to be developed by describing the problem faced by the enterprise and the problem that is engendered by this situation.

The actual ERP (Nodhos) permits to print a preparatory delivery bill (PDB) with the listing of the packages to be weighed. Then, this PDB has to be carried to the scale. Before weighing a package, some data have to be manually entered in the scale. These data are: product ID (PID), client ID (CID), tare, number of packages, use by date and batch number. It is obvious this is an interoperability issue. The data reentry task is a non-value-added task which presents a big risk of error.

After weighing the package, the values of weight and batch number have to be noted in the PDB (paper). The filled PDB is then carried back to the operator who

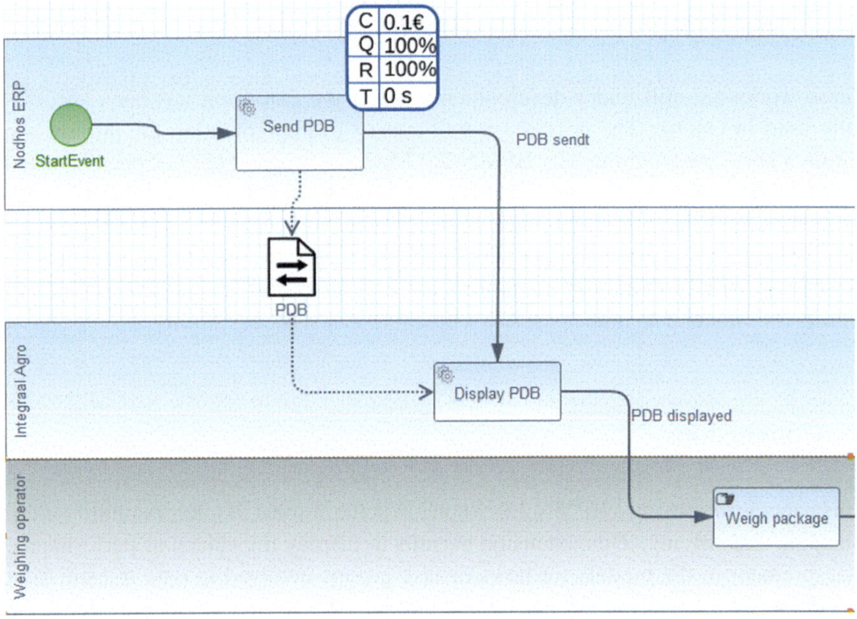

Fig. 8 TO BE model

enters in the ERP the values of weight and batch number, which presents a new possibility of error.

We grouped all the tasks involved in the data transmission, and we use the performanceMeasurement extension in each task as well as in each group. Of course, we used the reduction rules and the aggregation of performance measurement to calculate the values visible in the table coupled to the groups. We can now show the consequences of the interoperability issues to every participant of the project by displaying in the AS IS model the values of cost, quality, reliability and time of each task and each data transmission process. We will now be able to compare these values with those of the TO BE model which will solute the interoperability issues.

We use the extension we proposed to represent the data exchanges presenting an interoperability issue [4].

The TO BE model (Fig. 8) represents the solution that has to be implemented in order to solve the interoperability issues. We can see in this model that the interoperability issues have been solved. Consequently, there is no more non-value-added task and the risks of errors induced by the reentry tasks have disappeared. But, we can display the values of performance measurement of the data transmission tasks.

With these two models, we can quickly make a comparison of the performances between the existing and the future systems. This will be useful to show to the performance improvement.

7 Perspectives and Implementation

These works are still under development even if the extension has been successfully used in Onetik. The next phase will consist in implementing the proposition within a software solution. The BPMN 2.0 Modeler Project [15] has been selected for its capacity to integrate a model-driven approach, BPMN 2.0 models and simulation aspects. The conceptual phase has been achieved, and we are working on the development of this extension that will be the baseline for representing performance aspects on the BPMN model. A future work will be to plug a simulation engine to BPMN 2.0 modeler to make possible a machine-based processability.

8 Conclusions

This paper is proposing a BPMN 2.0 extension in the context of interoperability identification and solving. This extension permits to display the values of performance measurement in the models for tasks or task groups involved in data transmission presenting an interoperability barrier and problem solved thanks to a graphical icon added to the original task or group item of BPMN. The interest of this approach has been illustrated on a use case from industry. These new features of BPMN allow bringing to light to all participants of a CBP project (managers, IT technicians, processes owners, etc.) the presence of an interoperability barrier and its solution.

References

1. ATHENA. (2005). D.A2.1: Cross-organisational business process requirements and the state of the art in research, technology and standards version 2. ATHENA Project No. (507849).
2. OMG. (2003). OMG: MDA Guide Version 1.0.1, www.omg.org/mda/
3. Harmon, P. (2016). The state of business process management 2016. Business Process Trends.
4. Heguy, X., Zacharewicz, G., Ducq, Y., & Tazi, S. (2017). Interoperability markers for BPMN 2.0. Making Interoperability Issues Explicit, Advances in Engineering Research. In *2nd International Conference on Electrical, Automation and Mechanical Engineering (EAME 2017)* (Vol. 86, pp. 330–333).
5. Braun, R., & Esswein, W. (2014). Classification of domain-specific BPMN extensions. In *The practice of enterprise modeling*. Berlin: Springer.
6. Geraci, A., Katki, F., McMonegal, L., Meyer, B., Lane, J., & Wilson, P., et al. (1991). *IEEE standard computer dictionary: Compilation of IEEE standard computer glossaries.*
7. Konstantas, D., Bourrières, J. P., Léonard, M., & Boudjlida, N. (2005). *Interoperability of enterprise software and applications.*
8. Truptil, S., Bénaben, F., Pingaud, H. (2010). A mediation information system to help to coordinate the response to a crisis. In L. M. Camarinha-Matos, X. Boucher, H. Afsarmanesh (Eds.), Collaborative networks for a sustainable world (Vol. 336, pp. 173–180). Springer, IFIP Advances in Information and Communication Technology. 978-3-642-15960-2.
9. Ducq, Y. (1999). Contribution à une méthodologie d'analyse de la cohérence des Systèmes de Production dans le cadre du Modèle GRAI.

10. Pavlovski, C. J., Zou, J. (2008). Non-functional requirements in business process modeling. In *Proceedings of the fifth Asia-Pacific conference on Conceptual Modelling* (Vol. 79, pp. 103–112). Australian Computer Society, Inc.; Petriu, D., Bernardi, S., & Merseguer, J. (2008). An UML profile for dependability analysis and modeling of software systems. Technical report, Universidad de Zaragoza, Spain.
11. Lodhi, A., Köppen, V., Saake, G. (2011). An extension of BPMN Meta-model for evaluation of business processes.
12. Saeedi, K., Zhao, L., Sampaio, P. (2010). Extending BPMN for supporting customer-facing service quality requirements.
13. Cortes-Cornax, M., Dupuy-Chessa, S., Rieu, D., & Dumas, M. (2014). Evaluating choreographies in BPMN 2.0 using an extended quality framework.
14. Cardoso, J., Sheth, A., Miller, J., Arnold, J., & Kochut, K. (2004). Quality of service for workflows and web service processes.
15. Eclipse Foundation. BPMN 2.0 modeler project. https://www.eclipse.org/bpmn2-modeler

An Assessment Conceptual Framework for the Modernization of Legacy Systems

Abir M'baya, Jannik Laval and Nejib Moalla

Abstract Legacy systems modernization is one of the most common issues for enterprise organizations. As technology is changing rapidly, enterprises need to upgrade legacy systems and ensure suitable modernization. Evolving legacy systems to new ones can give new life to the existing applications; however, if the modernization process is not done correctly, it can lead to failure. Making valid decisions on the choice of modernization strategy and achieving successful evolutionary system require addressing the full modernization project life cycle. Accordingly, we propose in this paper an assessment and decisional framework for legacy systems modernization (ADF-LSM). ADF-LSM includes the main steps of migration strategy to guide the modernization process. Also it introduces quality metrics modeling at business, information systems and infrastructure levels, to assess the system and integrate the new needs of evolution. We validate the applicability of the quality metrics of the application level thorough a case study involving a control system for practically arbitrary automated vehicles. Our case study results show that the quality metrics identify well the current issues list and benefits needed to modernize a system.

Keywords Legacy system · Assessment framework · Legacy modernization · Quality metrics · System analysis

1 Introduction

A legacy system is the backbone of any enterprise as it refers to any business—critical software systems that significantly resist modification and are non-extensible when faced with new technologies; their failure can have a serious impact on its daily oper-

A. M'baya (✉) · J. Laval · N. Moalla
DISP Lab EA4570, Université Lumière Lyon 2, Lyon, France
e-mail: A.Mbaya@univ-lyon2.fr

J. Laval
e-mail: Jannik.Laval@univ-lyon2.fr

N. Moalla
e-mail: Nejib.Moalla@univ-lyon2.fr

K. Popplewell et al. (eds.), *Enterprise Interoperability VIII*, Proceedings of the I-ESA Conferences 9, https://doi.org/10.1007/978-3-030-13693-2_29

ation [1]. Accordingly, such systems need to be continuously adapted or enhanced to either satisfy changing user requirements or cope with the continuous evolution of the other systems around them [2]. Legacy systems evolution is the process of evolving existing software systems by replacing, redeveloping, reusing or migrating the software components and platforms, when traditional maintainer practices can no longer achieve the desired target system. The main aim of software modernization is to reduce maintenance cost and increase flexibility. This allows companies to handle market needs and chiefly ensure that the organization has a competitive advantage in the market. The evolution can be categorized into three categories: maintenance, modernization and replacement. When the business needs to start deviating from the initial intended usage, companies start maintenance activities to keep the system operational. When the software system becomes so outdated that maintenance activities become too challenging and expensive, modernization becomes inevitable. The modernization project might ultimately involve total replacement of the system when the old system can no longer be evolved. Despite the disadvantages of legacy applications such as the difficult to maintain, the lack of documentation and the difficulty to extend and integrate with other systems [3, 4], these systems are vitally important for the continuation of business in the enterprises and are still present in daily operation. It is estimated that more than 80% of the world's business runs on COBOL; 50–70% of the total IT costs are devoted in the maintenance of these systems [5]. To achieve the success of modernization projects is neither easy nor automatic; it is a very challenging task. Several factors influence the success or failure of these projects, such as the assessment process of the both premodernization and post-modernization situations, feasibility analysis, selection strategy, organization factors in terms of cost-benefit estimation and risk analysis [6].

Legacy system modernization is a major research domain, and a plethora of legacy system modernization methods has been proposed (cf. Sect. 3). However, a methodology that ensures the accompaniment of the modernization project from end to end and consolidates the aforementioned factors is still missing, as per our knowledge. In this paper, an assessment and decisional framework for legacy system modernization, here in after called ADF-LSM, is proposed that drives system evolution projects in full and combines the main factors to assure the success. ADF-LSM presents a strategy including a methodological approach to help maintainers in the evolution process. Further, it provides an analysis composed of five phases: (i) In the preprocessing phase, we aim to understand both legacy system and target system. (ii) In the assessment phase, we assess deeply the system to identify the key pains of the current system at different levels. (iii) Then, according to the results provided in the assessment phase, selection phase selects the most suitable strategy by performing cost, benefit and risk analysis. (iv) The transition phase aims to determine the evolution project plan. (v) Validation phase checks and validates all the changes made on the modernized system.

Our contributions are as follows: (1) a systematic framework for modernizing legacy systems and (2) a case study using, a control system for practically arbitrary automated vehicles, to demonstrate how the application quality metrics predict the issues of the system at the application level.

The rest of paper is organized as follows: In Sect. 2, we present the methodological approach of the framework ADF-LSM. Section 3 discusses the quality model used to qualify the system quality. In Sect. 4, we validate the quality metrics of the application layer through a case study on an openTCS system, a control system for practically arbitrary automated vehicles. Through this study, we assure that the quality metrics predict well the key pains in the application level. Finally, we conclude the paper and propose future research directions.

2 The Supporting Methodology of ADF-LSM for System Evolution

Transforming legacy applications to evolvable systems can be critical to many sectors as it involves risks and costs. Thus, it is important to carefully plan, document, transform and deploy the system. We describe in this section the proposed incremental approach ADF-LSM for performing legacy systems modernization. ADF-LSM method provides end-to-end guidance, from project planning through to system deployment, for managing evolution projects. In fact, it helps stakeholders to evaluate, analyze and recommend the best way to reach the desired target system that meets the new requirements and includes the new technologies. Then, it ensures the quality of the migration process during the transformation of the legacy system toward a modern paradigm while minimizing the risks and the additional costs associated with increased quality. The framework consists of five phases: preprocessing phase, assessment phase, strategy selection, transition phase and validation and verification phase. Figure 1 shows the different phases of ADF-LSM.

2.1 Preprocessing Phase

The phase starts with the start-up project activity that helps industrial practitioners to make a broad decision whether to evolve the system through an approximate qualitative and quantitative qualification of the system or not. This qualification leads to document the problem statement of the organization, understand the need for evolution, if it is required, and record the evolution requirements. To improve aspects of existing systems, it is appropriate to perform the activity of understanding the legacy system and the target system that is crucial to the success of any evolution [7]. Many legacy systems are difficult to understand as they are poorly documented and their structure is degraded that results from prolonged maintenance. Therefore, we need to capture a sufficient understanding of the main concepts, components and relationships of the system as it is a prerequisite for making an informed decision as to an appropriate evolution strategy. Evolution modeling addresses how to acquire knowledge with the aim of bringing these systems under control. Often, we

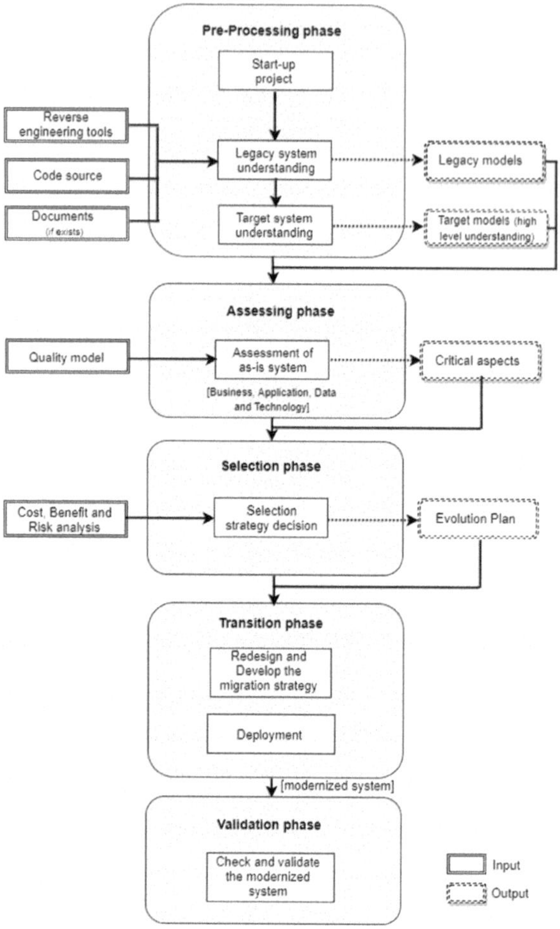

Fig. 1 Workflow of the framework ADF-LSM

use reverse engineering tools to construct system models from several viewpoints (business models, application models, data models and technology models). In the same way, we extend the legacy system's models with new business and technology constraints to build the target system's model. The benefit of modeling is to support the assessment activity. It therefore promotes an understanding of the system, which is vital for an effective assessment. In fact, where the assessment is based on a poor comprehension of the system, the results are at best loosely approximate and at worst wholly inaccurate and misleading.

2.2 Assessment Phase

Accurate assessment is an essential contribution to develop the right evolution strategy. This phase involves assessing the quality of the legacy system. It starts with the assessment activity that quantifies the current state of the system and understands the current issues list and benefits needed from the modernization. The business, application, data and technology quality of the system are evaluated and later used to determine the most appropriate modernization strategy to be applied to the system. This is achieved by applying the measurement model and evaluating the metrics it includes (detailed in the next section). The required information may be obtained through static analysis and interviews. Then, by analyzing the gap between the current system and the target desired system, we introduce a new control for the organization to control the critical aspects of the software system. In fact, it evaluates how the current system is mapped to the target business, application, data and technology quality for identifying gaps to be considered during the identification of the evolution approach. A critical aspect is made active if the gap assumed by a factor or attribute is not valid with regard to the determined thresholds, indicating the goodness level of the factor and attribute values. The assessment process is an incremental process. The assessment level depends on the level of understanding that we have reached in the preprocessing phase. In fact, we begin the evaluation through high-level sub-systems. Then, by decomposing these sub-systems, we evaluate the lower level components that result from the decomposition of the sub-systems.

2.3 Selection Phase

This phase aims to define the most suitable strategy for the evolution of the software system. Strategy selection acts as a decision point for deciding if migrating to a modern paradigm is feasible. There are several ways in which a particular system component can evolve and evolution strategies are considered as paths for reaching the target system. In this phase, we aim to select the best that satisfies the problem statement's aims and constraints by analyzing each strategy's costs, risks and benefits. The result of this phase constitutes the ultimate product of plan evolution. In fact, we can have a single strategy for the whole system or a hybrid strategy with a sub-strategy for each component.

2.4 Transition Phase

This phase involves the tasks, which must be performed to transform the legacy system to the target system according to the evolution strategy defined in selection phase. The solution transition phase includes activities such as updating the high-level migration strategy with information related the realization of the strategy. Then, it selects the appropriate approach that intends to take to deploy the system, analyzes the test effort and develops a clear test strategy and test data. It is important to select an appropriate transition technique to avoid any loss of the business during the transition.

2.5 *Validation Phase*

This phase is applied to assess and evaluate the modernized system and provide all the changes needed after the migration. Its aim is to check all the changes made on the system such as software updates and management changes and validate that the target system is very close to the legacy project requirements and needs of the business.

3 The Supporting Toolkit

A number of modernization approaches have been reported for managing legacy systems. Several authors have proposed decision frameworks to select from among the alternatives. This section describes the supporting toolkit of the framework. It includes the quality model. The quality model based on the metrics aims to support the assessment of the system. In the qualification phase, it evaluates the state of the current system and determines the critical aspects. It assesses the correctness of the migration process during the transformation of the software system. Finally, it validates whether or not the target system is achieved. The quality model is based on the identification of the business, application, data and technology quality profiles. Consequently, the framework considers the following high-level aspects:

- Business: represents the system value from the points of view of its direct and indirect users, where the direct and indirect users can be managers, end users, customers and so on
- Application: evaluates the quality of the information systems architectures of the organization, the structure, performance and interactions of the applications.
- Data: describe the structure and interaction of the organization's sources of data, logical data assets and data management resources.
- Technology: depicts the technology architecture layer of the organization. It represents the structure and interaction of the platform services and physical technology components.

The system is evaluated by applying the quality model presented in Table 1. The model includes qualitative and quantitative assessment. Although qualitative assessment cannot be graphed or displayed in terms of mathematical expressions, it is still an important indication in supporting the strategy making process. In the

Table 1 Quality model

Aspect	Factor	Attribute
Business	Economic	Time to market
		ROI
		Maintenance cost
		Usage frequency
		User satisfaction
	Organizational	Technical maturity
		Skill levels
		Response to change
Application	Flexibility	Modularity
		Simplicity
		Lack of technical documentation
	Application performance	Average responsiveness
		Average availability
	Reliability	Software reliability
	Maintainability	Correctness
		Testability
		Modularity
	Application obsolescence	SW obsolescence
	Interoperability	Degree of standardization
Data	Data integrity	Data consistency
		Data security
	Data usage	Data accessibility
		Data dependance
	Data obsolescence	DB obsolescence
Technology	Operational performance	Storage performance
	Obsolescence	OS obsolescence
		Hardware obsolescence
	Portability	Software system independence
		Machine independence

case of quantitative assessment, a quantification metrics need to be selected for measurement. To this purpose, the quality attributes are evaluated directly through a set of well-chosen metrics, while the factors are indirectly evaluated by exploiting the impact that attributes have on them. Let M_i a generic metric, X_i is the value assumed by $M_{i,}$ and $X_1, X_2, \ldots X_n$ are the measures normalized at 1, of the metrics $M_1, M_2, \ldots M_n$ related to the attribute At. At is computed as the weighted average of the corresponding metrics by using the following formula: $\mathrm{At} = \sum_{i=1}^{n} p_i X_i$, where p_i is the weight assigned to each metric M_i to express its relevance for evaluating At. The weight values are between 0 and 1 and are adjusted according to the expert feedback. The goodness level of the attribute value is indicated based on the range of the attribute, $[\mathrm{At}_{\mathrm{min}}, \mathrm{At}_{\mathrm{max}}]$, which can assume the attribute AT. $\mathrm{At}_{\mathrm{min}}$ and $\mathrm{At}_{\mathrm{max}}$ represent thresholds. For automating threshold derivation, we calculate the thresholds by applying statistical analysis techniques to a manually constructed benchmark set of software products (i.e., a benchmark repository). This is a technique widely used in the literature for thresholds derivation of source code measures. Threshold process is data driven so they are based on data and not on experts' opinions. This leads to a highly objective and a fully automatic assessment. The same formulas are used for aggregating the values of the attributes and evaluating the values of factors. Thus, the dimension values are assessed by aggregating their related factors values.

4 Case Study

In this section, we aim to validate the proposed quality model of ADF-LSM through an experiment on an open-source project. We focus on how the quality model is able to detect the issues of a system. Our case study is strictly limited to the application level. We selected an open-source project maintained by a community whose release notes are well described. Our attention was clearly focused on the open-source project Open Transportation Control System openTCS [8].

4.1 Background

OpenTCS is a control system for automatic vehicles. It was primarily developed for the coordination of automated guided vehicles (AGV). The code base is maintained and development continued by Fraunhoter institute for material Flow and Logistic (IML). OpenTCS itself is not a complete product that can be used "out of the box" to control an AGVS with, but it is a framework of the basic data structures and algorithms (routing, dispatching, scheduling) needed for running an AGVS with more than one vehicle. The software runs on the Java platform. All libraries required for compiling and/or using it are freely available, too. The openTCS project includes the following modules: (1) opentcs-api-base: contains the base API for clients and extensions. This is what most developer's probably want to use, (2) opentcs-api-injection: contains API interfaces and classes used for dependency injection within the kernel and plant overview applications, (3) opentcs-common: contains a collection of utility classes used by openTCS components, (4) opentcs-commadapter-loopback: contains a very basic vehicle driver simulating a virtual vehicle, (5) opentcs-strategies-default: contains the default implementations of strategies that are used by the kernel application, (6) opentcs-kernel: contains the kernel application, (7) opentcs-plantoverview: contains the plant overview application.

4.2 Experimentation and Results

To perform the metrics that the proposed quality model includes, we employ CKJM and PMD for the quantification of the metrics and findings, respectively. Metrics correspond to the source code metrics in the literature, such as cyclomatic complexity; findings, also called violations, represent the results provided by different types of static analysis tools such as bug pattern detectors and rule-based tools. OpenTCS provided 23 old versions between 2012 and 2017. As a first experimentation, we have detected the problems that have been properly corrected in version n through the release notes. This means that in version $n - 1$, some issues are present and our model should be able to identify them. The issues correspond to the unsatisfactory values of some attributes and to the gap between values n and $n - 1$. The unsatisfactory values are below the corresponding thresholds of the attributes, which are derived automatically in our model. The gap between versions is shown in Fig. 2. For example, we have identified in the release note version 4.3 some maintainability improvements in the dispatcher (included in the module strategies applying our model, we have also detected that the maintainability value is unsatisfactory in version 4.2 below the computed threshold. Then, the gap 4 proves that the maintainability quality improved in the version 4.3. As a second experiment, we look at the 8 m provided in 2017, ranging from 4.0.0 to 4.7 assessed each of the 8 modules previously presented. Thus, we have performed 64 evaluations for each factor. Due to the space limitation, we have displayed only values (Maintainability, Reliability, Performance) of the modules "opentcs-strategies-default" and "opentcs-common" (Figs. 2 and 3). The proposed quality model report consistency and continuity in the results. This is expressed in the following figures by a regular progression addition, the curves show gaps between two versions, which correspond to more or less significant improvement actions made by the team. This proves that our sensitivity is aligned with the evolution decisions.

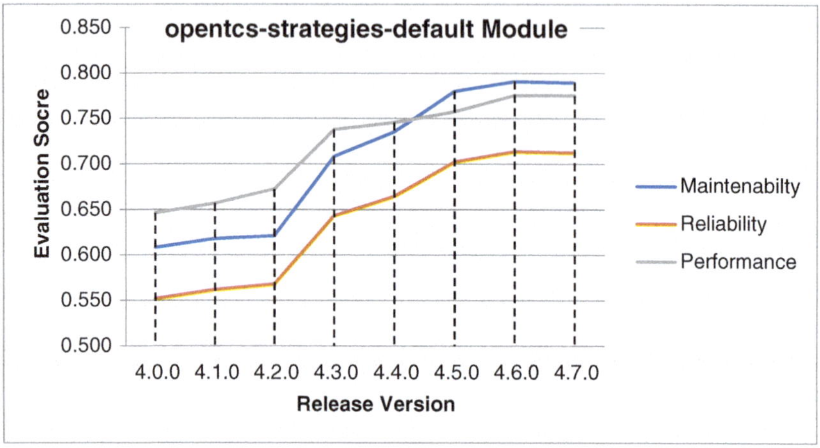

Fig. 2 Results provided by the quality model of ADF-LSM to evaluate the factors "maintainability, performance and reliability" of "opentcs-strategies-default" module in the openTCS system

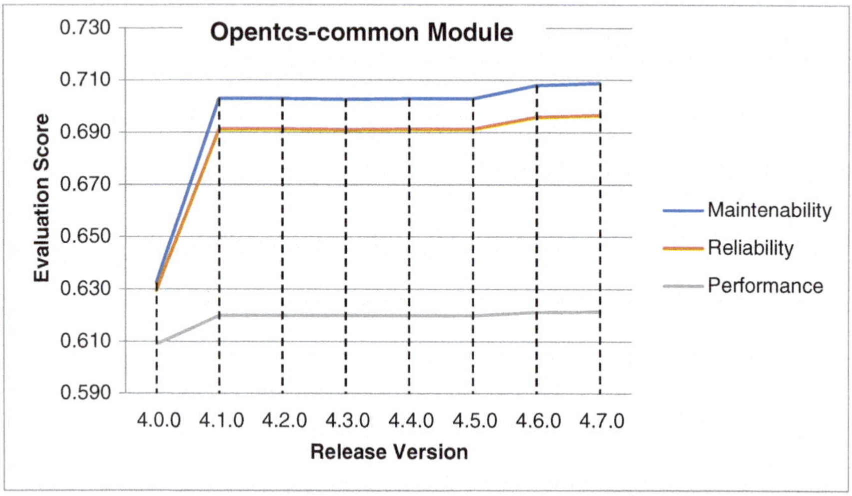

Fig. 3 Results provided by the quality model of ADF-LSM to evaluate the factors "maintainability, performance and reliability" of "opentcs-common" module in the openTCS system

5 Conclusion

Evolving software systems requires the assessment of their business, application, data and technology value. This paper presents a Systematic Legacy Assessment and Modernization Framework ADF-LSM to evaluate the current system by applying quality metrics. ADF-LSM includes an assessment process according to a predefined set of factors and attributes considering the requirements of the modernization project. The results of our use case validate the applicability of the quality metrics of the application level. As a future work, we aim to validate the whole model that includes qualitative and quantitative metrics on multiple case studies. Then, we would report a comparative results based on using this model to show what benefit it would provide to those teams dealing with legacy modernization issues. Also, to show what difference does using the model versus not using it makes from a practical perspective.

References

1. Brodie, M. L., & Stonebraker, M. (1995). Migrating legacy systems: Gateways, interfaces and the incremental approach. Morgan Kaufmann Publishers Inc.
2. Lehman, M. M., Perry, D. E., & Ramil, J. F. (1998, November). Implications of evolution metrics on software maintenance. In *International Conference on Software Maintenance, Proceedings* (pp. 208–217).
3. Khadka, R., Shrestha, P., Klein, B., Saeidi, A., Hage, J., Jansen, S., & Bruntink, M. (2015). Does software modernization deliver what it aimed for? A post modernization analysis of five software modernization case studies. In *Software Maintenance and Evolution (ICSME)* (pp. 477–486).
4. Khadka, R., Batlajery, B., Saeidi, A., Jansen, S., & Hage, J. (2014). How do professionals perceive legacy systems and software modernization? In *Proceedings of the 36th International Conference on Software Engineering* (pp. 36–47).
5. Sneed, H. M. (1995). Planning the reengineering of legacy systems. *IEEE Software, 12*(1), 24–34.
6. Khadka, R., Saeidi, A., Jansen, S., et al. (2013). A structured legacy to SOA migration process and its evaluation in practice. In *IEEE 7th International Symposium on the Maintenance and Evolution of Service-Oriented and Cloud-Based Systems (MESOCA). IEEE* (pp. 2–11).
7. Lewis, G., Morris, E., & Smith, D. (2005). Service-oriented migration and reuse technique (SMART). In *Proceedings of the 13th IEEE international Workshop on Software Technology and Engineering Practice*.
8. http://www.opentcs.org/en/index.html.

The European Union Landscape on Interoperability Standardisation: Status of European and National Interoperability Frameworks

Victoria Kalogirou and Yannis Charalabidis

Abstract Interoperability between public administrations was recognized by the European Commission (EC) with the launch of the Interchange of Data between Administrations (IDA). In 2004 the first version of the European Interoperability Framework for pan-European eGovernment Services EIF V1.0 was published followed by EIF V2.0 in 2010 and in 2017 the New EIF. This paper initially presents a summary in the landscape of interoperability, specifically a brief study on the EIF, NIFs (National Interoperability Frameworks), ISA and ISA^2 (interoperability solutions for public administrations, businesses and citizens) programmes with focus on the EIF and its implementation in MS' (member states) NIFs. How NIFs are currently utilized and what is beyond after the alignment or implementation with EIF. The description starts with some basic definitions and a review on the status of current EIF and NIFs in European countries and in several non-European countries. According to the New EIF's principles, layers, recommendations and model a comparison is made so as to recognize major commonalities and significant differences between EIFs and principles in the NIFs. The contribution after comparison and identification of gaps provides a short summary of outcomes, recommendations and new directions mainly in relation to the outburst of current trends.

Keywords European Interoperability Framework · Interoperability · Interoperability frameworks · Interoperability governance · eGovernance · eGovernment

---

V. Kalogirou (✉) · Y. Charalabidis
University of the Aegean, Samos, Greece
e-mail: vickalogirou@gmail.com

Y. Charalabidis
e-mail: yannisx@aegean.gr

K. Popplewell et al. (eds.), *Enterprise Interoperability VIII*, Proceedings of the I-ESA Conferences 9, https://doi.org/10.1007/978-3-030-13693-2_30

1 Introduction

1.1 The Meaning of Interoperability for Governments

Interoperability (IOP) is one of the most important aspects for government services [1]. IOP is not just a key enabler for an interconnected government and public services, but also a key enabler for productive and efficient public cross-border and cross-sector services and further an important asset for the progress on the Digital Single Market (DSM) [2, 3]. In the New EIF [4, 5] IOP is "the ability of organizations to interact towards mutually beneficial goals, involving the sharing of information and knowledge between these organizations, through the business processes they support, by means of the exchange of data between their ICT systems". IOP is of utmost importance for effective eGovernment (eGOV) [6]; it can minimize administrative burden, costs and time; foster innovation; facilitate reuse; improve security with an impact on growth and competitiveness; and enhance public and private policy goals for better decision-making [7]. Creating IOP and integration is a complex endeavour [6, 8].

The lack of IOP is challenging and a long-lasting problem [9]. However, as stated in the EIF strategy [4], IOP is a key factor and "MS are digitizing their public administrations (PAs) to save time, reduce costs, increase transparency, and improve both data quality and the delivery of public services". However, digital public services are not yet a reality in the European Union, as shown by the 2016 Digital Economy and Society Index (DESI) for eGOV. When it comes to cross-border services in the European Union, the situation is even more challenging as there are still barriers to access cross-border services. Therefore, a holistic, global, standardized and acceptable to all approach [10–12], coordinated on European, national and local levels, is required for interoperable interaction across borders and sectors so as to diminish fragmentation of data and services [4]. EIF is a commonly agreed approach to the delivery of European public services; MS should enforce a global approach to reach the required level of IOP with the help of EIF for common standards and requirements, as a very important topic for IOP is properly defined requirements and standardization [13–16].

1.2 Interoperability Frameworks in Europe

In Europe the need for standardization and interoperable systems started almost thirty years ago. The IOP concept in the EU started with the launch of European Commission (EC) programme CADDIA in 1985, the IDABC programme in 1995, the ISA programme in 2009 (Decision 2009/922/EC) and the establishment of the current Interoperability Solutions for European PAs (ISA^2) programme in 2016. Along with these there were also several EU directives, strategies, policies, initiatives, Ministerial Declarations and eGovernment Action Plans. A milestone for establishing a common European framework is the EIF v1 in 2004, followed by the EIF v2 COM (2010/7444/EC) which included EIS, to the current new EIF COM (2017/134/EC) which includes EIS and an Action Plan. Naming a few, the Manchester and the Lisbon Ministerial Declarations (2005), the eEUROPE 2005 initiative, the ModinisIDM programme (2005), the COM on IOP for Pan-European eGovernment Services (2006), the Service Directive (2006/123/EC), the Inspire Directive (2007/2/EC), the Malmo Ministerial Declaration (2009), the COM on Open Data (2011/882), the Digital Agenda and the Europe 2020 strategy, the Regulation on European Standardization (1025/2012), the Regulation establishing the Connecting Europe Facility (1316/2013), the Directive on the re-use of public information (2013/37/EU), the eIDAS Regulation (910/2014), the DSM Strategy (2015) and the General Data Protection Regulation (2016/679) and so on. The eGovernment Action Plan 2016–2020 [17] aims to accelerate the digital transformation of government and has IOP by default on setting up digital public services as a key principle, and EIF is a guideline to the digitalization of PAs in the 2017 eGOV Ministerial Declaration. Extending IOP for public services with the adoption of EIF and standards was part of DSM.

2 Status of ISA^2 Programme and the European Interoperability Framework (EIF)

The EC IOP policy in the EU is COM 2017/134 the European Interoperability Strategy (EIS), with the EIF and the Interoperability Action Plan as key documents to promote IOP. The EIS defines the governance, strategy, framework and principles, while the IOP Action Plan defines the implementation time frame. EIF is an action supported and monitored by the ISA^2 (2016–2020) programme. The programme supports actions and offers free digital solutions (frameworks, tools and services) for PAs, businesses and citizens in Europe to benefit from interoperable cross-border and cross-sector public services. NIFO National Interoperability Framework Observatory is an action for monitoring the implementation of National IOP Frameworks and the assessment of the compatibility with the EIF. EIF provides guidance to PAs with a set of common principles, models and recommendations and is the basis for the majority of the National IOP Frameworks (NIFs) and strategies. There are three versions of EIF: the 2004 EIF 1 (European IOP Framework for pan-European eGovernment Services), the 2010 EIF 2 (Towards IOP for European public services) and 2017 New EIF (Promoting seamless services and data flows for European public administrations).

The New EIF [4] is part of the EU COM/2017/134 and has taken into account the lessons learned by implementing EIF 2, the recent technological trends, the new information exchange needs and the recent EU policies, directives, political initiatives and the view of relevant stakeholders after a public consultation. The EIS and the Interoperability Action Plan have the priorities for EIF implementation until 2020,

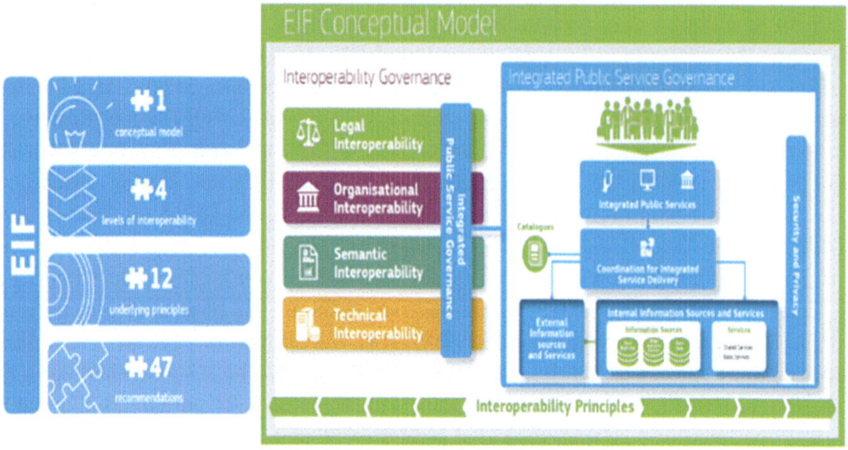

Fig. 1 New EIF basic elements and conceptual model [18]

and the NIFO is the tool for alignment and monitoring. In the New EIF, the focus is given on the actual application of principles and models that should be applied in practice with more specific recommendations to facilitate their implementation and with emphasis on openness and information management, data portability, IOP governance and integrated service delivery. Further to this, EIF provides the schema and the core components for interoperability elements to NIFs and Domain Interoperability Frameworks (DIFs). EIF operates at a higher level of abstraction, and it does not impose choices or obligations on the MS according to its subsidiarity principle. NIFs are more detailed and prescriptive than EIF. In the EIF–NIF alignment reports for IOP in Europe [18] full alignment is still not reached and a need for more guidance is stated. The New EIF has 47 recommendations for the basic elements of the principles, the model and the layers. The New EIF structure (see Fig. 1) has:

- Twelve principles grouped into four categories (C): C1. Principle setting the context for EU actions on interoperability Principle (P) P1. Subsidiarity and proportionality with 1 recommendation (1R), C2. Core interoperability principles for four (P): P2. Openness (3R), P3. Transparency (1R), and P4. Reusability (2R), P5. Technological neutrality and data portability (2R), C3. Principles related to generic user needs and expectations for four (P): P6. User-centricity (4R), P7. Inclusion and accessibility (1R), P8. Security and privacy (1R), P9. Multilingualism (1R), C4. Foundation principles for cooperation among public administrations for three (P): P10. Administrative simplification (1R), P11. Preservation of information (1R), P12. Assessment of Effectiveness and Efficiency (1R).
  The interoperability layers (L) with 14 (R): the horizontal interoperability governance with (5R), the integrated public service governance (2R), the legal interoperability (1R), the organizational (2R), the semantic (3R) and technical IOP (1R).

- The conceptual model for integrated services provision with 14 (R): the conceptual model for integrated public services (2R), the internal information sources and services (1R), the base registries (4R), the open data (3R), the catalogues (1R), the external information sources and services (1R), the security and privacy (2R).

3 Methodology

This research is focused on two areas: Firstly, a comparison of EIFs (part A) and, secondly, a comparison on the New EIF and the current NIFs that are available on NIFO, plus a few NIFs from non-EU countries (part B). In the first part, a very generic comparison is made among the versions of EIFs. Although the major differences and changes were depicted in the update of the version, this research focuses on examining the IOP issues still current from 2004 to today and the new changes that were taken into account. Initially, an abbreviation and numbering of the basic elements of the structure of the New EIF have been made: recommendations as (R), categories as (C), principles as (P), layers as (L), conceptual model for integrated public services as (M), interoperability governance as (LIG), integrated public service governance as (LPS), legal interoperability as (L1), organizational interoperability as (O2), semantic interoperability as (S), technical interoperability as (T), conceptual model for integrated public services: internal information sources and services as (MC1), conceptual model for integrated public services: base registries as (MC2), conceptual model for integrated public services: open data as (MC3), conceptual model for integrated public services: catalogues as (MC4), conceptual model for integrated public services: external information sources and services as (MC5), conceptual model for integrated public services: security and privacy as (MC6). Also, each (R) has the following abbreviation, e.g. 1 R1.P1.C1. In each of the elements "a change score" has been given according to the change made on the recommendations: 0 same R, 1 rephrase R, 2 new or major changes R. In some cases the score 2 was given although there was the same basis on the previous EIF recommendation because the change as a whole was affecting the change in the recommendation.

In the second part, a comparison was made with baseline the New EIF principles and the values (scores) from the previous comparison (part A) in relation to the differences between New EIF and EIF 2 principles' recommendations (19 of the 47). The same methodology as in part A was followed, with the matrix with abbreviations, numbering and scoring. The scoring values were taken from the NIFO analytical model to measure the alignment with EIF: "0 not aligned, 1 partially aligned, 2 aligned". In the countries with no NIFO model available estimation was proposed or in some cases an older version of the NIFO model was used to measure the alignment. Important references to this research were the reports from interoperability in Europe from 2016 and 2015 and the MS' eGOV 2017 factsheets [19] with the latest eGOV policies and initiatives. The examined countries are as follows: Austria, Belgium, Bulgaria, Cyprus, Czech, Croatia, Denmark, Estonia, Finland, France,

Germany, Greece, Hungary, Ireland, Iceland, Italy, Latvia, Lithuania, Luxemburg, Malta, Netherlands, Norway, Poland, Portugal, Romania, Slovakia, Slovenia, Spain, Sweden, the UK, Australia, New Zealand and Bhutan. To the extent of this research, the values from the 2016 NIFO alignment models were modified to fit the New EIF recommendations, except for Greece and Hungary where no model was available. Furthermore, for the non-EU countries an estimate was given to fit the comparison matrix with the alignment in their frameworks.

4 Findings and Discussion

In the first part of the preliminary results, the comparison between New EIF and EIF 2 shows an alignment with policy development, an alignment with the emerging technological trends, the interoperability governance, the integrated service delivery, the information management (big and open data), the European Interoperability Reference Architecture (EIRA) and the security and privacy. Although the alignment with the countries was 75% in 2016 still more concrete recommendations are needed to ensure implementation which is a challenge that this has to be accomplished by 2020 to be in accordance with the time frame of the Interoperability Action Plan. The EIF 2 has no time frame for the actions to be implemented. The EIF 2 has twenty-five (R), the same twelve (12) (P) which slightly differ in numbering and two new modifications: in (P) 5 the data portability and in (P) 12 the assessment of effectiveness and efficiency. The number of recommendations and principles are fewer, but most of the issues, policies or approaches were further elaborated and the technological development was taken into account. EIF 1 was focused on open standards, while EIF 2 on open specifications. Public administrations (PAs) can decide to use less open specifications if open specifications do not exist or do not meet functional interoperability needs and adapt NIFs. The services are referred to as e-services rather than specifying them. The terms of open data and openness were not viable at that time. The description of semantics and open standards and the need for multilingualism in technical aspects and the independency of proprietary technologies were recognized early. A summary of results is obtained by our proposed comparison, in comparison with the baseline of the New EIF recommendations and the previous framework's recommendations or references (see Fig. 2). According to the results, the majority of the recommendations were modified, and this shows that not only there is a need to follow the technological changes and ICT policies but also to take into account the actual alignment inputs and to frequent update EIF. Added to this, there are some basic recommendations, such as transparency, that are basic elements of the frameworks. In a nutshell, the same principles IOP identified as "problems" from 2004 are still vivid, and the underlying concepts are the same although they differ in wording or format.

The second part is the comparison of the baseline of New EIF recommendations and the alignment of NIF. The preliminary comparison results (see Fig. 3) indicate that what was proposed in the New EIF recommendations to the majority of the

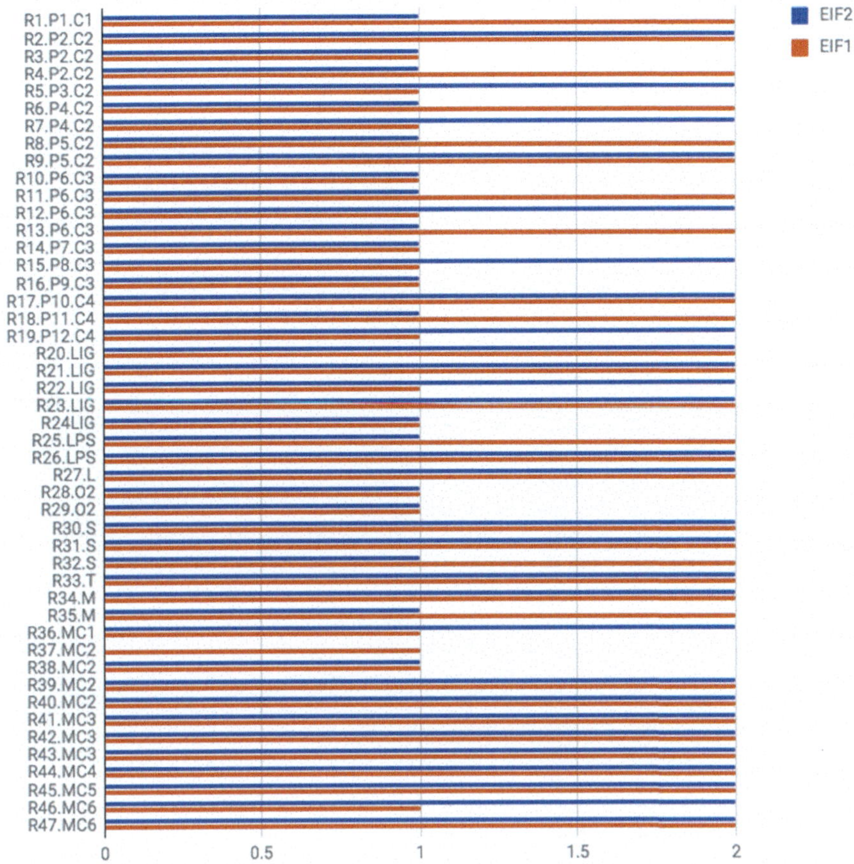

Fig. 2 EIFs summary comparison

NIFs not only needs to be aligned but to be actually implemented and within a very strict time frame (according to the IOP Action Plan). Some recommendations can be measurable for all countries e.g. open data while others are not e.g. preservation of information. For the non-European countries' score the alignment was measured to the principle recommendations without the European dimension; in case where common terminology was used, it was easier to measure for the matrix. Also, the principles that are still hard to reach from 2004 to today in most of the countries even for the ones more technologically advanced are the multilingualism, the administrative procedures, the preservation of information and the measurement of effectiveness and efficiency. This research needs to be extended to apply to all recommendations of EIF and further combine the alignment with the measurement of the actual implementation in each of the countries and then as a whole.

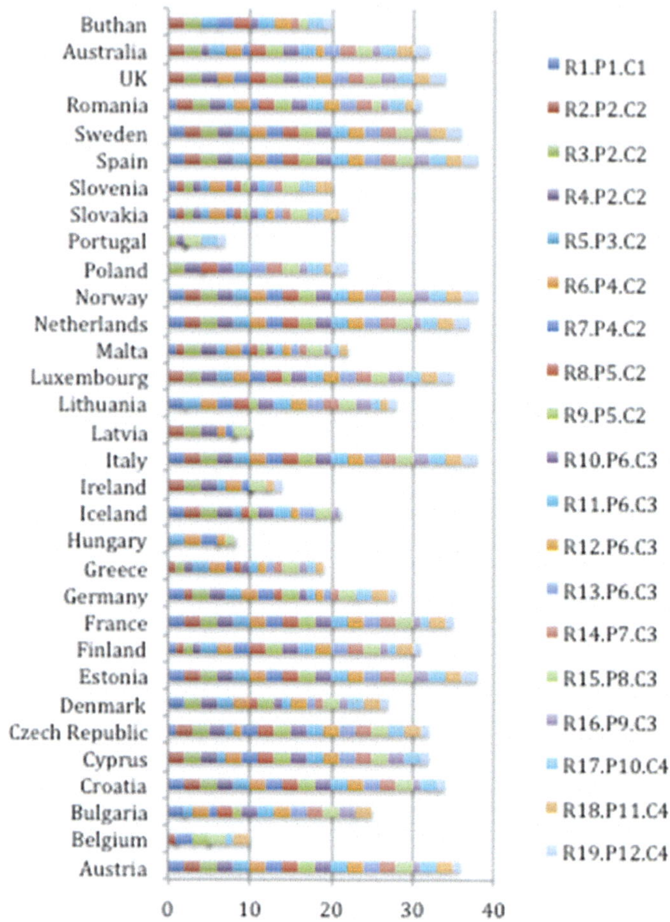

Fig. 3 Summary EIF–NIFs alignment

5 Conclusions

Most of the principles which were initially stated in EIF 1 in 2004 are still relevant to the New EIF in 2017, and most of the identified interoperability problems remain unsolved or partially fixed. Each version of EIFs differs in text and wording following the technological trends, but most of the guidance remains the same. The next version of EIFs needs to take into account the fast technological changes in ICT and consider as prerequisite the solving of technological barriers. In this way the full potential of interoperability will be unlocked nationally and across borders. It will help eliminate digital barriers and ensure DSM aims for free movement of goods, persons, services, capital and fair access to online goods and services, so that isolation and fragmentation will not be an issue. The digitalization of public administrations will offer better services to citizens and businesses and thus contribute to economic growth and making citizens' life easier. There is a need to regularly update EIFs and NIFs due to the rapid change in technological developments, and that was already recognized in 2004. Bearing that in mind, and the fact that the updates between EIFs took several years, and also that still there are MS that currently have an outdated version of NIF which in some cases reaches back to EIF 1 and that the actual implementation has fallen behind, there is a need to facilitate this issue at European level with close collaboration with MS. There is still the need for the identification of the implementation of EIF in NIFs and a process alignment that will minimize the gap on how to actually measure IOP in each MS public administration. Furthermore, there are still gaps between the readiness in reaching new trends and being on track with the current frameworks in MS. The alignment is a first step for concrete results based on the actual implementation and monitoring. Although, in some cases, it is easy to conclude, in some others the result was ambiguous due to the difference of the outcome and the framework schemes in each country. A combination of the NIFO analytical model and the IMM can perhaps adhere value to monitor MS' implementation by monitoring public services provided at national and local levels which have the maturity for eGOV services across borders and across sectors along with the EU stable mechanisms to ensure IOP.

References

1. Ordiyasa, W., Nugroho, L. E., Santosa, P.I., & Kumorotomo, W. (2016). Enhancing quality of service for eGovernment interoperability based on adaptive ontology. In *International Conference on Science and Technology-Computer (ICST)*, pp. 102–107.
2. European Commission, COM/2015/0192 final Communication from the Commission to the European Parliament, the Council, the European Economic and Social Committee and the Committee of the Regions, A Digital Single Market Strategy for Europe. http://eur-lex.europa.eu/legal-content/EN/TXT/?qid=1447773803386&uri=CELEX%3A52015DC0192. Last accessed November 27, 2017.
3. European Commission, EU Internal market: From crisis to opportunity: putting citizens and companies on the path to prosperity EU publications, ISBN 978-92-79-42535-6, https://publications.europa.eu/en/publication-detail/-/publication/f85c0e8f-4cdf-4859-be26-f9c17e7fbb6f. Last accessed November 27, 2017.
4. European Commission, COM/2017/134 Communication from the Commission to the European Parliament, the Council, the European Economic and Social Committee and the Committee of the Regions, European Interoperability Framework—Implementation Strategy {SWD(2017) 112 final} {SWD(2017) 113 final}. http://eur-lex.europa.eu/legal-content/EN/TXT/HTML/?uri=CELEX:52017DC0134&from=EN. Last accessed November 27, 2017.
5. European Commission, New European Interoperability Framework Promoting seamless services and data flows for European public administrations, EIF brochure, ISBN 978-92-79-63756-8. https://ec.europa.eu/isa2/sites/isa/files/eif_brochure_final.pdf. Last accessed November 27, 2017.
6. Scholl, H. J., & Klischewski, R. (2007). E-Government integration and interoperability: Framing the research agenda. *International Journal of Public Administration, 30*(8–9), 889–920.
7. Charalabidis, Y., Lampathaki, F., & Askounis, D. (2009). A comparative analysis of national interoperability frameworks. In *AMCIS 2009 Proceedings* (p. 694).

8. Weerakkody, V., Janssen, M., & Hjort-Madsen, K. (2007). Integration and enterprise architecture challenges in e-government: A European perspective. *International Journal of Cases on Electronic Commerce, 3*(2), 14–38.
9. Charalabidis, Y., & Askounis, D. (2008). Interoperability registries in eGovernment: Developing a semantically rich repository for electronic services and documents of the new public administration. In *Proceedings of the 41st Annual Hawaii International Conference on System Sciences, IEEE.*
10. Janssen, M., Charalabidis, Y., Kuk, G., & Cresswell, T. (2011). Guest editors' introduction: E-government interoperability, infrastructure and architecture: State-of-the-art and challenges. *Journal of Theoretical and Applied Electronic Commerce Research, 6*(1), I–VIII ISSN 0718-1876. (2011).
11. Charalabidis, Y. (2010). A framework for assessing interoperability in public administration at organisational, semantic and technical level. In *Beyond e -Government measuring performance: A global perspective* (pp. 77–98). Public Technology Institute (PTI).
12. Bovalis, K., Peristeras, V., Abecasis, M., Abril-Jimenez, R. M., Rodriguez, M. A., Gattegno, C., et al. (2014). Promoting interoperability in Europe's E-Government. *Computer, 47,* 25–33.
13. Hosseini, M., & Dixon, B. E. (2016). Syntactic interoperability and the role of standards (Chapter 8). In *Health Information Exchange* (pp. 123–136). Academic Press.
14. Shariatzadeh, N., Gurdur, D., El-Khoury, J., Lindberg, L., & Sivard, G. (2016). Using linked data with information standards for interoperability in production engineering. In *Procedia CIRP*, Vol. 41.
15. Daclin, N., Mallek Daclin, S., Chapurlat, V., & Vallespir, B. (2016). Writing and verifying interoperability requirements: Application to collaborative processes. *Computers in Industry, 82,* 1–18.
16. Cimander, R., Kubicek, H., & Scholl, H. J. (2011). *Organizational interoperability in E-Government—Lessons from 77 European good-practice cases.*
17. European Commission, COM(2016) 179 final, Communication from the Commission to the European Parliament, the Council, the European Economic and Social Committee and the Committee of the Regions, EU eGovernment Action Plan 2016–2020 Accelerating the digital transformation of government, {SWD(2016) 108 final} {SWD(2016) 109 final}. http://eur-lex.europa.eu/legal-content/EL/TXT/?uri=CELEX:52016SC0109. Last accessed November 27, 2017.
18. European Commission, ISA2, New European Interoperability Framework Promoting seamless services and data flows for European public administrations, ISBN 978-92-79-63756-8, (2017). https://ec.europa.eu/isa2/sites/isa/files/eif_brochure_final.pdf. Last accessed November 27, 2017.
19. European Commission, Joinup, NIFO Factsheets. https://joinup.ec.europa.eu/page/nifo-factsheets. Last accessed November 27, 2017.

Part XI
Entities in IoT

A Lightweight IoT Hub for SME Manufacturing Industries

Carlos Agostinho, Fabio Lopes, Jose Ferreira, Sudeep Ghimire and Maria Marques

Abstract With the advance in the Internet of Things (IoT), new ways of acquiring, processing, and managing collected data from electronic devices are being developed to provide support for more complex systems. This process of transforming the acquired data from the physical world, through the sensors, into viable information on which the applications can make decisions upon, must consider the various implementation scenarios and the business and technical requirements, such as security, privacy, and interoperability between heterogeneous devices (which often communicate using different protocols and require a common vocabulary). With the increasing complexity of these requirements, it becomes urgent to develop an infrastructure to handle the associated processes and provide a middle ground layer on which the physical and digital world are connected and translated into each other. This software layer, or middleware, can be described as a hub and aims to fill the gap between devices and information systems. This work contributes with a study of mechanisms and methodologies for the collection of data, interoperability of systems and data filtering, to optimize and automate, using a lightweight approach, and the collection and pre-analysis of the data to be used by the various applications of the IoT systems, such as the SME manufacturing industries.

Keywords Internet of things · Hub · Middleware · Data acquisition · Data filtering · Sensors · Interoperability

---

C. Agostinho (✉) · F. Lopes · J. Ferreira · S. Ghimire · M. Marques
Centre of Technology and Systems, CTS, UNINOVA, Campus Da Caparica, 2829-516 Caparica, Portugal
e-mail: ca@uninova.pt

F. Lopes
e-mail: fl@uninova.pt

J. Ferreira
e-mail: japf@uninova.pt

S. Ghimire
e-mail: sud@uninova.pt

M. Marques
e-mail: mcm@uninova.pt

K. Popplewell et al. (eds.), *Enterprise Interoperability VIII*, Proceedings of the I-ESA Conferences 9, https://doi.org/10.1007/978-3-030-13693-2_31

1 Introduction

The Internet of Things (IoT) is an emerging technological concept that aims to combine consumer products, industrial components, and other everyday objects with Internet connectivity and powerful data analytic capabilities that can transform the way we work and live [1]. The objective of this is to optimize the functionalities of each object, as every extractable information becomes a mean of analyzing and computing the functioning processes, to enhance its context functioning and to provide new purposes, that emerge when connecting a certain object to an intelligent network [2].

The IoT has the potential to change how people can be connected to objects, that can also be connected among themselves, creating intelligent technological environments, generating interest in both the business and scientific areas. Some of the addressed issues related to the devices include different manufacturing specifications and the lack of a common vocabulary, due to the different sources of models and semantics [3]. This may pose some barriers to interoperable and scalable IoT networks. The current research efforts involve connecting the machines, equipment, software, and devices using internet protocols for allowing communication without human intervention [4]. Hence, a methodology to realize homogeneous communication between applications and devices, and between the devices, should be applied.

One of the solutions that can be adopted is a middleware technology, a layer of software between the two systems that makes it easy for the two to communicate [5]. Figure 1 shows three different approaches for the data collection and integration of heterogenous devices where the proposed one is a middleware-based approach based on fog computing paradigm (b), providing an abstraction layer between devices and applications/services. The advantages consist in the possibility of implementing independent processes that enhance the quality of the methodology, providing more reliable systems, and also diminishing the required data processing in the applications. This solution is modular and applicable to a wider range of scenarios, due to its independency from the other layers (cloud and devices). It is appropriate for SMEs, representing a cheaper and easier solution due to the possibility of using low-cost equipment. Some of the useful applicable processes in this situation are data validation and filtering, event detection, and security checking.

The two solutions, provided in Fig. 1, are based on the processing being made directly at the devices (a) applying the edge computing paradigm, or being made in the data consumers (c) following the traditional cloud paradigm. The first, generally, does not allow complex processes due to the simplicity of the devices used in contemporary industry (and when it allows, they are very expensive), and the second, requires processing from the data consumers, increasing complexity in the applications and consuming processing time from other important application processes.

In this work, a solution for the creation of a middleware is presented, envisioning its industrial application in a real-world scenario related with the European C2NET project [6], to be explained further in this paper. The idea is to create a Lightweight

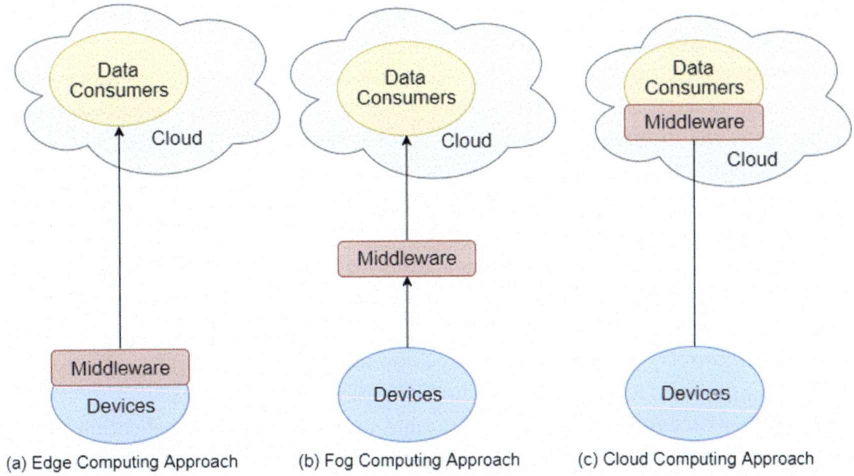

Fig. 1 Middleware-based approaches for data collection through sensor devices

IoT Hub which physically integrates the IoT network for communication purposes and provides middleware software processes.

2 Background

To specify and better understand the methodology proposed in this paper and how the solution contributes to IoT, some key concepts are mentioned and briefly explained.

Data Acquisition. This process consists on the collection of information and the system responsible for the regulation of data acquisition, defined as the monitor [7]. In IoT, this acquisition is heavily based on the low-layer devices from the network, as depicted in Fig. 1, and is based on finding, fetching, and transferring raw data to be processed and indexed by higher layers of the system [8].

Context Acquisition. An important aspect of the data collection is to clarify the meaning of data sources. Sensors are hardware components that measure environmental information such as temperature, location, and processing time and transform it into a digital signal. In larger networks, the amount of generated data is critically larger, and it is not feasible to process everything that is generated by the devices. Considering that, context-aware processes play a critical role in deciding what data need to be processed, which implies the understanding of sensor data, one of the main challenges of IoT. The concept of context management is also very important and essential for the software systems that apply this context modeling and reasoning. Data move from phase to phase, from the place where it is generated to where it is consumed, creating a data life cycle that starts with the acquisition [9]. The acquisition parameters, such as responsibility (data authoring), frequency, source, sensor type (software, virtual or physical sensor), are thoroughly explained in the works of [10, 11] and considered in this solution.

IoT Protocols. Communication protocols are formal descriptions of digital message formats and define rules that include: packet size, transmission speed, handshaking and synchronization techniques, address mapping, flow control, and other important communication aspects to achieve proper information exchange [12]. They are implemented in hardware (communication protocols) and software (message protocols) and used to exchange information between systems.

In IoT, there are various data protocols used for communication, that usually fall into three communication types: device-to-server (e.g., MQTT—http://mqtt.org/), server-to-server (e.g., AMQP—https://www.amqp.org), and device-to-device (e.g., DDS—http://portals.omg.org/dds/). Some communication/transport layer protocols, commonly used in IoT, are Wi-Fi, Bluetooth, ZigBee, and Ethernet.

Data Pre-processing. This is an important aspect of the data collection, using techniques to ignore noisy and unreliable data before the application layer. The objective is to transform raw data, often incomplete or inconsistent, into an understandable format that can be defined by the communication hub or the requesting data consumer. Some techniques used, mentioned in the work of [13], are data cleaning (identifying and resolving data inconsistencies), missing values (fill probable values that are missing), noisy data (corrects random error or variance in a measured variable), and inconsistent data (uses external references for correcting the information, based on aspects such as functional dependencies and attributes).

Middleware Solutions for IoT. The IoT environment holds a wide spectrum of applications, which make use of middleware layers to achieve the previously explained benefits. The main technical aspects to consider when designing middleware systems are the capability of interoperability, scalability, abstraction, interaction, multiplicity, security, and privacy. The existing solutions can be categorized according to the involved domains into three categories [14]: Semantic Web and Web Services, RFID and sensor networks, and robotics. Some of the most prominent approaches are Triple Space-Based [15], UbiWare [16], and TinyREST [17].

3 Lightweight IoT Hub Architecture

In this section, the developed architecture for the Lightweight IoT Hub is presented. It is designed to comply with the functionalities mentioned in the Introduction along with the specifications of the C2NET project, and considers the researched information provided in the Background section.

The Lightweight IoT Hub was designed to provide C2NET SME's with a low-processing, but robust, solution for data collection. This low processing requirement allows the use of a diverse range of readily available hardware to build IoT applications, not requiring a lot of processing capabilities. Examples of these hardware options are the single-board computers called *Raspberry Pi*[1] and the microcontrollers from the *Arduino*[2] series, which are used in the implementation of this solution for the Hub and sensor nodes, respectively. Hence, the Lightweight IoT Hub provides a solution that is platform independent and supports diverse communication protocols. It is also written in Java for portability purposes and targets Unix systems which provide device drivers for the majority of the ISO standards.

The Hub is composed by a set of submodules, presented in Fig. 2 and further detailed in Sect. 3.1, that focus on the system's communication, management of devices, and data handling. The **Communication Module** is responsible for the communication between the IoT systems and the C2NET platform, guaranteeing the external components' interoperability, protocol abstraction, and connection between consumers and the data providers. The **Device Management Module** manages the different types of IoT devices in the IoT network, ensuring their security and connectivity. The **Data Handling Module** allows the manipulation of data at the source and filtering and reducing the data flow before it reaches the application, the C2NET platform in this case. This improves performance and reduces inaccuracy in the retrieved data.

The IoT Hub is an external component relatively to the cloud platform. It communicates with the platform using a publish/subscribe (Pub/Sub) message queue, feeding the C2NET cloud platform with real-time data coming from the different IoT data sources (devices). The communication is bidirectional, enabling not only data input but also a configuration flow from the platform to the various possible existing Hubs, virtualizing the IoT network. Following this approach, the user (e.g., company IT) can register different devices and allocate specific configurations for how to receive the data.

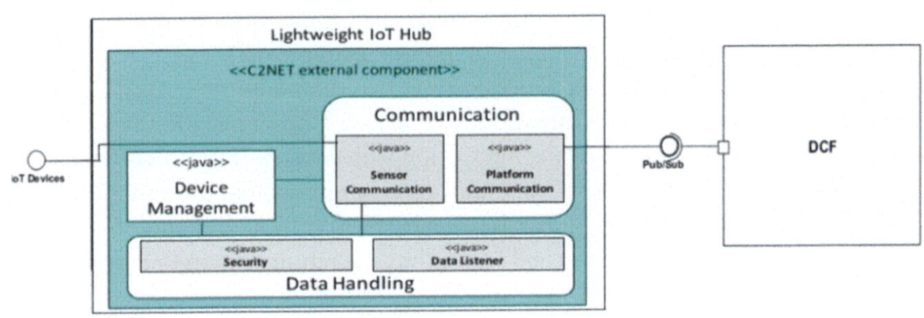

Fig. 2 Lightweight IoT Hub deployment view

[1] https://www.raspberrypi.org.

[2] https://www.arduino.cc.

3.1 Developed Architecture

The development of the architecture for the Lightweight IoT Hub started at the lower layer, with the scrutiny of the specifications for the contemporary sensor networks and the analysis of which protocols are best suited for the intended deployment of this work's solution. With this, considering the necessities of the C2NET project pilots, it was clear the need for event-driven processes, because these networks are typically intensively active for short periods of time and remain idle for the remaining periods. Hence, the Lightweight IoT Hub considers that sensors broadcast results immediately and the hub is responsible for managing thresholds, filter unwanted data, and forward the information to the C2NET platform in a pre-defined time frame. This allows for a larger variety of sensors to be used as it reduces the need for processing capability of each node while also reducing power consumption. Also, the volume of data does not require huge amounts of memory in the Hub, achieving one objective of the Hub that is to provide a reliable solution with low-memory provisions.

Infrastructural approaches based on communication layer protocols such as HTTP and ZigBee, with a shared medium and limited spectrum, are supported and fit perfectly the needs of home/office wireless sensor networks, while dedicated Can-Bus and serial communication approaches are more appropriate for heavy industrial environments (as used in the implementation of this work), enabling lower bit error rates.

Figure 2 shows a technical version of the Hub, where the submodules internal content is depicted. This subsection, the information from the previous subsection is, completed with the specifications of each submodule.

The **Device Management Module** allows the discovery and connection of several different sensors to the Hub, handling the different communication layer protocols according to the user configuration. It is responsible for monitoring the connectivity and availability of the devices, with the support of a simple database.

The **Communication Module** is divided into two submodules: **Sensor Communication** and **Platform Communication**. The Sensor Communication is related to the device management, enabling the interaction with devices coupled via one of the possible protocols. The Platform Communication is responsible for all interactions between the IoT Hub and the platform. It is used to receive the configuration of protocols and devices from the platform and forwarding data from sensors to the platform.

Each **Data Listener** of the **Data Handling Module** is an implementation of a protocol used to communicate with sensors. The Data Listener also performs data filtering activities when specified by the user during the resource network configuration. The Data Listener must be compliant with the requirements of the Hub and the platform, but developers can write their own Data Listeners according to their needs.

Finally, the **Security Module** is responsible for providing authentication and authorization for devices connected to the Lightweight IoT Hub. All communications with the C2NET platform must be authenticated, obtained via secured channels and immune to repetition and violation of data integrity. For this, HTTP over SSL with symmetric key cryptography is used and the Hub will request a public key via the Pub/Sub component. Within the platform, ACS will hold the private key that authenticates the Hub. This module provides features that allow the monitoring of the hub's hardware and informs the user about the global behavior of the system.

3.2 Communication

The communication of the Hub with the sensor nodes and the C2NET platform (Fig. 3) is a very important process of the Hub's functioning, allowing the essential interoperability that makes it an IoT system element, and also providing certain functionalities such as the management of sensors and data collecting procedures, configuration exchange, and data forwarding (to the cloud platform).

The sensors, pre-configured for communication, send data over the network, from the node to the hub, where the data are collected for further processing (on the hub or the platform). The hub is responsible for actuating on the devices, enabling them

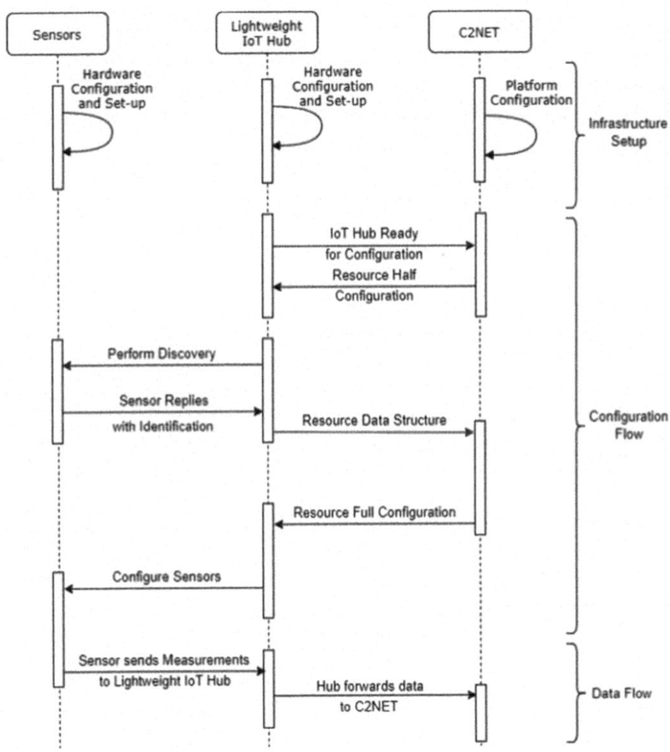

Fig. 3 Lightweight IoT Hub communication

or disabling them, and is responsible for the mechanisms to detect new devices, or nodes, automatically (as seen in Fig. 3). The data retrieved from the sensors are then forwarded to the platform for further analysis and display.

3.3 *Messaging Between Hub and Devices*

The messages that allow the management mentioned in the previous subsection are also important to consider. In this case, the structure of the messages is designed to be constituted by a maximum of 8 bytes, where each one of them contains specific data. The first byte is always used for the message type, which defines the use of the remaining bytes. Typically, the second byte denotes the node ID, the third defines the sensor ID (within each node), the fourth denotes the type of sensor (useful for messages from the node to the hub, where the hub acknowledges the existence of that sensor and what type of data it measures) and the remaining bytes are usually for values of measurements, thresholds and frequencies (for communication to happen). Those values may be very high and need to be represented by more than two bytes.

An example of this structure is presented in Fig. 4. It is a message intended for the actuation of a warning light. The actuator is in node 3 and its ID is 10. To activate it, the Hub should make the value 1 reach the correspondent node. So, the first byte has the pre-defined message type code “0x17” that denotes the message for actuation, “0x03” is the node ID and “0x10” is the actuator ID, and occupy the bytes 2 and 4, respectively. This type of actuator has a pre-determined value for its designation, “0x30,” and that information is inserted into the third byte of the message. The value for actuation purposes occupies the fifth byte and to activate this warning light “0x01” is written. The three remaining bytes are not used for this type of message.

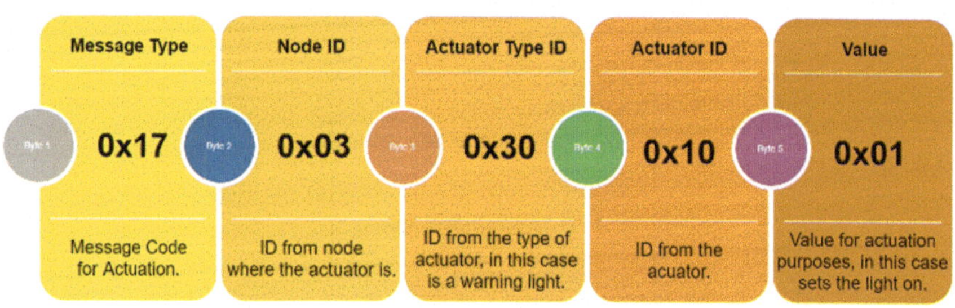

Fig. 4 Communication between Hub and node message example

4 Implementation of the Use Case Scenario

As it was mentioned, this work is related to the C2NET project and aims for the deployment of the Lightweight IoT Hub in the SME's manufacturing industries. The goal of the C2NET project is the creation of cloud-enabled tools for supporting the SME's supply network optimization of manufacturing and logistic assets based on real-time collaborative demand, production, and delivery plans [6]. Considering this, the Hub stands as the centralized element which provides the capability to communicate with the cloud-enabled tools and managing the factory's network.

4.1 Infrastructure Setup

For testing and validation purposes, the resultant solution of this work had a direct impact on the processes a metalworking factory in Portugal. The resultant IoT implementation of this deployment is illustrated in Fig. 5. The Hub is implemented in a *Raspberry Pi* single-board computer and the sensor nodes (C1, E1, 31, A1, and D1) are implemented using *Arduino* devices, with the exception of B1 which is connected to a thickness gauge sensor that requires more processing capabilities. Hence B1 is implemented in a *Raspberry Pi*.

4.2 Configuration

An example of the resulting configuration settings for a specific node, named "Painting Station," is presented in Fig. 6. These settings represent the message structure sent by the platform to the hub for defining from which sensors to start collecting data. For this to happen, the hub has already had to discover what sensors are present in the network and query for their information, such as type of sensor, data type, range, and data units. This information is passed to the hub and, only after that, the platform user sends the pre-defined settings for the data collection. It is important to mention that the Device Management module requires the information about the protocol to be used before performing the device discovery processes. This is possible by configuring the internal config.xml file and activating the appropriate Data Listener. The configuration presented in Fig. 6 is stored and monitored by the device management.

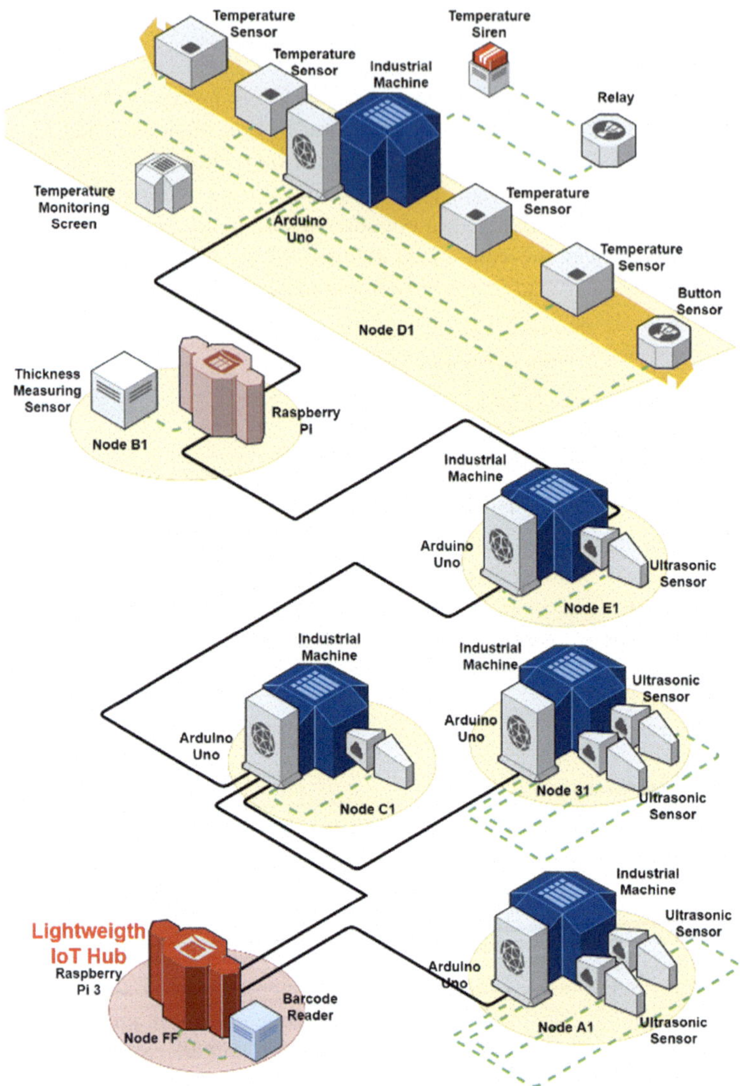

Fig. 5 IoT Implementation

4.3 *Data Collection*

After the configuration, the selected IoT devices start to transmit data. The sensor data are temporarily kept, with a timestamp, in memory storage. To forward the data to the C2NET cloud platform, as intended, the Platform Communication uses the message structure presented in Fig. 7. This information is sent to the platform, for displaying the collected data. The platform, accessible to the company's IT personnel, allows to

```
"title": "resource_full_configuration",
"hubID": 10000,
"body": {
    "id": 10000,
    "name": "AAMM_Hub",
    "properties": [],
    "resources": [
        {
            "id": 10001,
            "name": "Painting Station",
            "properties": [
            {   "name": "protocol", "type": "string", "value": "CanBus" },
            {   "name": "dataCollectionPeriod", "type": "integer", "value": 1440 },
            {
                "name": "sensing elements",
                "type": "JSON",
                "value": [
                    {"name": "temperature 1", "frequency": "100", "datatype": "CELSIUS",
                    "type": "double" },
                    {"name": "temperature 2", "frequency": "100", "datatype": "CELSIUS",
                    "type": "double" },
                    {"name": "temperature 3", "frequency": "100", "datatype": "CELSIUS",
                    "type": "double" }
```


Fig. 6 Lightweight IoT Hub message structure for the settings of data collecting


```
"title": "resource_data",
"hubID": 1000,
"resourceID": 1001,
"body":
[
    { "Painting Station.temperature1": "28.20",
        "Painting Station.temperature1.timestamp": "2017-11-05T13:15:22.000Z"
    },
    { "Painting Station.temperature2": "30.30",
        "Painting Station.temperature2.timestamp": "2017-11-05T13:14:43.000Z"
    },
    { "Painting Station.temperature3": "25.90",
        "Painting Station.temperature3.timestamp": "2017-11-05T13:13:30.000Z"
    },
    { "Painting Station.temperature1": "27.40",
        "Painting Station.temperature1.timestamp": "2017-11-05T13:14:25.000Z"
```


Fig. 7 Lightweight IoT Hub message structure for the settings of data collecting

perform analysis on the collected data and perform configurations on various aspects of the overall IoT process.

5 Conclusion and Future Work

In an IoT paradigm, the distributed scenarios prevail where the data sources are physically separated and are often directed to autonomous data consumers, which are often the higher-level applications with functionality for high-end data analytics and event detection. Still, this type of distributed scenarios, typically has issues regarding scalability and maintenance if the data sources and data consumers are tightly coupled. That means that if there are changes in the data sources or consumers, the overall systems need to also perform changes to deal with the new requirements. So, a viable industrial solution needs to have a system ready for seamless integration of data sources with a high level of abstraction between the data sources and consumers. Middleware aims to reduce the complexity of such systems and processes, providing a fog computation approach where the processing occurs closer to the devices and independently from the applications. The direction of IoT is to have an edge computation approach where the processing occurs in the devices.

In this work, a Lightweight IoT Hub middleware solution is presented, and a prototype has been developed and installed in the premises of a Portuguese metalworking company. The Hub is responsible for addressing interactions made between the C2NET cloud platform and the IoT devices, managing the real-world data collection process. It is a simple, robust, low-processing, and low-cost IoT middleware, for easy integration in SME's. It supports several protocols and implements some basic security mechanisms based on authentication and authorization.

One of the possible enhancements for this work is the integration more elaborate security methods, as for instance the ones described in [18]. Given the tendency to move computing to the edge, it is important to analyze if the middleware software and procedure proposed for the Lightweight IoT Hub fit within the capacity of contemporary smart devices. Finally, when analyzing the RAMI 4.0 architecture [19], it is also possible to identify other future research direction, namely the integration with the digitalization process (transition from the real to digital world) of the architecture, more specifically with the layers of integration and communication, allowing a contextualization in the Industry 4.0 paradigm.

Acknowledgements The research leading to these results has received funding from the EC HORIZON2020 Program under grant agreement n° C2NET 636909 (http://www.c2net-project.eu/) and n° VF-OS 723710 (http://vf-os.eu/), and from Faculty of Science and Technology of Nova University of Lisbon.

References

1. Rose, J., Eldridge, S., & Chaplin, L. (2015). The internet of things: An overview—Understanding the issues and challenges of a more connected world. Internet Soc.
2. Lopes, F., Ferreira, J., Jardim-Goncalves, R., & Agostinho, C. (2017). Semantic maps for IoT network reorganization in face of sensor malfunctioning.
3. Agostinho, C., Sarraipa, J., Goncalves, D., Jardim-Goncalves, R. (2011). Tuple-based semantic and structural mapping for a sustainable interoperability. In *IFIP Advances in Information and Communication Technology*.
4. Huang, Y., & Li, G. (2010). Descriptive models for Internet of things. In *Proceedings of 2010 International Conference on Intelligent Control and Information Processing*.
5. Rosenblum, D.S. (2001) *Interoperability and middleware* (pp. 1–5).
6. C2NET: C2NET Project. http://c2net-project.eu/.
7. Ojha, S., & Bandil, P. L. (2016). *IoT based data acquisition system using Raspberry Pi.*
8. Fersi, G. (2015). Middleware for internet of things: A study. In *Proceedings—IEEE International Conference on Distributed Computing in Sensor Systems*, DCOSS.
9. Fox, P. (2011). Data management considerations for the data life cycle. NRC STS.

10. Pietschmann, S., Mitschick, A., Winkler, R., & Meißner, K. (2008). *CROCO: ontology-based, cross-application context management*, SMAP 2008.
11. Chen, H., Finin, T., Joshi, A., Kagal, L., Perich, F., & Chakraborty, D. (2004). Meet the Semantic web in smart spaces. *IEEE Internet Comput, 8,* 69–79.
12. Salman, T., & Jain, R. (2017). Networking protocols and standards for internet of things. In *Internet of Things and Data Analytics Handbook* (pp. 215–238).
13. Han, J., Kamber, M., & Pei, J. (2012). *Data mining: Concepts and techniques.*
14. Chaqfeh, M. A., & Mohamed, N. (2012) Challenges in middleware solutions for the internet of things. In *Proceedings of 2012 International Conference on Collaboration Technologies Systems CTS 2012.*
15. Gómez-Goiri, A., & López-De-Ipiña, D. (2010). A triple space-based semantic distributed middleware for internet of things. In *Lecture Notes in Computer Science.*
16. Katasonov, A., Kaykova, O., Khriyenko, O., Nikitin, S., & Terziyan, V. (2008). Smart semantic middleware for the internet of things (pp. 169–178). ICINCO-ICSO.
17. Luckenbach, T., Gober, P., Arbanowski, S., Kotsopoulos, A., & Kim, K. (2005). TinyREST: A protocol for integrating sensor networks into the internet. In Proceedings of REALWSN.
18. Mozzaquatro, B. A., Jardim-Goncalves, R., Melo, R., & Agostinho, C. (2016). The application of security adaptive framework for sensor in industrial systems. In *SAS 2016.*
19. Adolphs, P., Bedenbender, H., Dirzus, D., Ehlich, M., Epple, U., Hankel, M., Heidel, R., Hoffmeister, M., Huhle, H., Kärcher, B., Koziolek, H., Pichler, R., Pollmeier, S., Schewe, F., Walter, A., Waser, B., & Wollschlaeger, M. (2015) *RAMI Industrie 4.0.*

Toward Service Orchestration Through Software Capability Profile

Abdelhadi Belfadel, Jannik Laval, Chantal Bonner Cherifi and Nejib Moalla

Abstract Open-source applications are a gold mine of potential functions for reuse. However, the difficulty lies in the identification and characterization of these functionalities to meet specific business needs. In this work, we aim to decompose and externalize functionalities from open-source business applications for service orchestration. We propose a solution for characterizing a functionality in a standardized and readable way. Then we propose an automated process based on ISO 16100 capability profile, to externalize the characterized functionalities as RESTful APIs. As a result, we expose reusable components to be used for service orchestration in response to a specific business need.

Keywords Open-source application · Software reuse · Service orchestration · RESTful API · ISO 16100 · Capability profile · Service orchestration

1 Introduction

Small- and medium-sized enterprises (SMEs) are the most common firms in many countries. In Europe (EU28), 23 million SMEs employ more than 90 million people. They represented, in 2015, 99.8% of all enterprises [1], and they are increasingly doing open innovation to bring ideas in the market to improve productivity, increase competitiveness and facilitate entrance to new markets [2]. In order to do so, SMEs need to explore new solutions and integrate new functionalities quickly. This results

A. Belfadel (✉) · J. Laval · C. B. Cherifi · N. Moalla
DISP Laboratory, University Lyon 2, Lyon, France
e-mail: abdelhadi.belfadel@univ-lyon2.fr

J. Laval
e-mail: jannik.laval@univ-lyon2.fr

C. B. Cherifi
e-mail: chantal.BonnerCherifi@univ-lyon2.fr

N. Moalla
e-mail: nejib.moalla@univ-lyon2.fr

K. Popplewell et al. (eds.), *Enterprise Interoperability VIII*, Proceedings of the I-ESA Conferences 9, https://doi.org/10.1007/978-3-030-13693-2_32

in prototyping new business needs in short time period without cost or engagement with a software vendor. But in most cases, SMEs operate under limited resources which restrict their innovativeness [2]. High number of open-source solutions results from the factories of the future (FoF) initiatives. This aims to develop the necessary key enabling technologies and help in manufacturing enterprises to adapt to global competitive pressures.

Currently, the development of software applications is based on the reuse of existing functionalities instead of developing them from scratch [3, 4]. Application programming interfaces (APIs) are considered as the most commonly used entities supporting software reuse [5, 6]. By providing implemented and tested functionalities, APIs increase software quality and reduce the effort spent on coding, testing and maintenance activities [7].

In this context, we provide a framework to facilitate the appropriation of the open-source applications and bring adequate solutions to SMEs. This framework consists of four steps: source code analysis, evaluation, servitization and orchestration model. In the source code analysis step, we analyze the source code to detect existing services and potential candidates (functionalities) to externalize. In the evaluation step, we qualify existing services of an open-source application and elect reliable and trustworthy candidates. In the servitization step, we characterize and servitize the elected candidates in an automated process. Finally, in the orchestration step, we propose an orchestration model for the externalized functionality to save the business logic offered by the open-source application. For this purpose, we present in this paper one part of the proposed framework with a proof of concept applied on an open-source application resulted from a FoF initiative (FITMAN[1]), starting from the characterization of the functionalities done in a standardized and readable way. Then we propose an automated process for the externalization of the characterized functionalities as REST (representational state transfer) APIs for service orchestration. Our approach is based on ISO 16100 capability profile, which proposes a standardized methodology for interoperability of manufacturing software. As a result, we externalize reusable components to be used for service orchestration in response to a specific business need. This paper is structured as follows. Section 2 focuses on the related work, and Sect. 3 presents an overview of some useful standards for our solving approach. Section 4 is dedicated to the proposed solution, and Sect. 5 presents an implementation of the proposed solution applied on an open-source application. Finally, conclusion and future works are drawn in Sect. 6.

2 Related Work

In this section, we discuss works related to source code analysis, software reuse and legacy to SOA migration.

[1] www.fitman-fi.eu

2.1 *Source Code Analysis and Software Reuse*

Metrics are strong support tools in software development and maintenance. They are used to measure software quality, to estimate cost, effort and complexity, and to improve processes [8]. The metrics that are important to calculate reusability are related to cohesion, inheritance and coupling. In [8], the authors measure the association between numbers of classes, check the direct and indirect dependencies, IO dependencies, number of in and out metrics in object-oriented programming. In [9], the authors propose a method to display dependencies between modules in reuse-based embedded software development and adding development management property data to each module in order to support developers to know which modules will be affected when some parts of the reused software are modified. Other authors in [10] identify components from object-oriented source code based on quality-centric metrics.

2.2 *Legacy to SOA Evolution*

In the literature, the concept of SOA is interpreted in many ways. Different approaches to SOA migration are proposed. A brief overview of legacy to SOA evolution is reported by [11] that divides the legacy to SOA evolution approaches into four categories: replacement, redevelopment, wrapping and migration. In [12], the authors report a systematic literature review of SOA migration approaches. They propose a reference model, called SOA migration frame of reference, that can be used for selecting and defining SOA migration approaches. In [13], the authors outline a semi-automated approach to migrate dynamic legacy Web applications to Web services-based SOA applications by using two technologies: service component architecture (SCA) and service data object (SDO). They used a manual approach to identify the potential service within each function. Other authors propose a framework and guidelines for the identification of specific services from legacy code [14]. Their approach focuses on defining the services based on a model-driven architecture approach.

3 Standards

In order to reach service orchestration using functionalities of open-source applications, we need to know more about the entities that have to interoperate. Two standards intend to provide this knowledge by offering a way to create profiles of the selected entities: ISO 15745 [15] (industrial automation systems and integration—open systems application integration frameworks) and ISO 16100 [16] (industrial automation systems and integration—manufacturing software capability profiling for interoperability). The ISO 15745 series defines an application integration framework (AIP),

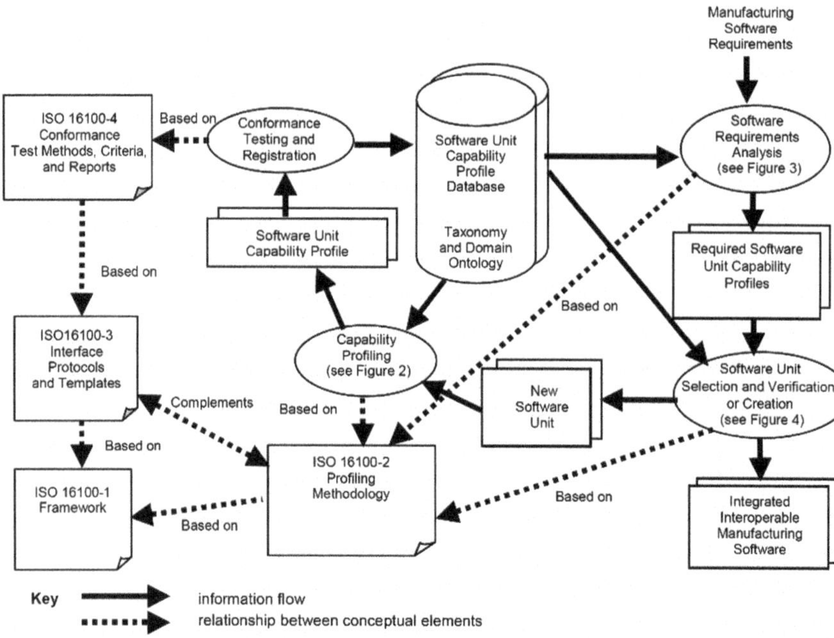

Fig. 1 Context of ISO 16100 [18]

which is a set of elements and rules for describing application interoperability profiles, communication network profiles and the communication-related aspects of device profiles based upon particular fieldbus technologies [15]. On the other hand, the standard ISO 16100 targets the representation of a software capability profile [16]. It specifies a framework for assessing the interoperability of a set of software products used in the manufacturing domain and provides a method which is independent of a particular system architecture or implementation platform for constructing profiles of manufacturing software capabilities [17]. Figure 1 Context of ISO 16100 [18] shows the concepts defined in the different parts of ISO 16100.

3.1 API Documentation

An API is the published interface, and a service is the concrete implementation of an API running in the back end. It is typically a black box, which means that source code is not publicly accessible. API documentation is very important for the successful adoption of an API. APIs expose services and therefore must be designed with an interface easily understood by API consumers. The documentation should help developers to learn the functionalities offered by an API and enable them to start using it quickly. The API documentation should provide all necessary information to API consumers or developers in a human-readable format and help them assess its suitability for use in their client app [19]. It should provide information about its usage requirements like input/output parameters, message format, licensing policy and more. The API documentation could serve as input to generate client and server-side skeleton code that can be further developed. To make API documentation effective, it should include the following aspects about the API: endpoint, title, method, message payload, header parameters, response code, error code. Several technologies and tools are available for API documentation. We can find RESTful API Modeling Language (RAML)[2], API Blueprint[3] and OpenAPI Specification[4] (originally known as the Swagger Specification).

4 Contribution

As outlined in the previous sections, our goal is to help companies select the most suitable open-source application and its reusable components to meet their business requirements which are composed by user- and service-type tasks. The final choice of reusing an open-source application depends on the precision of the expression of the needs. The more the needs are precise, the more the selection of the open-source application is easy. As a result, the company will choose the application which answers most in term of reuse to service tasks of the new business application. Figure 2 proposed Framework shows the proposed framework, composed by four main parts.

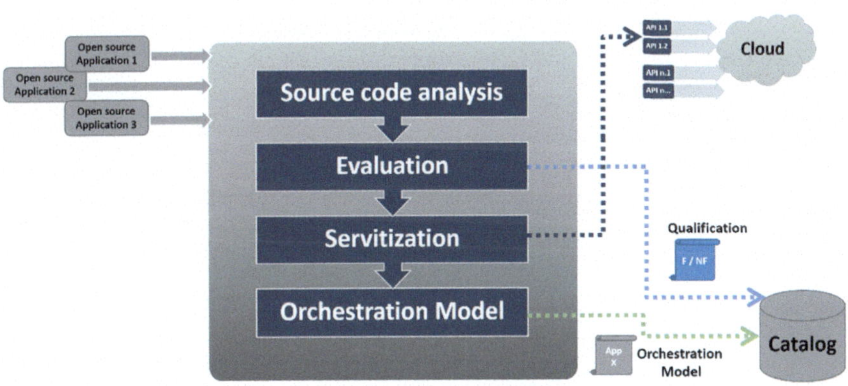

Fig. 2 Proposed framework

[2]http://raml.org/

[3]https://apiblueprint.org/

[4]https://www.openapis.org/

The first step is the code extraction and analysis. In this step, we transform the extracted source code into a model, allowing us, with the help of visualization techniques, to detect the existing services and identify potential candidates (methods) to expose as a service. The second step is the evaluation part. This step focuses on the technical aspect by applying some metrics helping to elect reliable candidate to externalize (next step process) and qualify existing services of the open-source application to explore their capabilities, system properties, conditions of use and limits. All this information is gathered in a single catalog for discovery, maintenance and reuse purpose. In the third step, we characterize the candidate functionality in a readable and standardized way, and an automatic process is applied to generate the API to be deployed on the cloud. The last part represents the orchestration model generation. In this step, we generate an orchestration template when the candidate method belonging to the core of the application and depend on another service. An orchestration template is generated to not modify the business logic offered by the application.

The originality of this work is to transform a servitized, semi-servitized or legacy application into SOA application. For (semi-)servitized applications, it allows to qualify, give visibility and secure existing services with APIs in order to ease the reuse. It provides also a way for semi-servitized applications to characterize the functionalities to expose, and automate the steps to reach the service. For non-servitized applications, the goal is to come out with a full-service application that facilitates its reuse.

In this paper, we focus only on the third step of the proposed framework which is composed by the characterization and servitization steps (detailed in Fig. 3 Servitization steps).

In order to reach our objective of the service orchestration, we have selected the framework offered by ISO 16100 for the characterization part because it focuses on the interfacing requirements for interoperability, instead of ISO 15745 standard

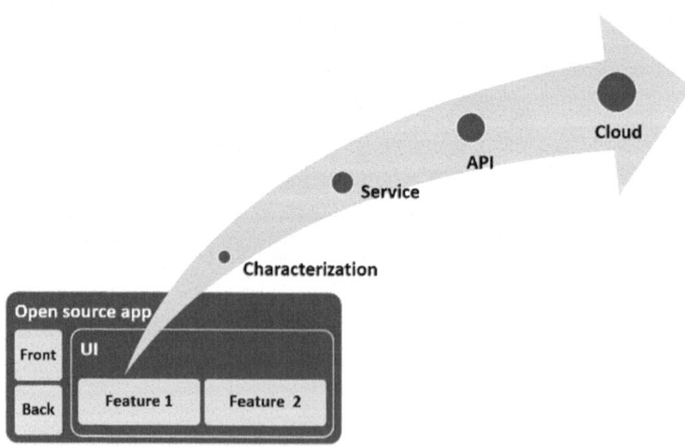

Fig. 3 Servitization steps

which identifies a larger set of profiles and elements needed to support interoperability between application components.

Our contribution is the proposal of an ISO 16100 capability template with its capability class described with XML Schema. We propose also an automated way to reach a reusable component based on OpenAPI Specification, which is one of the most popular API documentation frameworks. OpenAPI Specification provides standard way of defining REST API Interface [19]. It allows the consumer to understand the capabilities of the REST API without access to the service implementation code. We expose in the following (Sect. 4.1), the capability profile process as described in ISO 16100 standard. Then we propose in Sect. 4.2, the steps to reach the API documentation from the capability profile allowing to get the desired API.

4.1 Capability Profile Process

First, we have to identify candidate methods to externalize and expose as an API. In this paper, we do not present this step. We will deal with it during our future work. In the following, we take the output of the evaluation step, which are candidate methods, and we apply the next step of the framework which is the characterization step allowing to get the ISO 16100 capability profile of the method. The open-source application applications are represented as an activity tree structure that is hierarchical. And an activity is considered as a software unit in the ISO 16100 specification. The interoperability of software units can be described with its functionality, interface and structure. A capability profile for a software unit implies the generation of a concise statement of capabilities allowed by the software unit in terms of interfaces provided, the functions performed and the protocols used. A software unit to be profiled shall be analyzed and a template shall be filled to make a profile. In this work, we have formed a new capability class and a new capability template in XML Schema, helping to retrieve information needed to externalize a feature of an application. Figure 4 Example of a Specific part of capability profile shows an example of information contained in the specific part of a capability profile.

4.2 Capability Profile to an API

Once the common and the specific part of a capability profile are filled, the difficulty resides on how to reach a service from this formal structure (xml file). The solution that we propose is to transform the capability profile to an OpenAPI Specification, allowing to generate the interfaces of the new service. Table 1 shows the correspondence to be carried out, and a proof of concept is presented in Sect. 5 applied on an open-source application.

```
        <Specific>
            <ReferenceCapabilityClassStructure
                id="cea_1001" name="CAM_FunctionSample" version="001" url="">
                <Activities>
                    <Worker>
                        <Function ID="CreateAsset" action="create" level="4" />
                    </Worker>
                </Activities>
                <InformationExchange>
                    <InputDataTypes level="1" ID="inputDataTypesId">
                        <Numerical level="2" ID="numericalId">
                            <One-Dimensional level="3" ID="objId">
                                <string level="3" ID="name" />
                                <string level="3" ID="modelName" />
                                <string level="3" ID="ownerName" />
                            </One-Dimensional>
                        </Numerical>
                    </InputDataTypes>
                    <OutputDataTypes level="1" ID="outputDataTypes">
                        <Numerical level="2" ID="numericalId">
                            <One-Dimensional level="3" ID="objId">
                                <string level="3" ID="name" />
                                <string level="3" ID="modelName" />
                                <string level="3" ID="ownerName" />
                            </One-Dimensional>
                        </Numerical>
                    </OutputDataTypes>
                </InformationExchange>
            </ReferenceCapabilityClassStructure>
        </Specific>
    </CapabilityProfile>
</CapabilityProfiling>
```

Fig. 4 Example of a specific part of capability profile

Table 1 Capability profile to an API

Capability profile XML tag	OpenAPI specification property
<Owner> <ComputingFacilities> <Performance>	API description
<Function id="">	API path
<Function action="">	Method of API
<InformationExchange> <InputDataTypes>	Parameters of the API
<InformationExchange> <OutputDataTypes>	Response objects of parameters of API

5 Use Case

In order to validate our approach, we applied characterization and servitization process on an application called Collaborative Asset Manager (CAM) from FITMAN[5] Project. The FITMAN-CAM app is a Web-based, integrated platform for the management of virtualized Assets in the scope of service-oriented manufacturing ecosystems. (The term "asset" represents any item of economic value owned by an enterprise.) This application offers CRUD operations of virtualized assets (create, read, update and delete) using the user interface and exposes its own REST-based APIs to

[5] http://www.fitman-fi.eu/

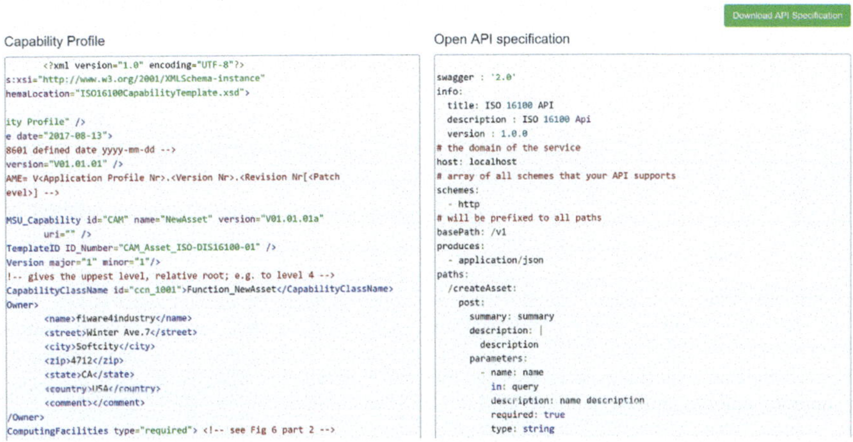

Fig. 5 Application screenshot: capability profile to Swagger

retrieve information about assets from database. For matter of reusing this application for new business requirements, there is a need to create assets without going through the user interface and orchestrate this action with other services. Our implementation of the servitization step helps to generate project skeleton in order to expose the REST-based API of any functionality already characterized by the proposed ISO 16100 capability profile. The characterization step (which is done manually for now) of the create asset method gives the profile in the left of Fig. 5 Application screenshot: Capability profile to Swagger.

The content of the capability profile reflect the signature of Java method *public void createAsset* (*String name, String modelName, String ownerName*). Once the profile filled, the next step is to upload the capability profile to our developed application, in order to transform it into an OpenAPI Specification following the steps described in Sect. 4.1. The generated OpenAPI Specification is described using YAML, a data serialization standard. From this stage, we generate the server stubs and client SDKs, by using Swagger Codegen process. The last step is the implementation of the business side and deployment of the generated API on an API Manager running on cloud to get a reusable component for service orchestration.

6 Conclusion and Future Work

In this paper, we have presented the characterization and servitization steps of the proposed framework with a proof of concept applied on a open-source application, starting from the characterization of a functionality done in a readable and standardized way, using a proposed ISO 16100 capability template with its capability class described with XML Schema. Then we provide an automated solution for the externalization of the characterized functionality as a REST-based application programing interface for service orchestration. As a result, we generate reusable components for service orchestration. As future work, we plan to focus on the source code analysis step. The aim of this step is to transform the source code of the open-source application into a queryable model and generate a visual to explore and qualify the existing services. The analysis step is followed by the evaluation step, where the use of metrics will help to elect reliable candidates to characterize and expose as presented in this work.

Acknowledgement This paper presents work developed in the scope of the project vf-OS. This project has received funding from the European Union's Horizon 2020 research and innovation program under grant agreement no. 723710. The content of this paper does not reflect the official opinion of the European Union. Responsibility for the information and views expressed in this paper lies entirely with the authors.

References

1. Almonaies, A. A. (2010). Legacy system evolution towards service-oriented architecture. In *International Workshop on SOA Migration and Evolution* (pp. 53–62).
2. Almonaies, A. A. (2011). Towards a framework for migrating web applications to web services. In *Proceedings of the 2011 Conference of the Center for Advanced Studies on Collaborative Research* (pp. 229–241).
3. Belfadel, A. (2017). Capability profile for enterprise application integration. In *2017 International Conference on Engineering, Technology and Innovation (IEEE/ICE/ITMC)*.
4. Boldrini, J.-C. (2017). *The journey in Open Innovation to develop a SME: A longitudinal case study in a French robotics company*.
5. Brajesh. (2017). *API management: An architect's guide to developing and managing APIs for your organization*. Apress.
6. Kambe, H. (2017). A method for analyzing and visualizing intermodule relations to support the reuse-based embedded software development. *Electronics and Communications in Japan, 7,* 18–31.
7. Kang, W. B. (2005). Software reuse research: status and future. *IEEE Transactions on Software Engineering, 31,* 529–536.
8. Kang, W. B. (2005). Software reuse research: status and future. *IEEE Transactions on Software Engineering, 31,* 529–536.
9. Kebir, S. (2012). Quality-centric approach for software component identification from object-oriented code. In *2012 Joint Working IEEE/IFIP Conference on Software Architecture (WICSA) and European Conference on Software Architecture (ECSA)* (pp. 181–190).
10. Muller, P (2016-11-24). Annual Report on European SMEs 2015/2016 SME recovery continues.
11. Patidar, K. (2013). Coupling and cohesion measures in object oriented programming. *International Journal of Advanced Research in Computer Science and Software Engineering, 3*(3).
12. Razavian, M. (2015). A systematic literature review on SOA migration. *Journal of Software: Evolution and Process, 27*(5), 337–372.
13. Roure, S. A. (2010). A service identification framework for legacy system migration into SOA. *IEEE International Conference on Services Computing, 2010,* 614–617.
14. Roy, M. F. (2011). Useful, but usable? Factors affecting the usability of APIs. In *2011 18th Working Conference on Reverse Engineering* (pp. 151–155).
15. Roy, M. F. (2011). Useful, but usable? Factors affecting the usability of APIs. In *2011 18th Working Conference on Reverse Engineering* (pp. 151–155).

16. Shatnawi, A. (2017). Reverse engineering reusable software components from object-oriented APIs. *Journal of Systems and Software, 131,* 442–460.
17. Standardization, I. O. (2000). ISO 15745, Industrial automation systems and integration. *Open systems application integration frameworks, ISO/TC/184/SC5.*
18. Standardization, I. O. (2005). ISO 16100-3:2005 Industrial automation systems and integration—Manufacturing software capability profiling for interoperability—Part 3: Interface services, protocols and capability templates.
19. Standardization, I. O. (2009). ISO 16100-1:2009 Industrial automation systems and integration—Manufacturing software capability profiling for interoperability—Part 1: Framework.

Empowering Product Co-creation Approaches Through Business Interoperability Concepts: The Toy Industry Case

A. Michalitsi-Psarrou, S. Koussouris, C. Kontzinos, O. Markaki, C. Ntanos, D. Panopoulos and J. Psarras

Abstract Although 99% of toy manufacturing companies are SMEs, the toy market is dominated by a limited number of massive global brands. Therefore, SMEs should enhance their current business model in a way that would give them a competitive advantage. This paper presents work from the European-funded research project ToyLabs that aims to build on this identified need to establish a new, collaborative value chain for product development in the toy industry, with business interoperability playing a key role in this quest. The methodology engages toy manufacturers, FabLabs, toy safety experts and end customers in the product design workflow to create new, safe customer-centred and market-targeted toys that are ready to be certified and put into production. For achieving the goals of this approach, the methodology behind the ToyLabs platform focuses on the incorporation of cross-sectorial business process interoperability capabilities that aim to create a common vocabulary, understanding and a communication gateway for the identification of potential partners even from sectors that were previously unrelated to the toy industry and the exchange of requirements and specifications between them.

Keywords Co-creation · Open innovation · Partner matching · Business process interoperability · Cross-sector interoperability

1 Introduction

A total of 99% of EU toy manufacturing companies are SMEs, accounting for 61% of the EU toy industry's employment. The current state of the EU toy market exhibits developments that may limit market growth prospects, especially for SMEs. In the context of an ever-changing demand, need for innovation, driven by the short product

A. Michalitsi-Psarrou (✉) · C. Kontzinos · O. Markaki · C. Ntanos · D. Panopoulos · J. Psarras
National Technical University of Athens, Iroon Polytechniou Str, 9, 15780 Zografou, Athens, Greece
e-mail: amichal@epu.ntua.gr

S. Koussouris
SingularLogic, Muntii Tatra Str, 4-10, 011078 Bucharest, Romania

K. Popplewell et al. (eds.), *Enterprise Interoperability VIII*, Proceedings of the I-ESA Conferences 9, https://doi.org/10.1007/978-3-030-13693-2_33

life cycle of toys, is widely acknowledged as necessary for a toy manufacturer, to maintain a competitive advantage. However, current R&D expenditures in the sector do not seem to respond effectively to this need, amounting to 0.6–2.6% of total turnover [1]. Although these observations concern all toy manufacturers, regardless of size, large companies may more easily survive in heavily competitive environments, since they have the capacity to offer a wide selection of toys, target multiple markets and enjoy scale advantages in purchasing, manufacturing, distributing, selling and marketing [2].

Nowadays, enterprises increasingly achieve a competitive advantage by forming innovative networks of value creation and bundling core competencies from different partners. Although the research community sees networked organisations as an undisputable reality, companies find it very time-consuming and difficult to establish electronic business relationships with a larger number of business partners [3]. A number of challenges that are even more persistent for SMEs, such as the lack of trust between partners [4] and the lack of IT capabilities [5] that make it difficult to integrate internal processes in a potentially cross-country business network, are making companies hesitant in pursuing the formation of collaborative networks. Specifically, in what concerns especially SMEs in the toy industry, strict toy safety regulations, governed by the Toy Safety Directive, lead toy manufacturers into isolation, to better control the conformity of their processes, materials, designs, etc. with EU toy safety standards.

These challenges have been also identified by the European research project ToyLabs that builds on the pillars of co-creation and open innovation and introduces a unified methodology that creates a multi-stakeholder network and consequently a multi-sided platform, where key players/stakeholders in the toy industry value network are brought together to collaborate closely, through standardised open processes to exchange designs, documentation and feedback, and come up with innovative toys and games. The platform that is being built on top of this methodology aims to bring to light an otherwise obscure collaboration potential with all relevant stakeholders to create products that may quickly enter the market, respond to a clear market demand, be cost effective and be customisable to meet multiple EU markets' demands. To achieve this, the platform is built to provide the means to facilitate the communication and the information exchange between actors that do not necessarily belong in the same industry, geographical market or even the same sector. This information exchange includes requirements, designs, manufacturing specifications (i.e. materials, resolution, strength, production capacity, etc.), manufacturing orders and others. This kind of cross-sectorial business process interoperability drives the need for the standardisation of both the search and the pre-selection of potential partners, as well as the interaction between collaborators after a partnership has been established.

2 Relation to Existing Theories

Two of the core value propositions being developed under ToyLabs concern the employment of co-creation/open innovation and partner matching methodologies in a reimagined product development environment, characterised by the diversity of its stakeholders. Although these concepts are well established on their own, especially in the cases of entities in similar or closely related domains, here they are combined to create a uniform environment between diverse actors that acts both as a collaboration platform and as a marketplace for very specific external competences that are integrally connected to the business processes involving new product development.

2.1 Co-creation and Open Innovation in New Product Development

Open innovation encourages companies to acquire outside sources of innovation to improve product lines and shorten the time required to bring products to market [6]. Co-creation involves the joint creation of value by a company and other stakeholders, especially consumers who intend to co-construct the service experience to suit their preferences [7]. Here, the focus is on co-creation in new product development (PD) that is also known as collaborative product development (CPD). In order for CPD to be effective, collaborators oftentimes need to modify their business activities/vocabulary/policies, etc. to participate in a joint framework that will help them achieve commonly agreed goals, a process known as business process interoperability (BPI) [8].

The concept of co-creation is applied in modern manufacturing through the development of collaborative platforms and ICT tools, whereas BPI is achieved by using models that include the most common transactions/interactions between specific businesses or business sectors. The importance of co-creation and co-innovation for creating business networks has been highlighted in the literature, proposing reference frameworks that can act as enablers in interlinking the participating organisations [9]. From a manufacturing point of view, there are examples of collaborative platforms based on cloud computing to support co-creation [10], while augmented/virtual reality is being used as a communication medium in other cases [11]. Ontologies are also considered to boost semantic interoperability in manufacturing related terms enabling reuse of knowledge resources [12].

2.2 Partner Matching in Manufacturing

Partner matching is among the core elements that drive business interoperability in business-to-business transaction systems as it relies on structures, and infrastructures which in most of the cases are of heterogeneous nature and therefore constitute a bottleneck of rapid production execution. The latter is even more evident in the case of collaborative and open-innovation-based manufacturing. The advantage of having large pools of collaborators and employing novel flexible production schedules with multitude collaboration suffers greatly from incompatibilities between the working characteristics of different potential partners and incompatible information flow and processing techniques. Research showcases that the absence of well-defined interoperable components in the different business transaction system layers results undoubtedly in situations where the larger a pool of collaborators is, the more difficult coordination and proper execution of orders becomes [13]. Such components are crucial to be present at the level of semantics that range from partner organisation and product descriptions to common vocabularies for automated process execution, at the technological level that concerns IT system interconnection, but also at the strategic level for governing manufacturing networks. Their absence is therefore severely thinning any advantage that a collaborative and open-innovation-driven approach can deliver [14].

In recent years, several approaches were formulated [15], offering solutions to such problems, ranging from simple business interoperability solutions, which address mostly peer-to-peer collaboration in business alliances and cover generally the lower interoperability layers, to dynamic manufacturing networks management methodologies that try to end-to-end orchestrate manufacturing networks at the semantics and technical levels [16, 17]. In fact, it is the concept of dynamic manufacturing networks, where extensive emphasis has been given to partner matching and selection [18], as these describe large pools of prospective collaborators, where the sharing of information about things, such as their capacities, schedules and cost structures, allows them to spontaneously engage into collaborations. This gives a significant advantage over traditional collaboration networks and allows novel supply chain methods to be deployed. Such solutions rely heavily on enhanced business interoperability, trying to embrace entities of different nature, size and technological infrastructure. This further promotes interoperability, making no discrimination between IT-rich and IT-poor organisations, suggesting in most of the cases low-cost, reusable components and providing solutions that are sector and operation agnostic.

3 Innovation Approach

Following closely the process by which a new product goes from inception all the way to production for a toy manufacturer, this paper describes a methodological framework that directly leads to a respective collaboration platform that can be used by any toy manufacturer and specifically SMEs for attaching more value to their design processes and new product development activities. The methodology also specifies how each stakeholder group is to be involved, according to their expertise or capacity on each of the discrete phases that takes an idea for a new toy to the market.

3.1 Toy Industry's Value Network and Involved Stakeholders

Potential users of ToyLabs have been grouped in meaningful categories with respect to their competences and projected role in the platform. In this value network, manufacturers play the most prominent role. However, following the ToyLabs approach, other, not prominent, stakeholders are also engaged early in the new toy development process, giving feedback and making proposals on the design, safety or marketability of the products. In that view, **safety experts** are invited to evaluate a product according to EU standards and provide early safety certification and consultancy for concepts, designs, prototypes and final products, thus rendering a complex and costly procedure more straightforward, avoiding expensive "back and forth". **FabLabs** (small-scale workshops offering digital fabrication) can offer short series and prototype manufacturing, handling small-scale production more cost-effectively than a toy manufacturer. These prototypes are then evaluated by dedicated groups of **childhood experts** (teachers, parents' associations, etc.) and **market representatives** (B2B customers, end customers, etc.). What is noteworthy is that due to the openness and innovative nature of the approach, these stakeholders may be called to step outside their core responsibilities and have a different or bigger role in various stages of the process, according to their stated competences.

3.2 Concept Description

Along the following lines, the framework that resulted to the ToyLabs collaboration platform is presented in brief, followed by the partner matching and selection methodology which is an essential part of the ToyLabs framework and respective value proposition.

Open Innovation and Co-creation Methodology

Apart from the identified stakeholder groups, an additional horizontal role is established, that of the "product owner", referring to the originator of a new toy idea, who can be any member of the network. Figure 1 shows the co-creation methodology that has been created specifically for the toy industry that builds on the idea of a V-Model [19], enriched with inspiration from the FITMAN Verification and Validation Methodology for the development of software for manufacturing enterprises [20].

The main steps of the typical development process of a toy are enriched with the foundations of this approach, as follows:

- The **immersion** phase refers to the inspiration process for the main characteristics, appearance and features of a product that could lead to a new or improved toy, passing through a better understanding of actual and potential market needs. Nowadays, social media—and in general Web 2.0 channels—is a valuable source of data, which, if analysed and processed accordingly, may offer invaluable insights. Therefore, market trends and social feedback analysis is the first added-value proposition of the methodology, where innovation for a new toy design emerges.

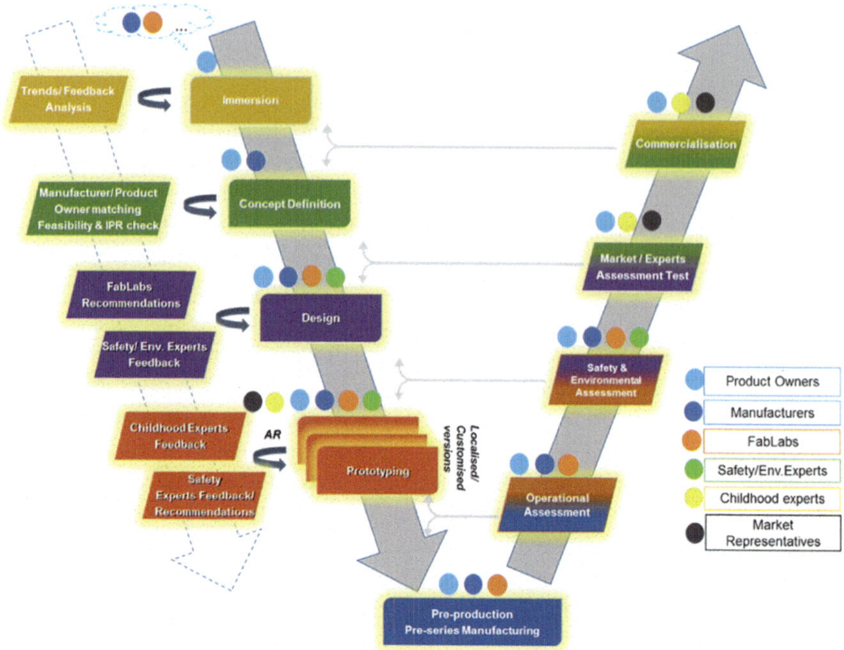

Fig. 1 Co-creation Methodological framework for the toy industry

- The **concept definition** phase begins from the decision on the main characteristics of a product and finishes with the conceptual design of the toy. The "product owner" selects the most appropriate manufacturer to collaborate, using the ToyLabs partner matching methodology and module, to support him in the realisation of the idea leading to a conceptual toy design and an initial feasibility assessment.
- The **design** phase refers to the creation of a detailed design for a new product (i.e. materials, required processing, etc.). The iterative design approach of ToyLabs is based on the continuous collaboration with external parties, selected using the partner matching methodology, presented below. Firstly, one or more FabLabs are engaged to give technical insights about the materials, the processing and the technical feasibility of the design and optionally make changes to the design or transform it from 2D to 3D, if the product owner lacks the expertise. Secondly, at least one safety expert is engaged, tasked with the provision of safety-related recommendations regarding safety standards applied to the specific toy category.
- The **prototyping** phase is where integrated, safe prototypes are produced for hands-on testing. The contribution of ToyLabs is threefold in this phase: (a) the partner matching methodology supports finding the most appropriate FabLabs in terms of experience and equipment for building the prototypes, (b) the cross-border collaboration potential with FabLabs enables the creation of different, localised versions of a prototype and (c) augmented reality technologies are used and com-

bined with the tangible prototype to offer an experience close to the final product. This phase runs iteratively until all parties reach a final version for the prototype that is then passed to selected experts for further assessment.
During the next phases, "**pre-production**", "**operational assessment**" and "**safety and environmental impact assessment**", depicted in Fig. 1, ToyLabs embraces the usual/traditional manufacturing process, so no further mention here is foreseen.

- **Expert/End-User Assessment**: Childhood experts, end-users and toy market experts are requested to give feedback from educational/psychological, user and marketing perspectives, to be used as input for the next phase, the product's commercialisation.

The **commercialisation** phase is the final phase of this workflow and incorporates the input from previous phases, while setting up a mechanism for feedback collection for the immersion phase of new toy development cycles.

Partner Matching Methodology

In ToyLabs, business interoperability is highly relevant when it comes to initiating a collaborative process, through the "partner matching and selection" methodology, as the overall process in identifying partners with the necessary capabilities/capacities relies profoundly on the ability of the system to retrieve and process data from different peers that are potential partners in the network. Since ToyLabs is a methodology to provide open collaboration opportunities, stakeholders are not limited to a singular role each, and their appropriateness for participation in the various processes has to be evaluated depending on the requirements of both the process and the product that is being designed. This is not a once-off project during toy manufacturing, as it is highly needed in various stages of the product's development, from the design to the certification phase, involving different stakeholder groups.

The component relies on a methodology having two parts:

(a) **partner matching** refers to the identification and classification of the parameters and factors (criteria) required for evaluating the suitability of different candidate partners to undertake certain steps of the proposed co-creation methodology.
(b) **partner selection** refers to the specification of the methods and algorithms required for making comparisons between candidate partners and selecting the ones that best cover the given requirements, which is out of the scope of the paper at hand.

For partner matching, a model-based approach (the ToyLabs blueprint model) allowed the management and interlinking of toy, partner, toy co-creation step and technical capabilities' information. The project's blueprint model, inspired by the IMAGINE blueprints [21], is an information meta-model that supports partner selection and improves decision-making in the toy co-creation process, thus minimising failure rates, shrinking development times for new toys, increasing adaptability to market trends and customisation needs, while ensuring increased compliance with safety and environmental regulations.

To come up with a blueprint model for the toy industry and the needs of the ToyLabs collaboration platform, that will be as exhaustive as possible to capture all the partnering or toy creation requirements, multiple, iterative discussions with the project's pilot partners—both FabLabs and toy manufacturers—as well as a survey where 43 toy industry representatives participated, were conducted. A set of key recommendations were thus generated and were used as requirements both for improving the structure and contents of the ToyLabs blueprints and for rendering the generated blueprints as analytical as possible.

Certain indicative derived requirements for the blueprint model were the following: The blueprint model should: (a) provide a low-cost method to collect partner and toy-related information, (b) allow fast and effective partner selection, (c) imprint the profile of different types of partners (e.g. companies/organisations, individuals), (d) enable partner authentication and certification, (e) include information on partners' reputation and inspire trust, (f) include historical information (e.g. track record), (g) include technical capabilities and equipment-related information and (h) associate partner information with toy requirements information. The list was though much more exhaustive.

The ToyLabs blueprint model is therefore a declarative meta-model that aggregates and modularises partner, toy and technical capabilities' data and information, by specifying two types of interrelated blueprints:

(a) The **partner artefact blueprint** capturing partners' unique skills and capabilities and making them available to potential contractors and
(b) The **toy artefact blueprint** that sets the constraints and preconditions up with respect to the creation of a toy that a partner should respond to.

The partner artefact blueprint is further split into two categories, the partnering organisation and the partnering individual blueprints, representing the "ecosystem" of the co-creation process. Table 1 illustrates the three derived blueprints in brief, avoiding, for the scope of this paper, to go into a deeper level of analysis for the various sub-elements. For example, the technical competencies are further analysed to more technical criteria and information for each "bullet", which is not presented in the table here.

4 Conclusions and Future Work

In today's competitive environment, identifying market trends effectively is not enough by itself. It needs to be followed by quickly reaching the market with new products. Business interoperability is the cornerstone of many of the solutions proposed, while calls are made for the integration of new ICT technologies and innovative techniques and materials. SMEs tend to have less opportunities to align with new manufacturing trends and ICT solutions. From a technical interoperability perspective, the innovation potential of ToyLabs focuses on a domain, where business procedures, information flow and communication among stakeholders can be intro-

Table 1 Blueprint model for the toy industry

Partner artefact blueprint		Toy artefact blueprint
Partnering organisation blueprint	Partnering individual blueprint	
Partner type: • Manufacturer • FabLabs • Experts' company	Partner type: • Safety expert • Environmental expert • Childhood expert • TSIG member	Artefact type: • Idea/concept • Design • Prototype
Toy category: • Dolls and soft-filled toys • Construction toys and puzzles • Activity toys	Area of expertise: • Safety • Quality • Environment • Children toys • Electronics	Toy category: • Dolls and soft-filled toys • Construction toys and puzzles • Activity toys
Company contact information	Expert contact information	ToyLabs methodology phase (i.e. immersion, design, etc.)
Locations and facilities	Location	Artefact version no.
Technical competencies/capabilities and related equipment properties: • 3D printing • 3D scanning • CAE/FEM and structural analysis simulation • Circuit production • CNC milling • Inkjet printing • Knitting machine • Laser cutting • Mould casting • Plastic transformation • Sewing machine • Soldering station • Vacuum forming • Vinyl cutting	Previous experience	Technical requirements
Certifications awarded	Expert certifications: • Safety certifications • Toy awards	Material requirements

(continued)

Table 1 (continued)

Partner artefact blueprint			Toy artefact blueprint
Partnering organisation blueprint		Partnering individual blueprint	
Products/services		Acting as representative of: • Educators/schools • Families • End-users	Time and cost requirements
Qualifications possessed:		Costs of the tests	Safety and environmental compliance requirements
• Novelty/innovation (e.g. patents) • Usability • Durability	• Accuracy • Productivity • Reputation		
Economic criteria (i.e. man-hour cost, product pricelist, way of charging, etc.)			Customisation parameters: • Text • Look • Style • Other
SLAs undertaken (e.g. terms and conditions, IPR, reward, capacity/availability, etc.)			Related open issues

duced or vastly improved. Disrupting any of these areas can substantially improve the competitive positioning of the EU toy industry's SMEs.

In this context, ToyLabs' contribution and innovation potential, as reflected by its progress beyond the state of the art, can be summarised in the following points:

- It introduces an innovative change in the toy manufacturers' business process, bringing together toy manufacturers, FabLabs, toy safety experts and childhood professionals to co-create and exchange value. ToyLabs value proposition encompasses a cross-sector and even cross-domain potential, which can also be applied in other domains.
- It introduces an innovation model that allows businesses to lower their production costs, helping toy industry's SMEs to compete against larger enterprises with access to interoperable business platforms, they wouldn't be able to afford otherwise.

From a business perspective, ToyLabs offers the manufacturers the opportunity to validate their concepts and designs in multiple remote locations, taking advantage of the rapid prototyping capabilities offered by FabLabs. This process aligns the manufacturers with their customers' needs and constructs rapid value chain operations based on the principles of dynamic manufacturing networks, both globally and nationally.

The approach presented in this paper is being implemented and tested in the EU co-funded project ToyLabs. ToyLabs is now in the process of completing the development of the ToyLabs collaboration platform that will drive the agile collaborative product design and creation process, bringing together all involved parties, springing this methodology into life. The overall approach will be validated, demonstrated and evaluated through two real-life pilot cases on mechanical puzzle toys and dolls and accessories. These pilots will exploit the ToyLabs co-creation process from conceptualisation to production and commercialisation, to reengineer their current new toy design processes. The methodology will be updated during the next months, to incorporate the feedback from the project's pilot partners and their connections with the toy industry.

Acknowledgements This work has been co-funded from the European Union's Horizon 2020 research and innovation programme under the ToyLabs project, grant agreement No. 732559.

References

1. ECSIP consortium: study on the competitiveness of the toy industry—final report (2013). (pp. 36–38). https://doi.org/10.2762/56910.
2. Toy & Game Manufacturing Report Summary. http://www.hoovers.com/industry-facts.toy-game-manufacturing.1207.html.
3. Legner, C., Wende, K. (2006). Towards an excellence framework for business interoperability. In *19th Bled eConference eValues - Conference Proceedings*, (pp. 1–16).
4. Hoyt, J., & Huq, F. (2000). From arms-length to collaborative relationships in the supply chain. An evolutionary process. *International Journal of Physical Distribution Logistics Management, 30,* 750–764. https://doi.org/10.1108/09600030010351453.
5. Dai, Q., Kauffman, R. J. (2001). Business models for Internet-based e-procurement systems and B2B electronic markets: an exploratory assessment. In *Proceedings of the 34th Annual Hawaii International Conference on System Sciences*, IEEE Computer Society (p. 10).
6. Chesbrough, H. W. (2006). Open innovation: the new imperative for creating and profiting from technology.
7. Prahalad, C. K., & Ramaswamy, V. (2004). Co-creation experiences: The next practice in value creation. *Journal of Interactive Marketing, 18,* 5–14. https://doi.org/10.1002/dir.20015.
8. Tsagkani, C. (2005). Inter-organizational collaboration on the process layer. In *Proceedings IFIP/ACM SIGAPP INTEROP-ESA Conference*.
9. Romero, D., & Molina, A. (2011). Collaborative networked organisations and customer communities: value co-creation and co-innovation in the networking era. *Production Planning Control, 22,* 447–472. https://doi.org/10.1080/09537287.2010.536619.
10. Vincent, Wang X., & Xu, X. W. (2013). An interoperable solution for Cloud manufacturing. *Robotics and Computer-Integrated Manufacture, 29,* 232–247. https://doi.org/10.1016/J.RCIM.2013.01.005.
11. Galambos, P., Csapó, Á., Zentay, P., Fülöp, I. M., Haidegger, T., Baranyi, P., et al. (2015). Design, programming and orchestration of heterogeneous manufacturing systems through VR-powered remote collaboration. *Robotics and Computer-Integrated Manufacture, 33,* 68–77. https://doi.org/10.1016/j.rcim.2014.08.012.
12. Lin, H. K., & Harding, J. A. (2007). A manufacturing system engineering ontology model on the semantic web for inter-enterprise collaboration. *Computers in Industry, 58,* 428–437. https://doi.org/10.1016/j.compind.2006.09.015.
13. European Commission: the new european interoperability framework. https://ec.europa.eu/isa2/eif_en.

14. Markaki, O., Kokkinakos, P., Panopoulos, D., Koussouris, S., Askounis, D. (2013). Benefits and risks in dynamic manufacturing networks. Presented at the (2013).
15. Zacharewicz, G., Diallo, S., Ducq, Y., Agostinho, C., Jardim-Goncalves, R., Bazoun, H., et al. (2017). Model-based approaches for interoperability of next generation enterprise information systems: state of the art and future challenges. *Information System E-bus. Management, 15,* 229–256. https://doi.org/10.1007/s10257-016-0317-8.
16. Papazoglou, M. P., van den Heuvel, W.-J., & Mascolo, J. E. (2015). A reference architecture and knowledge-based structures for smart manufacturing networks. *IEEE Software, 32,* 61–69. https://doi.org/10.1109/MS.2015.57.
17. Agostinho, C., Ducq, Y., Zacharewicz, G., Sarraipa, J., Lampathaki, F., Poler, R., et al. (2016). Towards a sustainable interoperability in networked enterprise information systems: Trends of knowledge and model-driven technology. *Computers in Industry, 79,* 64–76. https://doi.org/10.1016/j.compind.2015.07.001.
18. Viswanadham, N., & Gaonkar, R. S. (2003). Partner selection and synchronized planning in dynamic manufacturing networks. *IEEE Transactions on Robotics and Automation, 19,* 117–130. https://doi.org/10.1109/TRA.2002.805659.
19. ISTQB (2015). What is V-model- advantages, disadvantages and when to use it? ISTQB Exam.
20. Lampathaki, F., Panopoulos, D., Kokkinakos, P., Bompa, C., Koussouris, S., Askounis, D. (2014). Infusing verification and validation in ICT solutions in manufacturing: The FITMAN V&V Method. In *Enterprise interoperability VI*. (pp. 307–317). Springer, Cham.
21. Ferreira, Jose, de Beca, M. F., Agostinho, C., Nunez, M. J., & Jardim-Goncalves, R. (2013). Standard blueprints for interoperability in Factories of the Future (FoF). *IFAC Proceedings, 46,* 1322–1327. https://doi.org/10.3182/20130619-3-RU-3018.00427.

Part XII
Interoperability in M2M Interaction

Smart Process Communication for Small and Medium-Sized Enterprises

R. van de Sand, S. Schulz and J. Reiff-Stephan

Abstract The transformation process towards new industrial paradigms in the course of I4.0 places new demands on interoperable communication. Entities as part of cyber-physical production systems are required to interact autonomously with their environment. This can constitute a challenge for small- and medium-sized companies, since adapting existing manufacturing structures to the needs of I4.0 often involves major obstacles due to the high degree of innovation. Thus, the objective of this paper is to suggest approaches for the application of smart communication solutions to connect entities within the fully connected value network and show how components and be classified upon two criteria in order to determine whether these fulfil requirements on I4.0 applications. Especially, it provides an overview of the most forward-looking communication protocols that allow the share of information from the field level up to cloud applications and across company boundaries.

Keywords Industrial communication · Protocols · Smart entities · Assets

1 Introduction

Digitization plays an ever-increasing role in the industrial automation sector and is the key driver in the transformation process from conventional manufacturing strategies through cyber-physical production systems (CPPS). As illustrated in Fig. 1, the main focus of this process, called the Fourth Industrial Revolution (I4.0) [1], is placed on linking all production activities along the value chain both vertically and horizontally [2]. Consequently, this invokes increasing demands on data transparency and interoperability between technical entities. Since I4.0 stretches well beyond internal processes of an organization towards a fully connected value network, the amount and complexity of data produced during manufacturing processes increases rapidly. However, data must not only be managed and coordinated in future cyber-physical production systems, it has to be produced and made understandable

R. van de Sand (✉) · S. Schulz · J. Reiff-Stephan
Technische Hochschule Wildau, Hochschulring 1, 15745 Wildau, Germany
e-mail: van_de_sand@th-wildau.de

K. Popplewell et al. (eds.), *Enterprise Interoperability VIII*, Proceedings of the I-ESA Conferences 9, https://doi.org/10.1007/978-3-030-13693-2_34

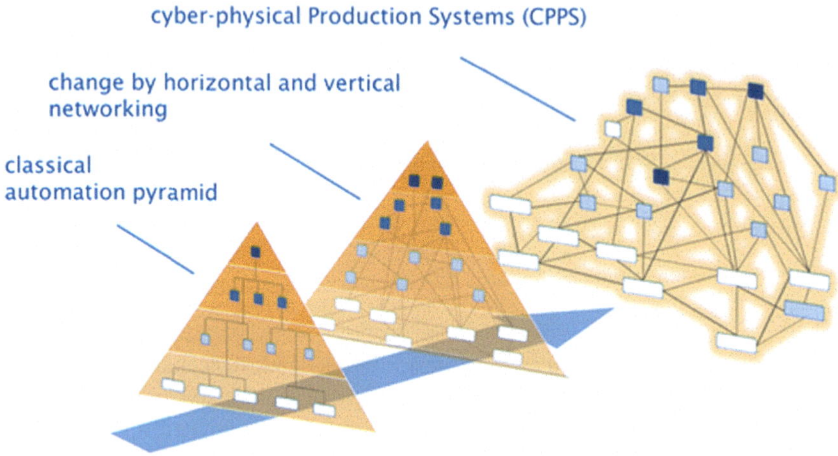

Fig. 1 Progress towards CPPS

for any participating party within a CPPS. This requires the digital transformation through innovation accelerators, such as the Industrial Internet of Things (IIoT) [1], which manifests itself as a global infrastructure for the information society in I4.0 transformation processes that provides advanced services by interconnecting things physically as well as virtually based on interoperable information and communication technologies [3, 4].

Considering the communication ability of industrial applications, the IIoT might unfold completely new opportunities regarding a new range of control systems. Although conventional automation strategies are mostly based on centralized control systems, a paradigm shift towards decentralized event-triggered control systems that operate at multiple scales and have the capability to reorganize and reconfigure [5] is aspired. Such decentralized control systems waive centralized logic devices by distributing computing, communication and decision-making tasks across the network. Consequently, new demands are placed on technical entities within CPPS where sensors and actuators can communicate via the Internet and are able to use Internet services [6] in order to enable interaction between an extensive number of devices.

However, enterprise structures, especially small- and medium-sized enterprises (SME) that have grown over decades might not provide appropriate conditions to adapt themselves to the I4.0-driven transformation process. Furthermore, the high degree of innovation constitutes a major challenge for SME which, therefore, require special attention. Although this transformation process brings major challenges, it can also be a benefit for such enterprises, since it serves to maintain and expand their own competitiveness [7]. Moreover, by applying I4.0 strategies, manufacturing companies can develop new business models and produce more effectively. Hence, the aim is to optimize the framework conditions and support structures so that as many SME as possible meet the challenge of Industry 4.0 [2]. Besides the implementation of cloud applications, big data analytics or security measures, which are indeed major obstacles, the networking of IT systems both within and beyond the company [2] plays the key role in this digital revolution. Thus, the contribution of this paper to the topic is to demonstrate how existing standards can be applied for the holistic process communication and suggests a concept for IT networking within CPPS.

After this brief introduction to the objective of this paper, the following section describes demands on I4.0 components and explains how classifications of assets can be made based upon two major criteria. Furthermore, currently available protocols for the communication within CPPS are reviewed. In Sect. 3, an approach for the introduction of smart communication solutions in a wide range of applications is suggested followed by a prototypical use case. Finally, Sect. 4 concludes the paper.

2 State-of-the-Art

2.1 I4.0 Compatible Components

The I4.0-driven transformation process places new demands, not only on standards regarding data exchange, but also on technical components. There are two major aspects in terms of I4.0 components in CPPS that must be considered independently. One is the level of awareness of assets in the information world that, according to [4], can be classified into four presentation degrees as indicated in Fig. 2. For example, a motor can be identified uniquely via an RFID tag during the production process and is, therefore, individually known in the IT system. If an asset assigns its own object in the information world for its administration [8], it can be considered as a self-managing entity providing various functionalities, such as life-cycle recording or quality control. Contrariwise, assets that cannot be identified by an unequivocal name but exist in a certain place are anonymously known. This, for instance, could be a bearing as part of a machine, which possesses particular properties that are known to the system. However, if it leaves the system, it becomes unknown and therefore no longer exists within the information world. Another feature of an I4.0 component is the ability to communicate within an IT system. Here, the classification ranges from assets that do not provide any communication interface to fully I4.0 conform communication interfaces supporting certain standards on information exchange, security and network identification [4]. Essentially, passive communication assumes that components cannot autonomously communicate with other entities but, however, are clearly identifiable due to information carriers, such as barcodes or RFID tags [9]. Components enabling active communication are capable of independently logging into the information network and participate in the autonomous exchange of information. On this basis, I4.0 conform communication expects that advanced requirements in terms of real time, semantics, and security are met [4]. Due to the communication and presentation (CP) classification, any asset can be assigned to one particular class indicating whether it is I4.0 conform or not. For example, a classical fieldbus device might be categorized as CP33, since it is individually known and can

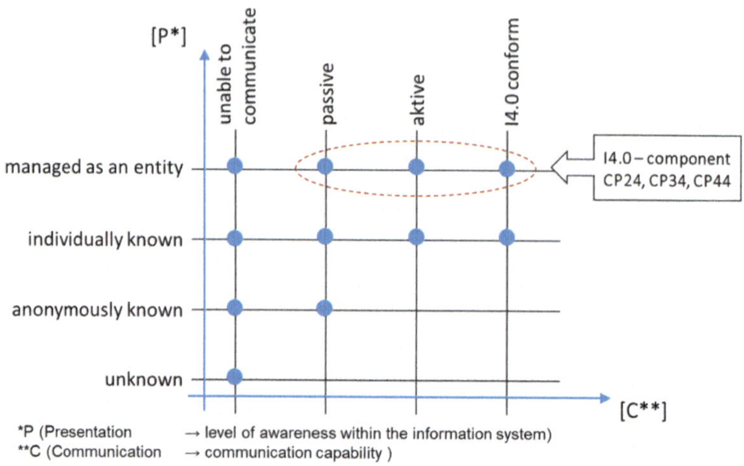

Fig. 2 CP classification according to [10]

communicate actively. Assets that would meet the requirements demanded by CPPS must provide certain structural preconditions in terms of communication abilities and hence are classified as a C24, CP34 or CP44 device [10].

According to [11], the idea of I4.0 components is to surround assets with an asset administration shell (AAS) which organizes the administration of assets within the digital IT system [10]. The AAS overtakes tasks regarding network-based communication, such as allocating a unique identification to the asset or processing data packages. Moreover, it exposes relevant properties and functions of an asset to its environment enabling other parties to make use of them without having to know details of networking technology and topology. It is composed of a 'header' and a 'body', where the header contains information for identification and designation of the AAS itself and all specific assets administrated by it [11]. In contrast, the body represents data for description of features belonging to an asset class with their respective disposition for the concrete asset [9].

It contains certain standardized submodules for an asset-specific characterization of the asset administration shell that can, amongst other things, outline process capabilities, such as drilling, energy efficiency or MES connection [11].

2.2 *Protocols*

Throughout the metamorphosis from classical production paradigms to intelligent, autonomous and, most of all, fully networked manufacturing strategies, certain standards must be determined in order to ensure a standardized communication basis. In order to meet this requirement, lightweighted protocols, based on the TCP/IP stack, have been developed which are better suited for scenarios in which many devices have to exchange data among each other and additionally consume the least possible bandwidth. The MQTT messaging protocol is such a lightweighted standard, which is based on the publish–subscribe principle [12]. It allows clients to connect to a broker and subscribe for a particular topic, where a client can be both a publisher and a subscriber. Due to the low communication overhead that this protocol entails, it is easy to implement on fieldbus systems with higher latency [13] and is, therefore, very suitable for applications where only limited computing performance is available. Another M2M communication protocol that is already widely used in the industrial automation sector is the OPC Unified Architecture (OPC UA). Combining the older classical OPC standards, such as data access or alarm and events, OPC UA provides an appropriate infrastructure in terms of interoperability between various, partly proprietary, systems in a I4.0 environment [4]. Currently, the OPC UA communication protocol reaches its limits when it comes to complex processes based on real-time information [14]. Therefore, industrial Ethernet protocols, such as Powerlink or Profinet, continue to represent powerful technologies on the field level, since these are capable of assuring hard real time, especially if low latency is required. However, the OPC Foundation has added the publish–subscribe pattern to the existing OPC UA specification, which was hitherto only based on the request–response principle, in order to fulfil demands on communication performance[15]. Another step in the direction of an overall I4.0 protocol, that the OPC foundation undertakes to assure a continuous and secure Ethernet communication from the ERP system to field level entities, is the combination of OPC UA and TSN (OPC UA TSN) [14]. Both the introductions of the publish–subscribe principle and TSN into OPC UA represent the most forward-looking technologies in terms of industrial deterministic real-time [1] networking in the changing environment of I4.0, and could eventually replace existing industrial Ethernet protocols.

3 Concept

3.1 Test Environment

Within the project ‘intelligent sensor systems for self-optimising production chains’(iSensPK), which is explained in greater detail in [2], the utilization of existing sensor technologies for CPPS is investigated. Based on a smart force/torque testing facility for various gripper systems in the area of fibre composite handling, the implementation of sensors with different properties and communication interfaces can be tested. The facility itself serves as an entity to simulate gripping scenarios, in which multiaxial stress states can be recorded during a test case. Each testing scenario or, more specifically, the resulting operational limits can be assigned to a specific gripper and consequently contribute to an autonomous decision-making with respect to handling processes within the automated production. Based on the data obtained from several test cases, entities can reject their use if the gripper does not fulfil certain parameters and consequently prevent misuse that could cause problems in the workflow. One current problem is the inaccuracy of measurement data recorded during test scenarios, as it depends on physical properties of the gripped object as well as on environmental parameters, such as temperature or humidity. Therefore, in order to exclude environmental influences on the result of the measurement, the testing facility shall be equipped with an environmental recognition system. This system enables the testing facility to perform autonomous self-calibration tasks and thus serves to compensate measurement inaccuracies. It is designed by linking various low-cost sensors that exchange data on the basis of the Ethernet standard by applying interoperable communication technologies. The following section, therefore, focuses on the communication ability of entities and, furthermore, proposes a communication architecture for the holistic process communication, which is freely scalable and transferable to other use cases.

3.2 Communication Architecture

According to the CP classification (see Sect. 2.1), only components rated as CP44 meet the requirements on I4.0 applications. However, if a device is at least capable of communicating passively, it can be considered as a potential I4.0 device [10]. In this section, the integration of a low CP classified temperature sensor as part of the aforementioned environment recognition system within a CPPS is described. The sensor provides a I^2C bus interface and can additionally be identified within the bus system through a register address. As a consequence, it can generally be considered as a CP33 device, as it is capable of active communication and is furthermore individually known in this subsystem.

The objective of this project is to surround the temperature sensor with an administration shell, as illustrated in Fig. 3, in order to ensure I4.0 conformity with communication and presentation. In this example, the AAS, or more precisely, its physical representation, is implemented on a single-board computer that is connected to the sensor. The AAS body contains several submodels that perform various tasks, such as connecting the asset to the I4.0 environment or acquiring data across the sensor's lifespan. As this asset only provides limited computing resources, which is therefore insufficient for the implementation of an OPC UA server, the MQTT protocol is used to share information in an I4.0 oriented manner. Here, the MQTT client provides a common M2M communication technology that can thus be applied to communicate with peripheral devices within the network by applying the publish–subscribe pattern. But as this protocol may not fulfil certain requirements on process reliability, the OPC UA technology shall provide the major interface to connect all entities directly into an information processing network by applying a hierarchical object model. The bridge between both technologies is realized through a client that passes data, distributed by the MQTT broker, to the OPC UA server. Generally, both protocols can be used to distribute data over the network, from field level devices up to cloud service applications.

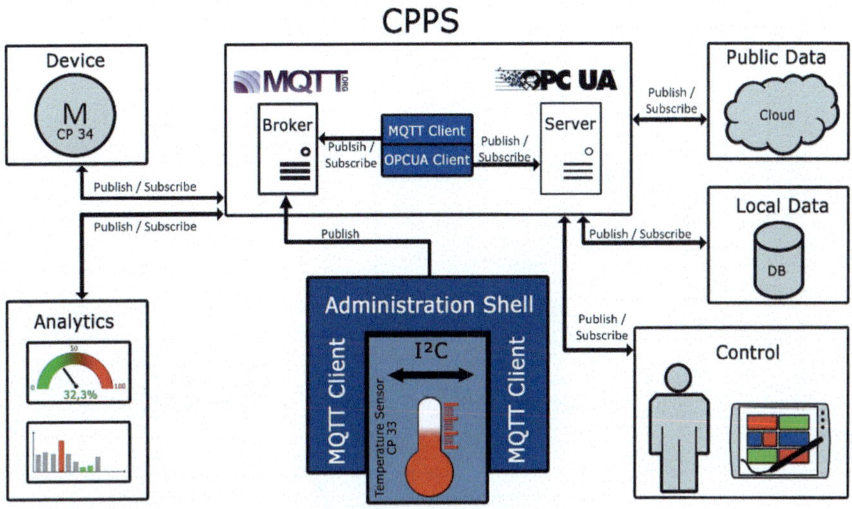

Fig. 3 Communication architecture, [16–18]

3.3 *Use Case—Holographic Process Visualization*

Besides technical entities within a cyber-physical production system, the human as a decision maker and problem solver plays the pivotal role in a constantly changing production environment. As part of a CPPS, he is required to collaborate effectively with the system, as he is the only actor being capable of detecting and eliminating malfunctions in the process flow. As a consequence, the human–machine interface (HMI) represents a key instrument for illustrating and influencing incidents within processes, and is thus an essential aspect to be taken into account as part of the holistic process communication.

Therefore, the project 'Holographic Process Visualization', started in 2017 at the TH-Wildau, addresses the utilization of augmented reality as a human–machine collaboration tool, for example in order to visualize incidents where they occur and thus enabling users to react quickly to changing circumstances. In contrast to conventional HMI tools, holograms that are projected via smart glasses are used for process monitoring and controlling tasks, supporting users to rapidly recognize location-dependent relations. In this context, a fully automated and decentrally controlled assembly line serves as an example. Here, each module is independently operated, whereby the information exchange amongst them is realized by use of the OPC UA protocol. Based on the presented communication architecture, which has been described in Sect. 3.2, the smart glasses are integrated within the system and thus serves as a human–machine interface. With the aim of interfacing with the assembly process, a MQTT client is implemented on the glasses for establishing a communication to a broker located within the network, as illustrated in Fig. 4. Furthermore, the data transfer between both protocols is realized through an independently operating

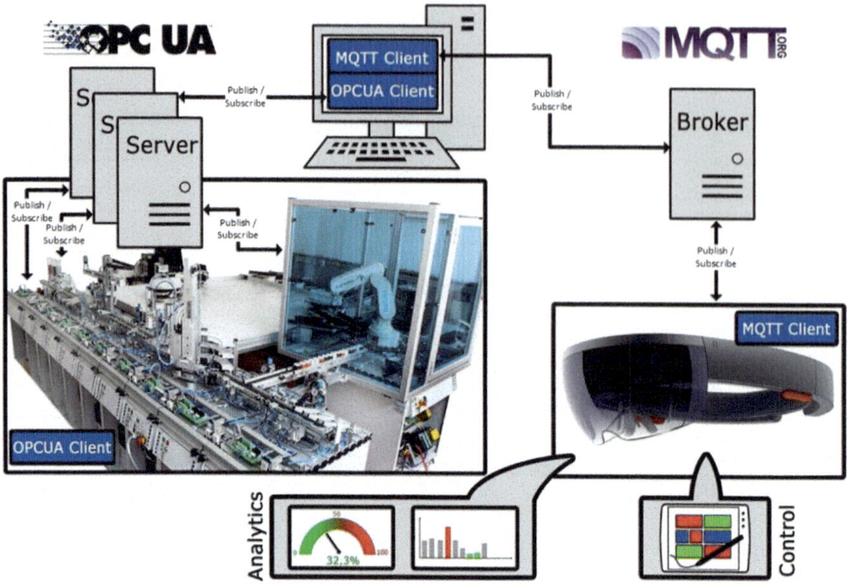

Fig. 4 Holographic process visualization, [16, 17]

application that allows to access process data by applying the MQTT protocol. The advantage of this approach resides in the small code footprint [12], which reduces the communication overhead and thus enables data exchange between the AR smart glasses and peripheral devices while using low computing resources. Moreover, this communication architecture allows a wide range of entities to interact with the production line by using one of the two protocols.

In the presented case, the AR smart glasses can be regarded as a CP44 device, since it manages itself within the IT System and deploys I4.0 conform communication technologies. It can directly access process information of the assembly line and subscribe for particular status messages, such as occurring alarms or malfunctions. Each module, in turn, provides certain functionalities to other network participants. For example, a module can be switched to set-up mode or deliver data in relation to its energy consumption. If a production station detects a malfunction or is running out of production resources, it can automatically request human intervention. Through the use of augmented reality, users can interact with the production line and respond to unexpected incidences.

4 Conclusions

The ability of interoperable communication between devices in the automation sector and beyond is becoming increasingly important. Consequently, certain standards must be determined and implemented within CPPS. The two communication standards OPC UA and MQTT represent very useful technologies in their respective areas but cannot replace each other, since each protocol serves different demands on intercommunication. OPC UA is already widely used in the industry as it provides decent security features as well as a hierarchal object model [15]. In contrast, the advantage of MQTT is the low communication overhead and the associated small code footprint, which enables devices with limited processing resources to interact with other entities in a simple and interoperable way. Furthermore, it improves the communication performance in unreliable network conditions. Even though both technologies reach their limits with respect to hard real-time communication under consideration of low-latency networking, they already cover a large spectrum of industrial applications. Just in the same manner, in which the communication capability of an asset is a significant factor in terms of I4.0 applications, its administration within the cyber-physical network is a further prerequisite in order to provide asset-specific data and functionalities to the information world. In this respect, the concept of the asset administration shell deserves more detailed consideration in terms of organizing functions and information of entities. Both the communication ability and the level of awareness of components are major aspects in the realization process of Industry 4.0 and play the key role in the implementation of cyber-physical production systems.

References

1. Watson, V., Tellabi, A., & Sassmannshausen, J. (2017). Interoperability and security challenges of Industrie 4.0. In *Informatik 2017*, Chemnitz.
2. Phillips, R., & Schröder, C. (2017). The challenges of industry 4.0 for small and medium-sized enterprises.
3. Wortmann, F., & Flüchter, K. (2015). Internet of things—technology and value added, Springer, Wiesbaden.
4. von Lipinski, R., Richter, M., & Reiff-Stephan, J. (2014). Intelligent sensor systems for self-optimising production chains. In *Proceedings of the 1st International Conference and Exhibition on Future RFID Technologies*, (pp. 115–125).
5. Guturu, P., & Bhargava, B. (2011). Cyber-physical systems: A confluence of cutting edge technological streams. In *International Conference on Advances In Computing and Communication*, (Vol. 138).
6. Bauernhansl, T., ten Hompel, M., & Vogel-Heuser, B. (2014). Industrie 4.0 in Produktion, Automatisierung und Logistik, Springer Viewig, Wiesbaden.
7. Weinert, N., & Plank, M. (2017). *Metamorphose zur Intelligenten vernetzten Fabrik*. Berlin: Springer Viewig.
8. DIN SPEC 40912—Core models—Specification and Examples (2014).
9. ZVEI (2016). Structure of the administration shell—Continuation of development of the reference model for the Industrie 4.0 component, ZVEI, Frankfurt a.M.

10. Epple, U. (2016). Status report, Industrie 4.0—technical assets, RWTH Aachen University, Düsseldorf.
11. ZVEI—Zentralverband Elektrotechnik und Elektroindustrie e.V (2017). Examples of the asset administration shell for Industrie 4.0 components, ZVEI, Frankfurt a.M.
12. Hillar, G. C. (2017). *MQTT essentials—A lightweighted IoT protocol*. UK, Birmingham: Packt Publishing Ltd.
13. Jeschke, S., Hess, F., & Isenhardt, I. (2014). *Automation, communication and cybernetics in science and engineering 2015/2016*. Irbid: Springer.
14. Lachello, L., Wratil, P., & Meindl, A. (2017). *Industrial ethernet facts* (3rd ed.). Fredersdorf: Ethernet Powerlink Standardization Group.
15. Deiretsbacher, K., & Mahnk, M. (2015). OPC-UA technology in detial. In *OPC Unified Architecture—Interoperability for Industrie 4.0 and the Internet of Things*, OPC Foundation, Scottsdale.
16. RTA automation: OPC UA Logo. Retrieved November 11, 2017 from https://www.rtaautomation.com/wp-content/uploads/opcualogo.png.
17. IOTLabs: MQTT Logo. Retrieved November 11, 2017 from http://iotlabs.org/wp-content/uploads/2016/12/mqttorg.png.
18. Microsoft: HoloLens. Retrieved December 1, 2017 from https://img-prod-cms-rt-microsoft-com.akamaized.net/cms/api/am/imageFileData/RE1r4TL, last accessed 2017/12/1.

Achieving Syntax Interoperability of Existing Traffic Controllers and Intelligent Transport Systems

Milos Ivanovic and Zoran Marjanovic

Abstract This paper presents a solution for achieving syntax interoperability between different systems. It is necessary to ensure that one system will accept data from the other systems and so to ensure efficient data exchange to work together, without changing the exchanged information. The goal of this research is to present a concrete solution of achieving syntax interoperability between existing traffic controllers and control traffic management system using DATEX II standard for Europe, based on Open Traffic Systems 2 protocol communication. The exchange of data and their intelligibility between the systems using the standard is the key for interoperability. The aim of this paper is to present how to convert binary format of data in any format which other traffic management system can understand. Syntax interoperability deals with the agreed data formats of communication so that machine-to-machine (system to system) communication can be parsed and understood by each machine (system). To achieve syntax interoperability with intelligent transport systems, it requires integrations at the following three levels: [1]

1. *Field to field* (roadside ITS station to roadside infrastructure)
2. *Field to vehicle* (roadside ITS station to vehicle ITS station)
3. *Field to command center* (roadside ITS station to back office)

In this paper, research will be focused on achieving syntax interoperability at the three levels previously mentioned. The solution is based on converting one data format (binary) to another data format (XML, etc.) which understood by systems.

Keywords ITS · Interoperability · Metamodeling · Traffic controller · Mediator

M. Ivanovic (✉) · Z. Marjanovic
Faculty of Organizational Sciences, Belgrade, Serbia
e-mail: smajhel.m@gmail.com

K. Popplewell et al. (eds.), *Enterprise Interoperability VIII*, Proceedings of the I-ESA Conferences 9, https://doi.org/10.1007/978-3-030-13693-2_35

1 Introduction

The study of this paper refers to the realization of communication between different transport systems. At one side, there are existing traffic controllers that regulate the flow of traffic at a certain intersection and, on the other side, are traffic management systems installed in the command center. The aim is to connect these two heterogeneous systems by safely and accurately exchanging information that is important for the functioning of the complete system. Intelligent transport systems (ITS) are now expanding and are striving to integrate new technologies into traffic management systems. However, there are already devices that regulate traffic at intersections around the world, and there are very large investments to replace devices with new ones. This is the most cost-effective solution which is to integrate the existing infrastructure with new technologies to realize intelligent transport systems. The way of this integration is to achieve interoperability between such systems that will be presented in this paper.

The research in this paper deals with problems of interoperability in traffic management systems. Specifically, the emphasis would be on intelligent transport systems (ITS). Interoperability in intelligent transport systems is a complex problem. Intelligent management of transport systems is still under development and will only be fully implemented in a few years, for the simple reason that existing systems should be integrated and adapted to new technologies applied in the transport infrastructure.

ITS interoperability should focus on effectively connecting devices and systems. Interoperability in such a traffic management system will be more important than ever since it seeks to connect communication between vehicles in traffic and communication of vehicles with transport infrastructure. Standards and architectures must continue to evolve in terms of technological advancements to maintain the interoperability and compatibility of an intelligent traffic management system [2].

First chapter of this paper is introduction to the research problem which will be presented in this paper. Second chapter of this paper focuses on four main aspects of ITS interoperability: technical, functional/logical, contractual/institutional and procedural. Third chapter of this paper presents the solution of achieving syntax interoperability existing traffic controllers and ITS. Solution is based on idea to make an engine which will convert one data format (e.g., Binary) of message to any other data format (e.g., XML).

2 Main Aspects of Intelligent Transport Systems Interoperability

2.1 Technical Interoperability

Technical interoperability is usually associated with hardware/software components, systems, and platforms that enable machine-to-machine communication to take place. This kind of interoperability is often centered on (communication) protocols and the infrastructure needed for those protocols to operate. Interoperability becomes an issue if a system is composed of both fixed and mobile subsystems. For example, onboard units in vehicles that travel across borders must be able to communicate with roadside equipment at different geographic locations.

So, the users want to use their transponders (tags) to pay toll road in every town at the national and transnational level. It should be achieved next tasks [3]:

1. Security mechanisms between RFID readers and tags (dedicated short-range communication protocol stack)
2. Use of appropriate standards (*The standardization is the key for interoperability*). If various systems use appropriate standards in communication, it will be easy to integrate increasingly systems and components.

2.2 Functional/Logical Interoperability

Functional reference system (functional spec, specs, functional specifications document (FSD), and functional requirements specification) in systems engineering and software development is a document that specifies the functions that a system or component must perform (often part of a requirements specification) (ISO/IEC/IEEE 24765:2010).

2.3 Contractual/Institutional Interoperability

Contractual/institutional interoperability requires agreements between network operators about data security, financial transactions, and service levels. This level of interoperability in ITS can be presented in tolling systems, and the aspect of contractual interoperability is being treated now within the EU co-funded project Pilot on Interoperable Systems for Tolling Applications (PISTA). It is the responsibility of each operator to ensure contractual interoperability through negotiations and contracts with other operators (COWI, ECOPLAN, UNIKARL, ISIS, 2003) [4].

2.4 Procedural Interoperability

Procedural interoperability is achieved when common procedures are used by all involved road network operators and by the users. Typical issues are harmonized data dictionaries (for example **NTCIP** in USA or **DATEX** in Europe or common human machine interfaces. [5]

In this paper, achieving syntax interoperability existing traffic controllers and intelligent transport system will be solved by using DATEX II data dictionary for Europe region, which is based on standard Open Traffic System 2 communication protocol (OTS2).

3 Solution for Achieving Syntax Interoperability in Intelligent Transport Systems

There are multiple definitions of interoperability, and Third Generation Partnership Project (3GPP) has defined interoperability as the ability of two or more systems to exchange data and use information. The categories of interoperability are technical, syntax, semantic, and organizational. The previously mentioned interoperability categories play a key role in achieving interoperability between two or more heterogeneous systems. Specifically, the technical point of view is technical interoperability. It represents the infrastructure (HW and SW) system that is needed to allow communication between multiple systems. It should pay attention to the communication protocols used by the system software, to clearly define and implement hardware infrastructure. The goal of this research is achieving syntax interoperability between existing traffic controllers and intelligent transport systems by using DATEX II data dictionary for Europe region, which is based on standard Open Traffic System 2 communication protocol.

Open Traffic Systems (OTS) is [6]:

1. *Successor* of **OCIT*;** origins: cities, municipal areas D/A/CH.*
2. *Open Communication Interface* for Road Traffic Control Systems.
3. *Supported* by the OTS City Association (OCA).
4. OTS main objective: **Interoperability.**

DATEX II—The standard for ITS in European Roads—CEN TS 16157

> DATEX II standard has been developed to provide a standardized way of communicating and exchanging traffic information between traffic centers, service providers, traffic operators and media partners. The specification provides for a harmonized way of exchanging data across boundaries, at a system level, to enable better management of the European road network. [7]

In Fig. 1, it is shown OTS2 system architecture, OTS2 + DATEX II. OTS2 is used as protocol to establish communication between central systems and between central station and roadside stations (e.g., traffic controller, car, VMS, etc.). DATEX II used as data model for exchanged traffic information [6].

A problem refers to the integration of multiple traffic controllers from different manufacturers with ITS traffic management system. Namely, a transport controller controls one intersection by light signals (red, yellow, and green) and with the use of additional symbols, such as arrows, pedestrian, and cyclist. The problem occurs when you are in a city of many different traffic controllers, who are primarily the product

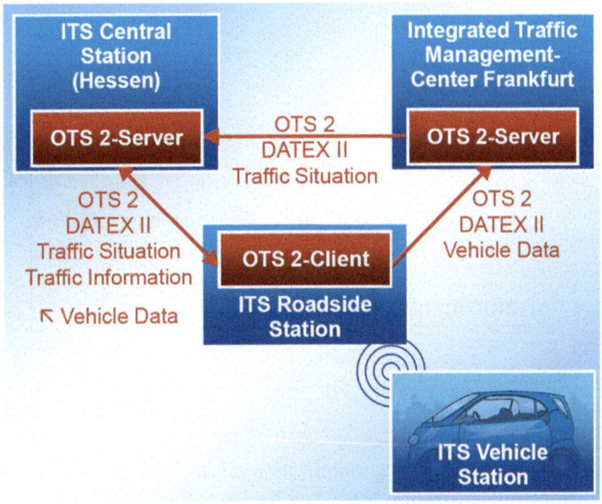

Fig. 1 Solution of OTS2 system architecture [6]

of different manufacturers and working under different protocols of communication. The aim is that all traffic controllers can communicate with each other and with the command center for the management of all controllers.

Benefits of interoperability traffic controllers are: reducing waiting time in traffic, achieving adaptive traffic management, relief roads, achieving the so-called green wave in the corridors, providing quality data for various reports, statistics, and knowledge base upon which they can make key business decisions, and so on.

The interoperability problem of integrating existing traffic controllers with ITS traffic management software (see Fig. 2) will be solved with the solution of achieving syntax interoperability, by using OTS2 standard communication protocol. Yellow circle (see Fig. 2) shows the place where standardization should be implemented as one of the solutions and key for interoperability.

The problem of syntax interoperability is that different existing traffic controllers have own (manufacturer) protocol of communication. Most of them are using serial communication RS232,485 by binary format of data. Intelligent transport systems are using Ethernet, USB, WiFi connection and XML, JSON, HTTP, etc. format of data. So, the aim of this paper is to present how to convert binary format of data in any format of other traffic management system understands. There are couple different messages which are exchanging between these two systems. In this paper, will be shown example of status message of traffic controller, as an example of solution. Every other message can be converted into any format of data on the same pattern as this one.

UML model (see Fig. 3) of status message will be used in this case. This UML model is created according to OTS2 protocol for message device status. OTS2 protocol is based on XML data format, and the aim of this paper is to show how we can convert status message from one data format to another. In this case, will be shown the conversion from binary data format to XML data format. The solution is based on metamodels of status messages. On the one side, we have metamodel of binary

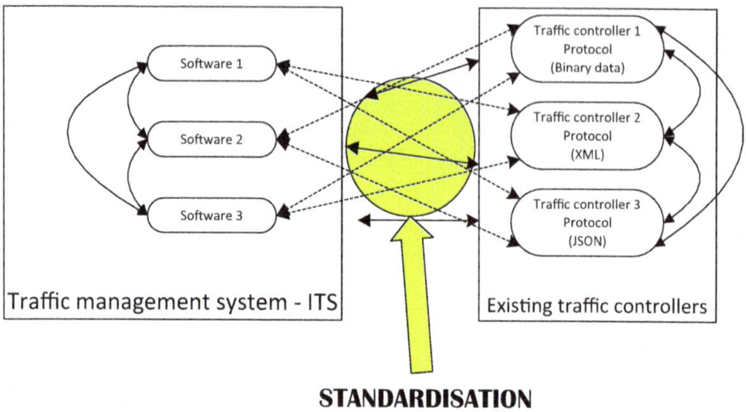

Fig. 2 Interoperability problem between existing traffic controllers and ITS traffic management system (standardization is the key for interoperability)

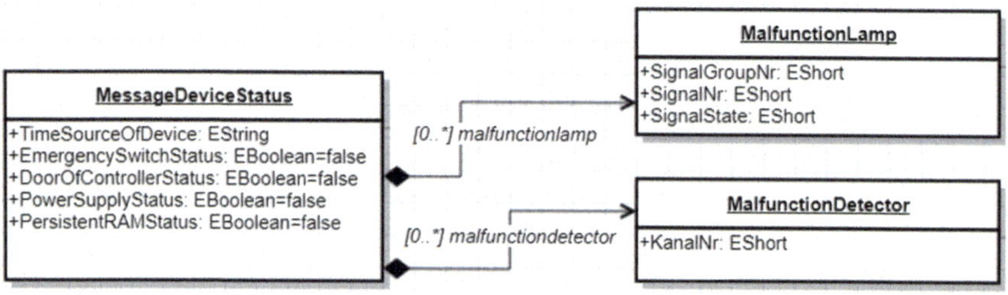

Fig. 3 UML model of status message (OTS2 protocol)

data format and on the other side metamodel of XML data format. With template engines frameworks (e.g., PHP symfony), we can make engine for model-to-model (M2M) transformation based on these metamodels. Key function in this engine is a recursive loop because of existing nodes in metamodels.

According to OTS2 system architecture (see Fig. 1), the architecture solution for problem of achieving syntax interoperability of existing traffic controllers and ITS is presented (see Fig. 4).

The idea is that OTS2 is a mediator component between traffic management softwares and existing traffic controllers. So, the whole logic of conversion messages from one system to another should be on OTS2 server. OTS2 server receives messages from all traffic controllers using appropriate way of link (Ethernet, serial communication (RS232,422,485), etc.) and converts them to OTS2 data protocol format (XML) per OTS2 protocol specification [8, 9], and proceeds to traffic management softwares. Template engine (e.g., Symfony PHP [10]) is a good framework solution for transforming one data format to another using metamodel structure of message.

Most of existing traffic controllers ("old fashion") have binary format of exchanging data. In example (see Table 1.) is presented the structure of binary data format

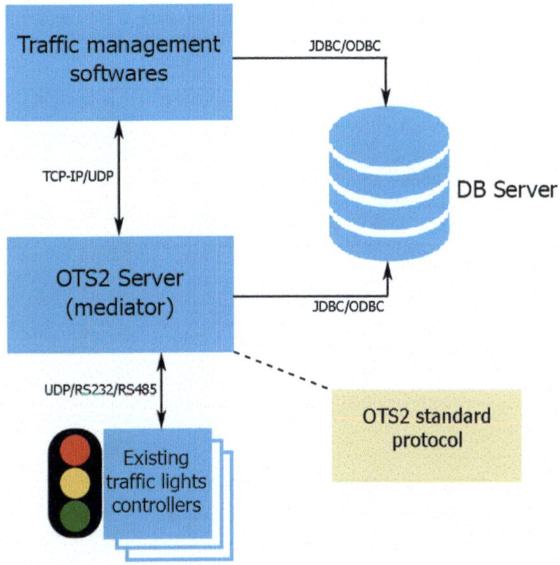

Fig. 4 Architecture solution for problem of achieving syntax interoperability of existing traffic controllers and ITS

with description of attributes and the length of them in bytes according to the type (Int, String, Boolean, list, etc.). Parsing binary data format is based on attribute positions and their lengths to pass through a series of bytes in the easiest way and take the appropriate values.

Example of status message in binary format is based on Table 1:

0x00 0x00 0x00 0x04 0x00 0x01 0x00 0x00 0x0D 0x00 0x01 0x00 0x03 0x00 0x01 0x00 0x01 0x00 0x03 0x00 0x02 0x03 0x00 0x01

Researching various references to find a solution in terms of transformation of messages from the initial format to a desired output format, I concluded that the classic model-to-model transformations (QVT, etc.) do not make much sense in this case of messaging. Because messages are not classic models presented through some nodes and connections. What makes sense is to introduce a model that would abstractly describe the messages and through which the transformations would be presented (see Fig. 5).

In this approach, based on the transformation specification, a code would be generated that would communicate between two heterogeneous systems, i.e., the transformation would be compiled and executed from a component (mediator). This implementation will be implemented in next six phases.

Table 1 Binary message data format of existing traffic controller

Name	Type	Length	Params
Time source	Int32	4 bytes	**0x03 – LAN** **0x04 - GPS**
Emergency Switch status	Bool	1 byte	**0x00 – true** **0x01 - false**
Door status	Bool	1 byte	**0x00 – true** **0x01 - false**
Power Supply status	Bool	1 byte	**0x00 – true** **0x01 - false**
RAM status	Bool	1 byte	**0x00 – true** **0x01 - false**
Malfunction lamps	List	1+(n* Lamp)	**Byte 1: 1+n*6** **Byte 2,3:** (short16) Lamp n – SignalGroupNr **Byte 4,5:** (short16) Lamp n – SignalNr **Byte 6,7:** (short16) Lamp n - SignalState
Malfunction detectors	List	1 +(n*Det)	**Byte 1: 1+n*2** **Byte 2,3:** (short16) KanalNr

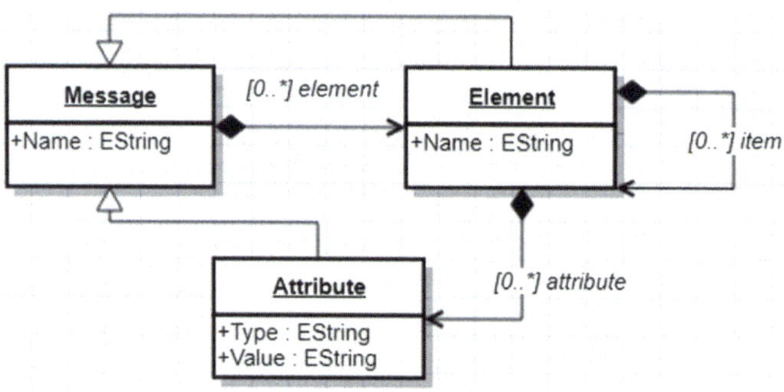

Fig. 5 Metamodel for abstract data structure (message)

3.1 I Phase: Defining the Mapping of the Abstract Structure of Messages to a Concrete Structure

Abstract data structure entities (Status message, Lamp, Detector):

Malfunction lamp (signal): DEF_TREE MFL (SignalGroupNumber: INT, SignalNumber: INT, SignalState: INT); *Malfunction detector*: DEF_TREE MFD (KanalNumber: INT);	*Status message traffic controller*: DEF_TREE SPSU (TimeSourceOfDevice: INT, EmergencySwitchStatus: BOOL, DoorOfControllerStatus: BOOL, PowerSupplyStatus: BOOL, PersistentRAMStatus: BOOL, MalfunctionLamps: {TREE MFL}, MalfunctionDetectors: {TREE MFD});

3.2 II Phase: Creating Metamodels for Abstract Data Structures (Message)

In Fig. 5, is shown metamodel for abstract data structure of message which is result from preview I phase, where has been defined mapping of the abstract structure of message to a concrete structure.

3.3 III Phase: Instantiation of Metamodel for Example of Status Message of Traffic Controller in the Eclipse Modeling Framework

In Fig. 6, is shown instantiation of metamodel for example of status message of traffic controller in Eclipse Modeling Framework. Elements and attributes for this instanced metamodel have been used from first phase for status message.

3.4 IV Phase: Defining Metamodel for Mapping an Abstract Structure into a Physical Structure (Conceptual Solution)

In Fig. 7, is shown the conceptual solution of defining metamodel for mapping an abstract structure into physical structure. Tree node is on high level and other nodes are on lower levels which is included in tree node or extend it.

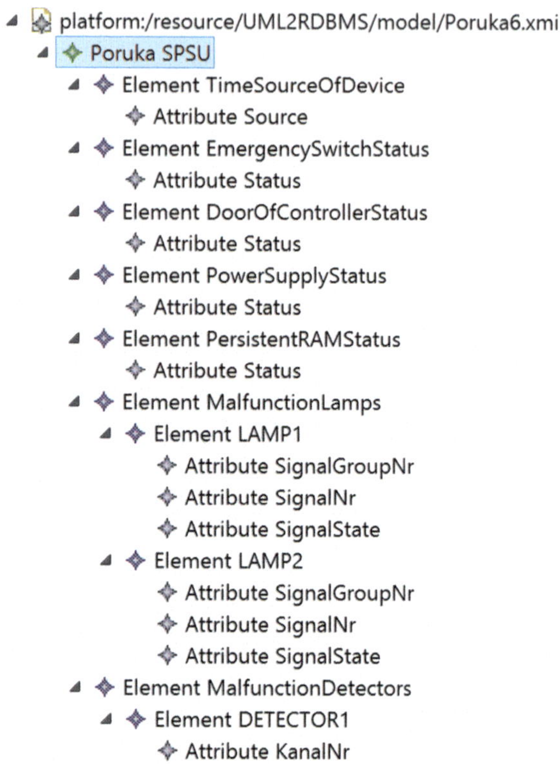

Fig. 6 Instanced metamodel for status message in Eclipse Modeling Framework

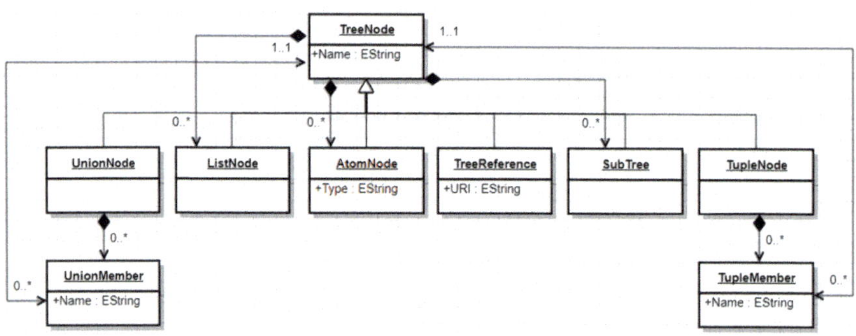

Fig. 7 Metamodel for mapping abstract structure to physical (conceptual solution)

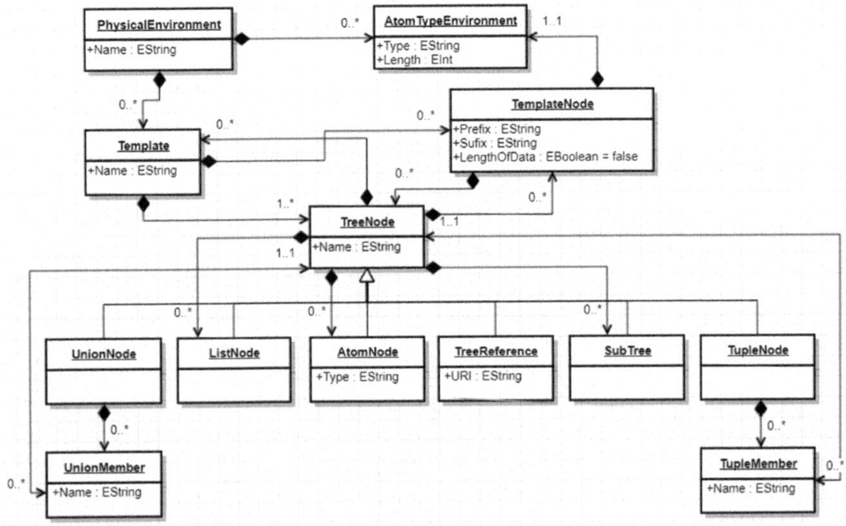

Fig. 8 Metamodel for mapping abstract structure to physical

3.5 V Phase: Creating Metamodel for Mapping an Abstract Structure to a Physical Structure

In Fig. 8, is shown metamodel for mapping abstract structure to physical structure. The conceptual solution from preview chapter has been extended with Template node which contains key attributes of prefix, sufix and length of data. With these attributes we can create the elements in structure with appropriate data types of each one.

3.6 VI Phase: Generating Application Which Will Convert Message from One Data Format to Another Data Format

Based on metamodel schema (see Fig. 8), is created Twig template engine in Symfony PHP. This template is parsing the message binary array and goes through all nodes using the recursion function. The input in this engine is message in binary format, in which this engine parses creating the output code. The output of this code is PHP code which creates all entities (objects) from binary array. At the same way, we can create any type of programming code like PHP in this example. From this output code is easy to make output data format (XML, JSON, etc.).

Output PHP code or other programming language from Twig template engine in symfony PHP is the result of making programmer code which can be compiled and run on the selected platform. This engine can make the code from other programming language OTS2 server needs. This code makes output data format, for example XML data format, or other data format needs.

4 Conclusion

This paper presents selected problems of intelligent transport systems (ITS) interoperability. The ITS interoperability is a complex problem. The solution of achieving syntax interoperability is given through converting one data format of message to another (binary data format to OTS2 XML data format) using Twig template engine in Symfony PHP [10]. Benefit of this research is to present the way how to achieve interoperability by using the DATEX II standard with OTS 2 protocol because the standardization is the key of interoperability. Complex problem of interoperability between existing systems and other heterogeneous system is that they do not have the same format of message in communication. This paper presents the solution of how to make an engine to produce the programmer code which will compile and convert one data format to another. Using standardization (DATEX II standard) in way of achieving interoperability is an important action in this implementation. We do not need to replace the existing systems with newer, and the idea is to keep the existing systems and equipment and to integrate with newer systems.

References

1. Austroads, Research report AP-R458-14, C-ITS Interoperability with existing ITS infrastructure (April 2014).
2. Luras, M., Zelm, M., Archimede, B., Benaben, F., & Doumeingts, G. (2015). *Enterprise interoperability*. London: ISTE Ltd.
3. Festag, A. (2015). Standards for vehicular communication—from IEEE 802.11p to 5G.
4. Mertner J., & Skov, K. M. (2005). Harmonisation and interoperability of national tolling schemes.
5. World Road Association mondiale de la route. https://rno-its.piarc.org/en/rno-basics-integrated-operations-systems-integration/interoperability. Last visited in August 2017.
6. Schön, T. (2010). OTS 2 Standardisation, Datex II Forum Berlin.
7. Datex II standard for ITS in Europe. http://www.datex2.eu/sites/www.datex2.eu/files/Datex_Brochure_2011.pdf. Last visited in August 2017.
8. Schön T. (2010). OTS 2 Protocol specification, Bultmann.
9. Schön, T., & Haftmann A. (2013). OTS data types for traffic controllers.
10. Symphony PHP. https://symfony.com/. Last visited in August 2017.

An Ontology Framework for Multisided Platform Interoperability

Quan Deng, Suat Gönül, Yildiray Kabak, Nicola Gessa, Dietmar Glachs, Fernando Gigante-Valencia, Violeta Damjanovic-Behrendt, Karl Hribernik and Klaus-Dieter Thoben

Abstract A successful B2B marketplace must ensure that suppliers and producers in a supply chain can find each other, communicate and negotiate in an effective way, while performing business processes. To this, we present an approach that involves two core ontology modules, e.g., the Catalogue Ontology and the Business Process Ontology, which can be extended by adding specific domain ontologies. For the representation of certain business aspects, the Catalogue Ontology exploits the Universal Business Language (UBL), while for the description of product characteristics related to different domains, this ontology makes use of the relevant industrial standards (e.g., the furniture ontology is based on the FunStep ISO 10303-236 standard and the eClass ontology is based on eCl@ss standard). The Business Process Ontology encompasses machine readable vocabularies for the semantic description of business processes and could be extended by adding new ontologies or data schemas. Finally, we validated the design and functionality of the ontology framework by defining and performing a set of queries related to product and services retrieval.

Keywords Interoperability · Multisided platforms · Ontologies · Taxonomies

---

Q. Deng (✉) · K. Hribernik · K.-D. Thoben
BIBA—Bremer Institut für Produktion und Logistik GmbH, Hochschulring 20, 28359 Bremen, Germany
e-mail: dqu@biba.uni-bremen.de

University of Bremen, Bibliothekstraße 1, 28359 Bremen, Germany

S. Gönül · Y. Kabak
SRDC Yazilim Arastirma, Gelistirme ve Danismanlik Anon. Sti, Ankara, Turkey

N. Gessa
ENEA, Via Martiri di Monte Sole 4, 40129 Bologna, Italy

D. Glachs · V. Damjanovic-Behrendt
Salzburg Research, Jakob Haringer Str. 5/II, 5020 Salzburg, Austria

F. Gigante-Valencia
AIDIMME, Benjamin Franklin Str. 13, 46980 Paterna, Valencia, Spain

K. Popplewell et al. (eds.), *Enterprise Interoperability VIII*, Proceedings of the I-ESA Conferences 9, https://doi.org/10.1007/978-3-030-13693-2_36

1 Introduction

Multisided platforms are characterized by interactions and interdependence between multiple groups of surrounding organizations [1]. The main aim of this paper is to design an ontology framework for enhancing semantic interoperability of cloud-based, multisided platforms and their instances operating under various regulatory norms, business rules and in heterogeneous working domains. Our use case study is built on practical experiences gained during the implementation of a European research project NIMBLE (Grant Agreement No. 723810), whose aim is to create multisided platform for collaboration, inter-sectoral and cross-sectoral interoperability in the EU. In NIMBLE, we advance conventional enterprise paradigms and allow marketplace participants to communicate with each other and to perform business interactions of various complexities over the platform. Although, in NIMBLE, various marketplace participants operate over the same multisided platform, the problem of heterogeneity in the description of resources, business transactions, documents and contracts to be exchanged between participants remains present on the platform. Hence, this paper addresses an approach for the improvement of the enterprise interoperability by raising the abstraction from domain data models and taxonomies to standardized metamodels. In that context, we discuss here an extensible ontology framework, which consists of two core ontologies: the Catalogue Ontology and the Business Process Ontology. For the semantic representation of business features in the Catalogue Ontology, we use the Universal Business Language (UBL) standard, while the description of business transactions in the Business Process Ontology is based on Moda-ML (Middleware Tools and Documents to enhance the Textile/Clothing supply chain through XML) framework [2]. The presented ontology framework is extensible for the description of products, services and business transactions in various sectors, e.g., in the furniture sector. A comprehensive report on the NIMBLE Ontology Framework is given in [3].

Paper organization. Section 2 describes our main motivation for the design of the NIMBLE Ontology Framework. Related works are discussed in Sect. 3. Section 4 presents the structure of our ontology framework and its modules. Section 5 demonstrates a specific use case in furniture sector, for which we demonstrate an advanced query-based reasoning system, combining SPARQL query filters and the results obtained via Apache Marmotta Linked Data platform. Finally, Sect. 6 draws conclusions and states overall future steps.

2 Motivation

Our motivation for the design of the NIMBLE Ontology Framework relies on an attempt to create a scalable knowledge network for ensuring interoperability, information integration and information exchange through business processes. A typical supply chain scenario consists of the following steps:

1. Supplier A publishes a catalogue of one or more products via the platform;
2. Supplier A creates new process model or reuses an established process model that specifies the documents to be exchanged through business interactions;
3. Producer B searches for products via the platform, e.g., *"Who in Spain can deliver product X which has a feature F, in the next Y days, at price P?"*;
4. Producer B obtains the search results and selects supplier A as the best match;
5. Producer B initiates a business transaction with supplier A and exchanges specified business documents.

It is common that companies describe their products using different data models and vocabularies that relates to a specific sectoral knowledge. However, the lack of common structures and/or vocabularies for product description results in interoperability and findability issues. One possible solution for the improvement is to use XML Schemas Definitions (XSD) and define common data structures with user-preferred vocabulary. However, XSD cannot enable semantic interoperability, which calls for ontologies and their formal specifications of common vocabularies to be employed for the description of enterprise domains. The NIMBLE Ontology Framework addresses search issues with an extendable Catalogue Ontology and improves interoperability in business interactions using a specifically created Business Process Ontology. The Catalogue Ontology supports publishing and searching of products with fine-grained technical (e.g., percentage of volatile organic compound in furniture sector) and commercial details (e.g., delivery, pricing). The Business Process Ontology allows the description of business transactions for different enterprise sectors from various aspects, e.g., *behavioral* (the order of execution of activities), *organizational* (business roles and entities in the business process) and *document consumption aspects* (data exchanged in business activities) [4].

3 Related Work

For the description of offered resources in an e-commerce platform, various attempts were undertaken so far to provide syntactic and semantic interoperability for B2B systems and services, for example, the international product and service classification standard eCl@ss with its transformation into eClassOWL ontology [5]; the lightweight ontology GoodRelations, initially used for describing offerings of goods and commodity services on the Web [6] which today covers many B2B aspects, including Web resources, offers, prices, terms and conditions, etc. However, many important B2B concepts, such as business entity, delivery, warranty and payment, are not sufficiently detailed yet to effectively support enterprise interoperability. Meanwhile, the UBL standard (ISO/IEC 19845:2015) provides a free library of standard XML business documents for e-commerce [7]. UBL covers concepts such as `Address`, `Item`, `Payment`, `Delivery`, `Warranty`, which are used to describe offers of various resources in an e-commerce platform. As UBL is defined in XSD format, it cannot express semantic relationships among business concepts.

In addition, UBL contains many elements which are not needed for the resource description. Some efforts have been taken to tailor UBL schemas to UBL catalogue [8] or transform UBL schemas to OWL format [9]. However, little has been done to derive a practical Catalogue Ontology from UBL, which can be extended with additional product taxonomies for the description of various offered resources.

For the purpose of business process descriptions, business process modeling languages have been developed, e.g., Business Process Model and Notation (BPMN) [10], Petri Nets [11]. Different business process modeling ontologies have been proposed for bridging semantic gaps in various business process models [12–14]. These ontologies are too generic and have little concentration on the description of business transactions in B2B marketplace for different enterprise sectors. Some efforts have been made in specific sectors. For example, in textile/clothing sector, Moda-ML [2] is developed as a vertical standard for data exchange. The Moda-ML Business Process Ontology defines concepts (vocabulary terms) and exchange documents related to activities in textile sector [15]. The defined shared concepts support the development of business transactions models, which should be understandable to all entities interacting via the platform.

In short, there is a lack of practical extensible ontology framework, which can provide sufficient support for resource discovery and business transactions in B2B marketplace for different enterprise sectors, while performing business processes.

4 NIMBLE Ontology Framework

In context-aware systems, business processes can be automated as long as there is a common agreement on knowledge behind those processes and their context. The NIMBLE Ontology Framework is a semantic collection of domain-specific concepts, e.g., furniture taxonomy, Moda-ML Business Process Ontology. As illustrated in Fig. 1, the core NIMBLE Ontology Framework consists of Catalogue Ontology and Business Process Ontology. Catalogue Ontology enables semantic publishing and searching of products and services, and business transaction executions are controlled by the concepts defined in Business Process Ontology.

The NIMBLE Ontology Framework can be extended by introducing new ontologies encompassing various domains. For example, eClass can be used to specify the description of individual resources, while the furniture sector taxonomy enables description of resources in the furniture industry. While the core ontology must be pre-filled to drive the main functionality of the NIMBLE platform, new domain-specific extensions can be added at any time by means of the NIMBLE platform.

The relationships between the ontologies in the NIMBLE Ontology Framework are illustrated in Fig. 2. The ontologies will be populated through the provision in the NIMBLE platform of services that allow the ingestion of product catalogues offered by the participants to the platform.

Relevant concepts, their properties and relationships in the ontology framework are maintained in the semantic metadata repository. In NIMBLE, this repository is

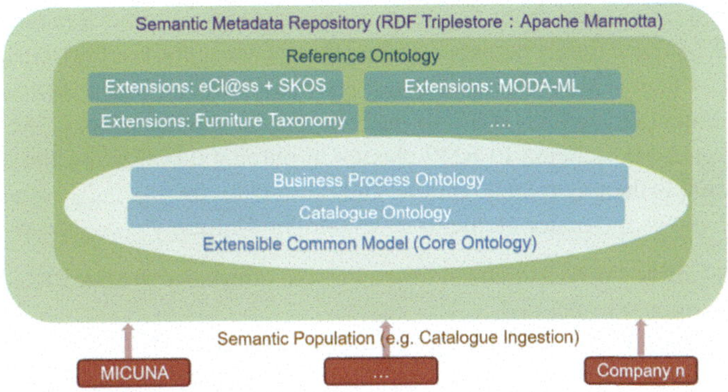

Fig. 1 Structure of the NIMBLE Ontology Framework

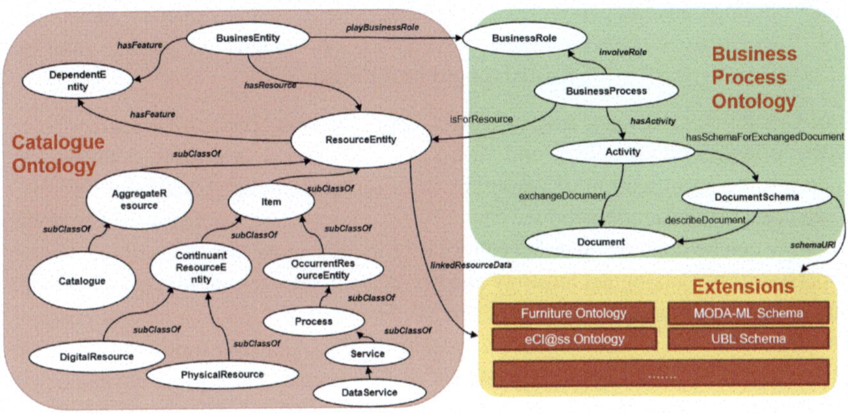

Fig. 2 NIMBLE ontology framework: classes and properties

driven by Apache Marmotta (http://marmotta.apache.org), which is an open source implementation of the Linked Data Platform (LDP) (https://www.w3.org/TR/ldp/).

4.1 Catalogue Ontology

Catalogue Ontology in NIMBLE is based on UBL, a worldwide standard providing a royalty-free library of XML business documents used in supply chain operations [16]. UBL covers various concepts in cross-sector use cases, including concepts for the description of companies, persons, catalogues, products, product properties, delivery terms, trading terms, etc.

In order to have Catalogue Ontology, we firstly presented a mapping between the UBL concepts and the concept in use cases in the NIMBLE project. Secondly, we adopted the concepts from GoodRelations ontology [6] and Svekatalog UBL Catalogue 2.1 [8] and finally categorized the relevant concepts. For the transformation of the UBL schema into Catalogue Ontology, we used Ontmalizer (https://github.com/srdc/ontmalizer), which is a tool that transforms concepts (e.g., `Catalogue` and `Item`) in XML schema to RDF classes in ontology. Following the automatic transformation is then manual adaptions and optimizations.

Figure 2 shows the major elements of Catalogue Ontology:

- **Business Entity** can be a legal Party or a Person, offering some resources or taking part in business transactions.
- **Resource Entity** is a product/service that is held by a Business Entity. Each Resource Entity has resource-specific characteristics or properties, e.g., price.
- **Dependent Entity** is derived from `Dependent Continuant Entity` of the Basic Formal Ontology (BFO) [17], which includes entities that are ontologically dependent on `Independent Continuant`. Examples of `Dependent Continuant Entity` are weight or color, and an `Independent Continuant` could be a tomato.

In Catalogue Ontology, `Dependent Entity` concepts can be used to specify the `Business Entity` and `Resource Entity` in a more detailed way. To enrich the specification of resources in different domains, Catalogue Ontology is extended with domain product category taxonomy. The extension of Catalogue Ontology can be done using either inheritance mechanisms, which combine product properties from Catalogue Ontology and product category taxonomy, or by using the Linked Data mechanisms (shown in Fig. 3). Here, RDF triples connect the *subject* from Catalogue Ontology with the *object* from the extension modules. This way, the specification of the resource instance in Catalogue Ontology is enriched with the instance descriptions in the product category taxonomy. Furthermore, a resource instance may be linked to multiple instances in product category taxonomies.

Beside Linked Data mechanisms, NIMBLE supports abstracting the details of target concepts from users and presenting these details as a flat data struc-

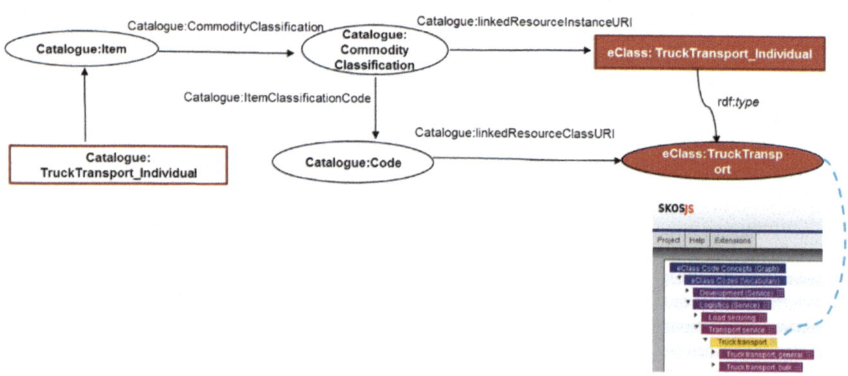

Fig. 3 Example extension with Linked Data mechanisms

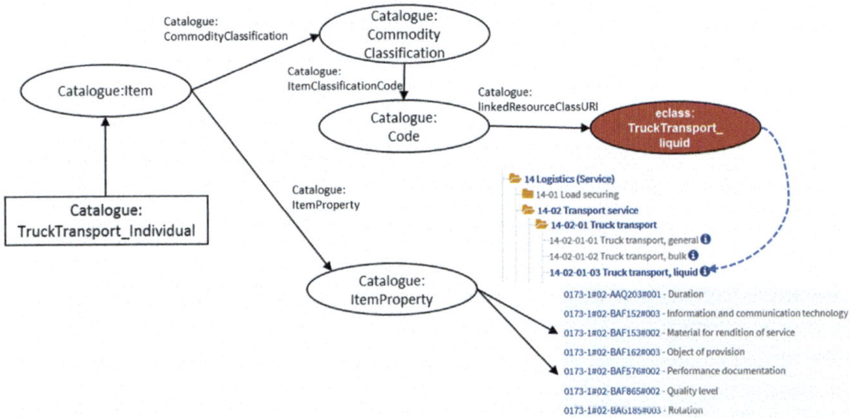

Fig. 4 Example extension based on modified Linked Data extension mechanism

tures so that users need to specify only values for the presented attributes. Figure 4 illustrates this modified approach that links product categories from external taxonomies, e.g., a `Commodity Classification` to the resource instance `Item`. RDF resources describing product categories can be accessed through `linkedResourceClassURI` of `Code` class. Finally, we create `ItemProperties` for those properties specified by all categories that represent the product. Subsequently, each `ItemProperty` has a link to the corresponding property definition.

The current version of the NIMBLE Ontology Framework includes the following extension modules:

1. eClass taxonomy, which is an ISO/IEC compliant industry standard for cross-industry product and service classification;
2. Furniture sector taxonomy, which merges an old release of furniture taxonomy (capturing industrial processes, machinery, techniques and materials used in manufacturing processes, components and product catalogues in furniture industry), and the furniture ontology, which is based on the funStep ISO standard (FunStep ISO 10303-236, see: http://www.funstep.org).

4.2 *Business Process Ontology*

For the description of business transactions in NIMBLE, we reuse the concepts in Moda-ML Business Process Ontology. Moda-ML Business Process Ontology is formalized in OWL (Web Ontology Language) and built around the following concepts: `Process, Actor, Activity` and `Document`. Moda-ML Business Process Ontology contains more than 100 classes of various documents and about 30 types of business transaction activities in textile sector. For the representation of metadata of exchanged documents, the `DocumentSchema` concept has been derived from the `DataSchema` concept in the DRM (Data Reference Model) ontology (http://vocab.data.gov/def/drm).

As shown in Fig. 2, Business Process Ontology in the NIMBLE ontology framework includes the following main concepts:

- **Business Entity** is shared with Catalogue Ontology and has a specific Business Role.
- **Business Role** defines a specific role that can be played in a Business Process, by a Business Entity.
- **Business Process** is a set of structured activities with logical behavior that produce a specific service or product [14].
- **Activity Entity** is a specification of an activity that is carried out with the aim to perform a Business Process. An Activity Entity can be atomic (cannot be split into further detailed activities) or compound (consists of several atomic activities).

In single atomic activity, one and only one **Document** can be exchanged; for example, a purchase order is a document that can be exchanged between the buyer and the seller in a business transaction activity. Documents are described by **Document Schema** with metadata that provides common understanding to different actors. The entities in the document schema are categorized into different types, following the UBL standard; for example, `InvoiceDocumentSchema` contains metadata of invoice documents.

With respect to the high variability of vocabularies and document schemas in specific business transactions, sector-specific business process ontologies should be introduced to extend the current Business Process Ontology. Its extensions can be achieved either using the inheritance mechanisms or Linked Data mechanisms. For example, `SubcontractedFabricManufacturing` process in Moda-ML Business Process Ontology can extend `ProductionProcess` in our ontology using the inheritance mechanisms. Linked Data mechanisms enable reuse of the document schemas that are defined in other standards (e.g., UBL) or by other business entities. For example, document schema `TextileOrderStatusReport` in textile ontology extends the Business Process Ontology using a data property `schemaURI` that specifies the exchange documents in textile transactions.

5 Use Case Study

In the following, we present an example of the NIMBLE Ontology Framework and its use for search and discovery of products and business transactions as a first step toward the complete validation of the project. Here, Catalogue Ontology is extended with a sector-specific furniture ontology, and Business Process Ontology is linked with the products from the furniture ontology.

The following example demonstrates the use of ontology modules for performing a single search request. The data and ontologies used in the following example are available online from: http://nimble-ldp.salzburgresarch.at.

Search request. *"Who in Spain can deliver an un-foldable product from furniture category Cradle, in the next 8 days, at price less or equal to 200 Euro, and can generate the Certification of Origin document as a part of business transactions?"*

By using predefined RDFPath, the search request could be automatically transformed into a SPARQL query. At present, SPARQL query is manually created, defining a search filter with the information on delivery period and delivery unit (e.g., 8 days), price (e.g., less or equal to 200 Euro), country (e.g., Spain) and feature of product is "unfoldable," etc. The search filter in the following SPARQL query produces the output by looking in Catalogue Ontology, sector-specific extension modules, Business Process Ontology and an extension document schema.

```
SELECT  ?supplierName ?supplierTel ?furnitureEAN ?furnitureColor
(concat(str(?deliveryPeriodDecimal), " ", str(?deliveryPeriodUnit)) as ?deliveryTime)
(concat(str(?priceDecimal), " ", str(?currencyID)) as ?price)
(concat(str(?minOrderQuantiyDecimal), " ", str(?minOrderQuantiyUnit)) as ?minOrderQuantity) WHERE {
  ?party nimble_ubl_aggregate:Contact/nimble_ubl_basic:Telephone ?supplierTel;
    nimble_ubl_aggregate:PostalAddress/nimble_ubl_aggregate:Country/nimble_ubl_basic:Name ?countryName;
    nimble_ubl_basic:Name ?supplierName;
    nimble_ubl_catalogue:Catalogue/nimble_ubl_aggregate:CatalogueLine ?catalogueLine;
    nimble_ubl_aggregate:DeliveryTerms/nimble_ubl_aggregate:EstimatedDeliveryPeriod
    /nimble_ubl_basic:DurationMeasure ?deliveryPeriod.
  ?deliveryPeriod nimble_ubl_basic:ValueDecimal ?deliveryPeriodDecimal;
    nimble_ubl_basic:unitCode ?deliveryPeriodUnit.
  ?catalogueLine nimble_ubl_aggregate:GoodsItem/nimble_ubl_aggregate:Item ?item;
    nimble_ubl_aggregate:RequiredItemLocationQuantity ?requiredItemLocationQuantity.
  ?requiredItemLocationQuantity nimble_ubl_aggregate:Price/nimble_ubl_basic:PriceAmount ?priceAmount;
    nimble_ubl_basic:MinimumOrderQuantity/nimble_ubl_basic:ValueDecimal ?minOrderQuantiyDecimal;
    nimble_ubl_basic:MinimumOrderQuantity/nimble_ubl_basic:unitCode ?minOrderQuantiyUnit.
  ?priceAmount nimble_ubl_basic:ValueDecimal ?priceDecimal;
    nimble_ubl_basic:currencyID ?currencyID.
  ?item nimble_ubl_aggregate:CommodityClassification ?externalItemLinke.
  ?externalItemLinke nimble_ubl_basic:ItemClassificationCode
    /catalogue:linkedDataURI/rdfs:subClassOf*  furniture:Cradle;
    catalogue:ItemInstanceURI ?linkedFurnitureItem.
  ?linkedFurnitureItem furniture:isFoldable ?isFurnitureFoldable;
    furniture:hasEAN ?furnitureEAN; furniture:hasColour ?furnitureColor.
 ?bp bp:isForResource furniture:Cradle; bp:hasActivity ?activity.
 ?activity bp:hasDocumentExchangedFromRole/^bp:playBusinessRole ?party;
   bp:hasSchemaForExchangedDocument/bp:schemaURI ?schemaURI.
 FILTER (?deliveryPeriodDecimal <= "8"^^xsd:decimal && ?deliveryPeriodUnit = "DAY"^^xsd:string
    &&  ?priceDecimal <= "200"^^xsd:decimal && ?currencyID = "EUR"^^xsd:string
    &&  ?isFurnitureFoldable = "false"^^xsd:boolean  && ?countryName = "Spain"^^xsd:string
    &&  regex(str(?schemaURI), "CertificateOfOrigin"))}
```

Sample features in catalogue ontology

Sample product features in sector specific extension module

Link to business process ontology

Link to extension document schema

By using sample data, provided in http://nimble-ldp.salzburgresarch.at, the above-presented SPARQL query creates the output, as shown in Fig. 5. It finds that the company "MICUNA S.L." in Spain (this information is based on international calling code, +34) can deliver unfoldable white Cradle in next 7 days, at price 150 Euro, with a minimum quantity of order that is 201 packages.

supplierName	supplierTel	furnitureEAN	furnitureColor	deliveryTime	price	minOrderQuantity
"MICUNA S.L."	"+34 [redacted]	"8431830130773"	furniture White	"7.0 DAY"	"150 EUR"	"201 PK"

Fig. 5 Sample output from the example query

6 Conclusion

The ontology framework presented in this paper is designed to enhance interoperability of activities and transactions performed via the NIMBLE multisided platform. Here, we firstly considered several existing enterprise ontologies, arguing that they are not sufficient for supporting knowledge exchange, interoperability and multisidedness of current and future platform solutions. Starting from this analysis, we designed our ontology framework to support information retrieval and enhance search and negotiation activities via the NIMBLE platform.

The presented ontology framework consists of two ontology modules: Catalogue Ontology and Business Process Ontology, which can be further extended to cover new enterprise domains. Both modules are based on standards for facilitating semantic and cross-domain interoperability. For example, the current knowledge extension of the NIMBLE Ontology Framework includes Furniture Ontology (based on FunStep ISO 10303-236), the eClass ontology (based on eCl@ss) and textile taxonomy (based on Moda-ML standard).

We preliminarily tested and validated the design and functionality of the NIMBLE Ontology Framework through the definition of queries that demonstrated its searching potential. Future work on this ontology framework will focus on the inclusion of other product and domain ontologies. The diversity of enterprise sectors, their business models and opportunities to collaborate via the platform will further drive the design and the extensions of the NIMBLE Ontology Framework. As next steps, the ontology framework will be extended and validated in further industrial use cases to ensure interoperability between suppliers and producers in a supply chain, while performing business processes.

Acknowledgements This research has been funded by the European Commission within the H2020 project NIMBLE (Collaborative Network for Industry, Manufacturing, Business and Logistics in Europe), No. 723810, for the period between October 2016 andSeptember 2019.

References

1. Boudreau, K., & Hagiu, A. (2009). Platform rules: Multi-sided platforms as regulators. In Annabelle Gawer (Ed.), *Platforms, markets and innovation*. Cheltenham, UK: Edward Elgar Publishing.
2. Gessa, N., De Sabbata, P., Marzocchi, M., Vitali, F. (2004). Moda-ML: Building a collaborative sectoral framework based on ebXML. In *Proceedings of CAISE'04—INTEROP 2004 Workshop on "Enterprise Modelling and Ontologies for Interoperability" EMOI*, Riga, Latvia. Retrieved from June 7–11, 2004, edited by Grundspenkis/Kirikova, 2004, ISBN-9984-9767-3-4.

3. NIMBLE D2.2 (2017). Semantic modelling of manufacturing collaboration assets.
4. Axenath, B., Kindler, E., and Rubin, V. (2005). The aspects of business processes: An open and formalism independent ontology. TR-RE-05-256, University of Paderborn. http://wwwcs.uni-paderborn.de/cs/kindler/Publikationen/copies/AKR05.pdf.
5. Hepp, M. (2005). eClassOWL: A fully-fledged products and services ontology in OWL. In *Poster Proceedings of the 4th International Semantic Web Conference* (*ISWC2005*), Retrieved from November 7–11, 2005. Galway, Ireland.
6. Hepp, M. (2008). GoodRelations: An ontology for describing products and services offers on the web. In *Proceedings of the 16th International Conference on Knowledge Engineering and Knowledge Management,* Acitrezza, Italy.
7. ISO/IEC 19845:2015 (2015). https://www.iso.org/standard/66370.html. Last visited September 20, 2017.
8. Forsberg, M. (2011). Svekatalog UBL Catalogue 2.1. http://www.sftiverifiering.se/svekatalog/XSD/1_1.htm. Last visited September 20, 2017.
9. Yarimagan, Y., & Dogac, A. (2009). A semantic-based solution for UBL schema interoperability. *IEEE Internet Computing, 13*(3), 64–71. https://doi.org/10.1109/MIC.2009.50.
10. White, S. A., Miers, D. (2008). BPMN modeling and reference guide: understanding and using BPMN, future strategies Inc.
11. Reisig, W. (1985). *Petri nets: An introduction*. New York: Springer.
12. Hepp, M., & Roman, D. (2007). An ontology framework for semantic business process management. In *Proceedings of Wirtschaftsinformatik 2007*, Karlsruhe.
13. Rospocher, M., Ghidini, C., & Serafini, L. (2014). An ontology for the business process modelling notation. In *8th International Conference on Formal Ontology in Information Systems (FOIS)*.
14. Von Rosing, M., Laurier, W. & Polovina, S. (2015). The BPM ontology. In *The complete business process handbook* (pp. 101–121). Elsevier.
15. Gessa, N., Busanelli, M., De Sabbata, P., & Vitali, F. (2006). Extracting a semantic view from an ebusiness vocabulary. In *Proceedings of IEEE CEC 2006*. San Francisco, 2006, Published by IEEE Computer Society, ISBN 0-7695-2511-3.
16. Universal Business Language Version 2.1. 04 November 2013. OASIS Standard. http://docs.oasis-open.org/ubl/os-UBL-2.1/UBL-2.1.html.
17. BFO (Basic Formal Ontology). Online http://ontology.buffalo.edu/bfo/. Last visited September 20, 2017.

The manufacturer's authorised representative in the EU is Springer Nature Customer Service Centre GmbH, Europaplatz 3, 69115 Heidelberg, Germany. If you have any concerns regarding our products, please contact ProductSafety@springernature.com

Printed and bound by CPI Group (UK) Ltd, Croydon, CR0 4YY
23/09/2026
02221533-0002